Ius Comparatum – Global Studies in Comparative Law

Volume 25

More information about this series at http://www.springer.com/series/11943

Académie Internationale de Droit Comparé
International Academy of Comparative Law

Graeme B. Dinwoodie
Editor

Secondary Liability of Internet Service Providers

 Springer

Editor
Graeme B. Dinwoodie
Faculty of Law
University of Oxford
Oxford, UK

IIT Chicago-Kent College of Law
Oxford, IL, USA

ISSN 2214-6881   ISSN 2214-689X   (electronic)
Ius Comparatum – Global Studies in Comparative Law
ISBN 978-3-319-55028-2   ISBN 978-3-319-55030-5   (eBook)
DOI 10.1007/978-3-319-55030-5

Library of Congress Control Number: 2017941537

Printed on acid-free paper

This Springer imprint is published by Springer Nature
The registered company is Springer International Publishing AG
The registered company address is: Gewerbestrasse 11, 6330 Cham, Switzerland

Contents

1   A Comparative Analysis of the Secondary Liability
    of Online Service Providers..   1
    Graeme B. Dinwoodie

2   Secondary Liability of Internet Service Providers in Poland   73
    Xawery Konarski and Tomasz Targosz

3   Secondary Liability of Internet Service Providers
    in the United States: General Principles and Fragmentation.............   93
    Salil K. Mehra and Marketa Trimble

4   ISP Secondary Liability: A Portuguese Perspective
    on Omissions as the Basis for Secondary Liability   109
    João Fachana

5   The Legal Framework Governing Online Service
    Providers in Cyprus...   125
    Tatiana Eleni Synodinou and Philippe Jougleux

6   Analysis of ISP Regulation Under Italian Law   141
    Elisa Bertolini, Vincenzo Franceschelli, and Oreste Pollicino

7   Secondary Liability of Service Providers in Brazil:
    The Effect of the Civil Rights Framework...   171
    Caitlin Sampaio Mulholland

8   Internet Commerce and Law   185
    Katja Lindroos

9   Common Law Pragmatism: New Zealand's Approach
    to Secondary Liability of Internet Service Providers   213
    Graeme W. Austin

10 Secondary Liability of Internet Intermediaries and Safe
   Harbours Under Croatian Law 229
   Ivana Kunda and Jasmina Mutabžija

11 Information Society Between Orwell and Zapata:
   A Czech Perspective on Safe Harbours.. 255
   Radim Polčák

12 Website Blocking Injunctions under United Kingdom
   and European Law... 275
   Jaani Riordan

13 The Liability of Internet Intermediaries
   and Disclosure Obligations in Greece 317
   Georgios N. Yannopoulos

14 Internet Service Provider Copyright Infringement in Taiwan 339
   Lung-Sheng Chen

15 Secondary Liability for Open Wireless Networks
   in Germany: Balancing Regulation and Innovation
   in the Digital Economy 361
   Christoph Busch

International Academy of Comparative Law... 383

About the Authors

Graeme W. Austin is chair in private law at Victoria University of Wellington and professor of law at the University of Melbourne. With a doctorate from Columbia Law School, Professor Austin returned to Australasia in 2010 after spending 10 years at the University of Arizona, where he was the J. Byron McCormick Professor of law. He is an elected member of the American Law Institute and has served as a member of the New Zealand Copyright Tribunal. His publications include *Human Rights and Intellectual Property* (Cambridge University Press), coauthored with Professor Larry Helfer.

Elisa Bertolini is an assistant professor of comparative public law at the Law School "Angelo Sraffa", Bocconi University, Milan, Italy. Her research interests include rights protection and constitutional adjudication in the Japanese and Chinese legal systems, Internet law, and the economic crisis in the EU context. Recently, she has published "Censoring the past? Suggestions on the German, Italian and Japanese approach to the totalitarian past," in the *Bulletin of the Nanzan Center for European Studies*, and "Corte Suprema e potere democratico: la particolare interazione nel sistema giuridico nipponico," is forthcoming in D. Butturini & M. Nicolini (Eds.), Giurisdizione costituzionale e potere democraticamente legittimato, ESI.

Christoph Busch is professor of European business law and private international law at the University of Osnabrück (Germany) and speaker of the Center on European Services Law at Osnabrück's European Legal Studies Institute. Christoph holds graduate degrees from the Universities of Münster and Paris Nanterre and a Dr. jur. from Bielefeld University. He is a cofounder and editor of the *Journal of European Consumer and Market Law* (*EuCML*) and a fellow of the European Law Institute. His research interests focus on European consumer and contract law, Internet law, and new regulatory techniques such as standardization of services.

Lung-Sheng Chen is associate professor of law at the Department of Law, National Chung Hsing University, Taiwan. Professor Chen obtained his LLM and JD from the School of Law, Washington University, in St. Louis, United States. Prior to his US study, Professor Chen was a practicing lawyer specializing in litigation and IP-related matters. Professor Chen's academic interests focus on intellectual property law and biotechnology law. He has published research papers in many prestigious law reviews and authored a patent law textbook offering a concise and deep analysis of Taiwan patent law.

Graeme B. Dinwoodie is the professor of intellectual property and information technology law at Oxford University, the director of the Oxford Intellectual Property Research Centre, and a university professor at Chicago-Kent College of Law. He has previously taught at the National University of Singapore (as the Yong Shook Lin Professor in intellectual property law), the New York University School of Law (as a global visiting professor of law), the University of Pennsylvania School of Law, and the University of Cincinnati College of Law. Immediately prior to taking up the IP chair at Oxford, Professor Graeme B. Dinwoodie was for several years a professor of law at Chicago-Kent College of Law and, from 2005 to 2009, also held a chair in intellectual property law at Queen Mary College, University of London. Professor Graeme B. Dinwoodie holds law degrees from the University of Glasgow, Harvard Law School (where he was a John F. Kennedy Scholar), and Columbia Law School (where he was a Burton Fellow). He is an elected member of the American Law Institute and served as president of the International Association for the Advancement of Teaching and Research in Intellectual Property from 2011 to 2013. In 2008, the International Trademark Association awarded Professor Graeme B. Dinwoodie the Pattishall Medal for Teaching Excellence in Trademark Law. He is the author of numerous articles and books on trade mark law and on international and comparative intellectual property law.

João Fachana has a Master of Laws in private law from Faculdade de Direito da Univesidade do Porto and a LL.M. in international business law from Católica Global School of Law of Universidade Católica Portuguesa. He is a member of the Portuguese Association of Intellectual Property Law (Associação Portuguesa de Direito Intelectual) and of the Internet Society (ISOC) - Portugal Chapter. He is an in-house lawyer, working in the fields of TMT, Intellectual Property and Data Privacy.

Vincenzo Franceschelli is professor of law at the University of Milano-Bicocca, School of Economics. Born in Milan, Italy, on October 11, 1947, he attained the Classical High School Diploma at the Liceo statale Giovanni Berchet, Milano, and his law degree at the University of Milan School of Law, magna cum laude. A professor since 1980, Vincenzo Franceschelli has taught at the University of Trieste (comparative private law, private law, and civil law), at the University of Siena (civil law), and at the University of Parma (private law).

He has been a Fulbright Scholar at the Academy of American and International Law in Dallas, Texas.

He has been a visiting professor at Seton Hall University School of Law (1985 Winter semester, 1995 Summer semester) and guest professor of the Japan Society for the Promotion of Science in Tokyo and Kyoto. He is coeditor of the *Rivista di Diritto Industriale*, Milano, Giuffrè. He has practiced law in his Milan law firm. Professor Franceschelli's research interests are in the areas of the law of contract, intellectual property law, computer law, communications law, and tourism law.

Philippe Jougleux is an associate professor of private law at the Faculty of Law in European University Cyprus and attorney-at-law at the Thessaloniki Bar Association. He graduated from the Faculty of Law of Lille II (France) in 1998. After completing a master's degree in the University of Aix-Marseille III in MME Law in 1999, he worked in the same university as an assistant from 1999 to 2002.

Xawery Konarski is a senior partner at Traple Konarski Podrecki & Partners and legal expert of the Polish IT and Telecom Chamber of Commerce and Internet Advertising Bureau Poland. He is the author of numerous publications on new technologies and personal data protection laws, including among others "Commentary to the Polish Act on rendering e-services."

Ivana Kunda is the head of the International and European Private Law Department at the Faculty of Law of the University of Rijeka. She is regularly invited as a guest professor abroad, such as at the University of Navarra (Spain) or the IULM (Italy). Her personal research projects have been supported by institutions such as the Fulbright Postdoctoral Research Program at Columbia University and the Gewerblicher Rechtsschutz und Urheberrecht at the Max Planck Institute for Innovation and Competition. She is currently an active researcher in three EU-funded international projects. She has authored papers and book chapters published in Croatia and abroad and a monograph on overriding mandatory provisions.

Salil K. Mehra is a professor of law at Temple University's Beasley School of Law, whose faculty he joined in 2000. His research focuses on antitrust/competition law and technology. He has published articles in a variety of journals, including the *Minnesota Law Review*, the *Emory Law Journal*, and the *Virginia Journal of International Law*. Professor Mehra is a past chair of the AALS Section on Antitrust and Economic Regulation and is a nongovernmental advisor to the International Competition Network. He is a former Abe Fellow of Japan's Center for Global Partnership and the Social Science Research Center. He graduated with honors, Order of the Coif, from the University of Chicago Law School, where he was on the University of Chicago Law Review. In 2016, Professor Mehra won a Temple University Lindback Award for Distinguished Teaching.

Caitlin Sampaio Mulholland (PhD from the State University of Rio de Janeiro, 2006) is an assistant professor of civil law of the Department of Law at the Pontifical Catholic University of Rio de Janeiro (PUC-Rio), where she currently coordinates the undergraduate program. The most frequent subjects of her scientific production are fundamental civil rights, civil liability, and contracts. She is the author of the books *Presumption of Causality in Civil Liability* and *Internet and Contracts*.

Jasmina Mutabžija has more than a decade of experience in managing companies providing intellectual property and Internet hosting services. In parallel, she is pursuing her teaching and academic career. She is a lecturer at Business School PAR in Rijeka (Croatia) and currently acting as a team member and consultant to two Croatian scientific projects. She regularly presents at scientific and professional conferences and publishes papers in the field of intellectual property and Internet law. She received her LLM in intellectual property law from the University of Turin and her PhD in commercial and company law from the University of Zagreb.

Radim Polčák is the head of the Institute of Law and Technology at the Faculty of Law, Masaryk University. His professional interests include legal philosophy, cyberlaw, and energy law. He is the general chair of the Cyberspace Conference, the editor in chief of the *Masaryk University Journal of Law and Technology* (MUJLT), and a founding fellow of the European Law Institute and the European Academy of Law and ICT.

Oreste Pollicino is a full professor of constitutional law at Bocconi University and the director of the series "Law and Policy of the New Media." He is the author of 5 books, 13 edited books, and more than 100 articles in the field of European constitutional law, media law, human rights law, and antidiscrimination law. His research areas include European and comparative constitutional law, media law, Internet law, and law and cinema. He is regularly invited as a speaker in national and international conferences and has been a visiting scholar at Oxford University (Institute of European and Comparative Law, and Programme in Comparative Media Law and Policy), Central European University (Budapest), and the University of Haifa Global Law Program, and has been a judge at the final rounds of the Monroe Price Moot Court Competition in Media Law, Oxford University.

Jaani Riordan is a barrister at 8 New Square in London. His practice spans all areas of intellectual property and technology litigation, with particular expertise in disputes involving internet and computer technology. He advises clients in the telecommunications, internet, broadcasting, media, biotechnology, fashion, and creative industries. Jaani holds a doctorate in law from the University of Oxford and degrees in computer science and law. Before being called to the English bar, he practised as a commercial solicitor in Australia. Jaani is the author of *The Liability of Internet Intermediaries* (2016, Oxford University Press).

Tatiana Eleni Synodinou is an associate professor of private and commercial law at the Law Department of the University of Cyprus. She obtained her law degree from the Aristotle University of Thessaloniki. She completed her postgraduate studies at the University of Aix-Marseille III and her doctoral and postdoctoral studies at the Aristotle University of Thessaloniki receiving fellowships from the Greek Scholarship Foundation and by the Committee of Research of the Aristotle University of Thessaloniki.

Tomasz Targosz is an assistant professor and the chair of intellectual property law at the Jagiellonian University in Kraków and partner at Traple Konarski Podrecki & Partners, Poland's leading IP law firm. He studied law at the Jagiellonian University in Kraków. In 2004–2006, he was a member of the Graduate College of the Universities of Krakow, Heidelberg, and Mainz. Tomasz specializes in copyright law, patent law, trademark law, unfair competition and antitrust law, as well as civil law, which was his original field of research.

Marketa Trimble is the Samuel S. Lionel Professor of Intellectual Property Law at the William S. Boyd School of Law of the University of Nevada, Las Vegas. She specializes in international intellectual property law and publishes extensively on matters at the intersection of conflict of laws/private international law and intellectual property law, particularly patent law and copyright law. She has authored several works in the area of cyberlaw, particularly regarding the legal issues of geoblocking and the circumvention of geoblocking. She is the coauthor of a respected international intellectual property law casebook.

Katja Lindroos (nee Weckstrom) is professor of commercial law at UEF Law School. She is also vice head of UEF Law School and responsible for international affairs and director of the MDP in International Economic and Resources Law. She frequently gives lectures to expert audiences in various fields, including continuing legal education for attorneys in Finland and abroad. Lindroos taught at the Chicago-Kent College of Law in the United States as an IP Fellow and Fulbright Scholar in 2006–2007 and served as a visiting assistant professor at the University of Louisville Law School in 2008. Lindroos (Weckström) has published broadly on issues regarding international and European intellectual property law and Internet commerce law. Her doctoral thesis "A Contextual Approach to Limits in EU Trademark Law" was published in 2011. The book *Internet Commerce and Law* was published in 2015. Her research tackles current topics relating to the protection of intellectual property on the Internet and in virtual worlds and focuses in particular on the liability of Internet service providers.

Georgios N. Yannopoulos is assistant professor at the Law School of the National and Kapodistrian University of Athens, teaching IT law and legal informatics. He studied law (LLB) at the University of Athens and has a PhD from the University of London, where he also conducted postdoctoral research. In addition, he has professional qualifications in computer programming and system analysis. He is a qualified lawyer registered to the Athens Bar Association. He has several publications including the books *Modelling the Legal Decision Process for IT Applications in Law*, 1998 (PhD thesis); *Information Flows in the Internet, Technology & Legal Regulation*, 2002 (in Greek); and *The Liability of Internet Intermediaries*, 2012 (in Greek).

Chapter 1
A Comparative Analysis of the Secondary Liability of Online Service Providers

Graeme B. Dinwoodie

Introduction

Any comparative analysis of the "secondary liability" of online service providers (OSPs) confronts a set of threshold definitional questions: most importantly, who is an "online service provider" and what is meant by "secondary liability"? Determining who is an online "service provider" primarily involves interpretation of definitions contained in various pieces of relatively recent legislation. The latter question— what is "secondary liability"—presents far more fundamental challenges, though

This Chapter draws heavily on national reports prepared by national reporters in advance of the Congress of the International Academy of Comparative Law in Vienna in July 2014, some of which have in turn been revised for publication as chapters in *Secondary Liability of Internet Service Providers* (Graeme B. Dinwoodie ed., Springer 2017) to which this is the Introductory Chapter. I am extremely grateful to the various National Reporters for that Congress on whose work this Chapter builds: Graeme W. Austin; Christoph Busch; Lung-Sheng Chen; João Fachana; Vincenzo Franceschelli, Oreste Pollicino, and Elisa Bertolini; Xawery Konarski and Tomasz Targosz; Ivana Kunda and Jasmina Mutabžija; David Lametti and Abby Shepard; Caitlin Mulholland; Clement Petersen; Radim Polčák; Cyrill Rigamonti; Jaani Riordan; Tatiana Synodinou and Philippe Jougleux; Marketa Trimble and Salil K. Mehra; Pierre Trudel; Katja Weckstrom; and Georgios N. Yannopoulos. These reports were authored in 2013–2014, although this Chapter takes into account subsequent developments. This Chapter also draws in part on Graeme B. Dinwoodie, *Secondary Liability for Online Trademark Infringement: The International Landscape,* 36 COLUMBIA. J. L. & ARTS 463–501 (2014). Thanks also to Matthew Kruger for editorial assistance with the book in which this Chapter appears.

G.B. Dinwoodie (✉)
Faculty of Law, University of Oxford, Oxford, UK

IIT Chicago-Kent College of Law, Chicago, IL, USA
e-mail: Graeme.dinwoodie@law.ox.ac.uk

© Springer International Publishing AG 2017
G.B. Dinwoodie (ed.), *Secondary Liability of Internet Service Providers,*
Ius Comparatum – Global Studies in Comparative Law 25,
DOI 10.1007/978-3-319-55030-5_1

these are of types not unknown to comparative law scholars generally. Some of these challenges are terminological—by what *name(s)* do we call the different bases on which an online service provider might be held liable for some way enabling third parties to engage in unlawful activity? But others are more conceptual: for example, to what extent is "secondary liability" dependent upon actual or notional proof of primary liability; is there a horizontal legal concept of "secondary liability" that delineates liability independently of the particular nature of (or policies animating) the alleged primary liability; does the term "secondary liability" encompass obligations that the law may impose upon service providers without suggesting that they might be monetarily "liable" for the unlawful conduct of the third party users of their services; and, is *online* service provider liability simply an application of traditional principles long applied in the offline environment? This Chapter seeks to address these conceptual challenges while also adumbrating the underlying doctrine now developing at the national level.

The Chapter proceeds as follows. Part I addresses definitional questions (both terminological and conceptual). Part II explains the reasons for the rise of secondary liability claims and the policy concerns that are implicated by the imposition of secondary liability. Part III discusses a range of standards adopted in different countries (in different contexts) to delineate the secondary liability of intermediaries. There, I highlight two different approaches to establishing the circumstances when an intermediary might be liable: a "positive" or "negative" definition of the scope of liability. The former flows from the standards for establishing liability; the latter grows out of the different safe harbour provisions that immunize intermediaries operating in particular ways, although there can obviously be connections between the standard for liability and the conditions for immunity.

Part IV considers the mechanism ("Notice and Takedown") that in practice has come in many countries to mediate the responsibilities of right owners and service providers for a range of unlawful conduct that occurs using the facilities of the service providers, as well as noting some variants (such as "Notice and Notice" in Canada). These mechanisms typically reflect OSP responses to potential secondary liability, and have developed both in contexts when that liability is defined positively and when it is framed in negative terms. But regardless of the varying impetus for the mechanisms, they are largely implemented through private ordering (with some of the concerns that attends any such activity) that is subject to differing level of public structuring and scrutiny.[1]

---

[1] "Public structuring" refers to the legal framework that might be established to shape (whether by compulsion or incentive) private ordering and thus give some greater comfort regarding the form the latter takes. *See generally* Graeme B. Dinwoodie, *Private Ordering and the Creation of International Copyright Norms: The Role of Public Structuring*, 160 J. Instit. & Theor. Econ. 161 (2004). And, separately, private actors (largely, OSPs) are increasingly making public on an ex post basis aggregated information concerning this private activity, which affords some level of public scrutiny. *See, e.g.,* Google, Inc., Transparency Report, at https://www.google.com/transparencyreport/?authuser=1 (listing 40 major internet companies that release such data). Between 4 January 2016 and 4 January 2017, Google removed 916 million URLs as a result of takedown requests from copyright owners.

Part V focuses on the concept of (judicially-enforceable) "responsibility without liability", a growing feature of the landscape in this area, especially but not exclusively in the European Union (EU).[2] Service providers in several fields, most notably intellectual property law, are being required actively to assist in preventing wrongdoing by third parties regardless of their own fault (but for example, engaging in so-called "web-blocking" of allegedly infringing sites). These mechanisms, found in several legislative instruments but developed in greater detail by courts through applications in private litigation, operate to create a quasi-regulatory network of obligations without imposition of full monetary liability. It is not clear that the obligations that might be ordered against intermediaries under this rubric should truly be conceptualised as instances of secondary liability; indeed, they are clearly something different in juridical character.[3] But these obligations are of great importance to the scope of secondary liability, and they do impose costs on intermediaries. Efforts to internationalise such mechanisms such as via the Anti-Counterfeiting Trade Agreement (ACTA)[4] or the Transpacific Partnership Agreement (TPP)[5] have met with fierce resistance, as have several pieces of legislation in the United States that critics saw as mimicking some of the developments in the European Union.[6]

Reflecting the bulk of both legislation and case law in which the secondary liability of OSPs has been addressed, this Chapter focuses in large part on claims under copyright law, trade mark law, and defamation law.[7] Part VI concludes, however, by considering briefly whether generally applicable principles can be derived from,

---

[2] The most recent jurisdiction to adopt such an approach is Australia. *See* Roadshow Films Pty Ltd. v Telstra Corporation [2016] FCA 1503 (Dec 15, 2016); § 115A of the Copyright Act 1968 (Cth); *see also* § 193DDA of Copyright Act (Sing.) (orders to disable access to flagrantly infringing online locations).

[3] *See* Jaani Riordan, The Liability of Internet Intermediaries at §§ 1.49–1.60, at 12–14 (Oxford Univ. Press 2016) (using terminology of "primary", "secondary" and "injunctive" liability, and describing these mechanisms as "injunctions without wrongdoing"); Martin Husovec, Injunctions Against Intermediaries in the European Union: Accountable But Not Liable? (Camb. Univ. Press 2017) (forthcoming).

[4] Anti-Counterfeiting Trade Agreement, Dec. 3, 2010 [hereinafter ACTA], *available at* http://trade.ec.europa.eu/doclib/docs/2010/december/tradoc_147079.pdf.

[5] Trans-Pacific Partnership Agreement, Chapter 18, Feb. 4, 2016, [hereinafter TPP] *available at* https://ustr.gov/sites/default/files/TPP-Final-Text-Intellectual-Property.pdf; *see also* Neha Mishra, *The Role of the Trans-Pacific Partnership Agreement in the Internet Ecosystem: Uneasy Liaison or Synergistic Alliance?* 20 J. Int'l Econ. L 31, 55–56 (2017).

[6] *See, e.g.,* Stop Online Piracy Act ("SOPA") H.R. 3261, 112th Cong. (2011); Preventing Real Online Threats to Economic Creativity and Theft of Intellectual Property Act, S. 968, 112th Cong. (2011) ("PIPA" or "Protect IP Act"); *see generally* Mark Lemley, David S. Levine & David G. Post, *Don't Break the Internet,* 64 Stan. L. Rev. Online 34 (2011).

[7] *See* Sea Shepherd v. Fish & Fish [2015] UKSC 10 at [40] (Lord Sumption, dissenting) ("In both England and the United States, the principles have been worked out mainly in the context of allegations of accessory liability for the tortious infringement of intellectual property rights. There is, however, nothing in these principles which is peculiar to the infringement of intellectual property rights. The cases depend on ordinary principles of the law of tort"). These contexts involve tortious liability. For a much fuller analysis, encompassing for example claims in contract, see Paul S. Davies, Accessory Liability (Hart, 2015).

and extended beyond, the specific context in which they first arose. This analysis leads to two central propositions, which it is argued hold true descriptively and warrant endorsement prescriptively. First, an assessment of secondary liability cannot be divorced from (and indeed must be informed by) the scope of primary liability or other legal devices by which the conduct of service providers or their customers is regulated.[8] And, second, despite the claims that secondary liability is simply the application of general principles of tort law, secondary liability is rarely a subject-neutral allocation of responsibility among different potential defendants according to autonomous principles of fault; rather, it maps in part to the policy objectives of the different bodies of law where the claim of (secondary) liability arises.[9]

I. Definitional Questions

The scope and content of this Chapter depends in large part on the meaning of the terms "service provider" and "secondary liability." Each is discussed below.

A. Online Service Providers

The term "service provider" (let alone "online service provider", or "internet service provider" or "ISP", with which "OSP" is sometimes used interchangeably both in this Chapter and more generally) has no consistent meaning across borders; indeed, defining "service providers" operating in the bricks-and-mortar world is also a difficult task. An equally (perhaps more) common usage in the literature is "intermediary". But, like "online service provider", this term lacks a single, common and consistent usage.

Each of these terms has in recent years received some legislative definition (along with yet other synonyms) in provisions creating safe harbours, or *immunity* from liability, for such actors.[10] The terms make more fleeting appearances outside this context, primarily in defining which actors are subject to certain disclosure obligations vis-à-vis customers and enforcement authorities.[11] For example, the UK's Digital Economy Act 2010 required an "internet service provider" to

---

[8] See infra text accompanying notes 58–75.

[9] See National Report of the United Kingdom, at 5 ("although the connecting factors for secondary liability are uniform, the different elements of primary wrongdoing will mean that attaching secondary liability is attended by very different practical considerations depending on the cause of action.").

[10] See, e.g., National Report of New Zealand, at 3 (citing section 2(1) of the Copyright Act 1994 (NZ)).

[11] See National Report of the United Kingdom, at 6 ("there is no general concept of a 'service provider' that is used specifically to impose secondary liability. However, a small number of English statutes do create secondary liability for online services in particular contexts.")

participate in a subscriber notification regime, and for that purpose defined an "internet service provider."[12]

But even in supposedly harmonised schemes of immunity (such as in the implementation of the E-Commerce Directive in the Member States of the European Union), there has been variation in interpretation of who comes within the definition (or the safe harbour, which will be a more fact-specific question and vary from case to case). For example, the Court of Justice of the European Union has found that both the eBay auction site and the Google search engine are *potentially* entitled to the benefit of safe harbours offered an information society service provider by the E-Commerce Directive.[13] But national application of the principles to those entities (and their subsidiaries, sometimes performing different online roles, such as YouTube) has varied across Europe.[14]

To some extent, the apparent variation in decisions at the national level may be because the *definition* of those actors who are potentially immune may implicitly contain some of the *conditions* for availing oneself of immunity.[15] (Indeed, in some jurisdictions this occurs quite explicitly[16]). Strictly, these conditions do not define who is a service provider but rather whether the service provider is acting a way that will allow them to take advantage of the immunities conferred.[17] Thus, for example,

---

[12] *See* National Report of the United Kingdom, at 6 (defined as "a person who provides an 'internet access service', which in turn means an electronic communications service consisting wholly or mainly of access to the internet, where an IP address is allocated to each subscriber to enable access"); *see also* Digital Economy Act 2010, §§ 3–4.

[13] *See* Joined Cases C-236/08–C-238/08, Google France SARL v. Louis Vuitton Malletier SA, 2010 E.C.R. I-2417, ¶¶ 106–20; Case C-324/09, L'Oréal SA v. eBay Int'l AG, 2011 E.C.R. I-6011; Directive 2000/31 of the European Parliament and of the Council of 8 June 2000 on Certain Legal Aspects of Information Society Services, in Particular Electronic Commerce, in the Internal Market, 2000 O.J. (L 178) 1, 11 (EC) [hereinafter E-Commerce Directive].

[14] *See infra* text accompanying notes 185–188 (discussing liability of eBay and YouTube under French, Spanish and Italian law).

[15] Even within a single jurisdiction, different safe harbours may also contain different specific conditions. *See* National Report of the United Kingdom, at 16 ("The scope of the safe harbours and limitations ... is delimited by two criteria: first, the defendant must be a "service provider" ... Second, the activity of the defendant said to give rise to liability must fall within one of the protected activities. The specific nature of these activities is usually a more significant limitation than the identity of the service provider."). Thus, a service provider may be protected under one safe harbour but not another. *See* National Report of the United Kingdom, at 17 (discussing immunity of search engines under the different safe harbours of the E-Commerce Directive).

[16] *See* National Report of Taiwan, at 7 & 12–13 (defining internet service provider in terms of the different types of service that might bring them within a safe harbour); *cf.* National Report of New Zealand, at 2 & 5–6 (quoting section 2(1) of the Copyright Act 1994, which does the same but with a much smaller set of conditions; instead, conditions appear outside the definition of the safe harbours themselves).

[17] The same approach can be found in provisions defining service provider for the purpose of imposing regulatory obligations. Thus, the UK Digital Economy Act only required "qualifying ISPs" to participate in its subscriber notification regime. *See* National Report of the United Kingdom, at 6.

the Italian courts have developed a distinction between passive and active intermediaries that has its roots in the conditions under which the protections of the E-Commerce Directive will be available.[18] One could read this case law as refining the notion of service provider (for these purposes) or simply imposing conditions on when immunity will be available. The latter is probably the better reading because a service provider may be active in one scenario but passive in another. Likewise, and relatedly, statutes providing immunity for different kinds of service provider performing different online roles will frequently define the term "service provider" in varying ways to accommodate those differences. For example, an "access provider" will inevitably be defined differently from a host provider at a certain level of detail.[19]

In the context of the E-Commerce Directive of the European Union, one of the most influential legislative instruments in the field, a "service provider" is a person providing an information society service. This basically means "any service normally provided for remuneration, at a distance, by electronic means and at the individual request of a recipient of services."[20] This is a broad definition but does exclude some commercial actors such as internet cafés (because not provided remotely), and broadcasters (who, rather than the user, determines when and what transmissions occur).[21] And "although the safe harbours [to which such providers can have recourse] apply only to economic operators rather than non-commercial services, English courts have held unequivocally that personal websites, such as blogs and discussion fora, which have no profit motive or revenue model, may qualify for protection."[22]

This Chapter adopts a broad view of the term "service provider". But it might also be helpful to set out briefly the types of cases and defendants that are considered; these leading examples, as much as any formal definition, establish the parameters of the analysis in this Chapter.

The principal bodies of law where secondary liability of online service providers has been addressed are copyright law, trade mark law, defamation law, and privacy law.[23] The principal cases have revolved around relatively similar fact patterns in

---

[18] *See* National Report of Italy, at 5.

[19] *See* 17 U.S.C. 512(k)(1)(A)-(B) (providing narrower definition of providers able to come within the scope of the copyright infringement immunity conferred by Section 512(a) on access providers).

[20] *See* E-Commerce Directive, recital 17.

[21] *See* National Report of the United Kingdom, at 16–17.

[22] *See id.* at 17 (citing *Kaschke v Gray* [2010] EWHC 690 (QB), [43]); *see also* National Report of New Zealand, at 2 (commenting that the definition of a "hosting" service provider for the purposes of the copyright safe harbour "would extend the concept of 'Internet service providers' to 'Web 2.0' platforms, bulletin boards, blogs, or even websites operated by firms, public entities, and private parties"); Case C-484/14, Tobias McFadden v Sony Music, ECLI:EU:C:2016:689 (CJEU 2016) (discussing free public wifi offered by small business in the vicinity of the business under Article 12 of the E-Commerce Directive).

[23] *Cf.* Sea Shepherd v. Fish & Fish [2015] UKSC 10 at [40] (Lord Sumption dissenting).

different jurisdictions. For example, copyright owners have sued manufacturers of copying technologies for infringements caused by those who use their equipment,[24] purveyors of peer-to-peer file sharing software for the activities of those who download material without rights holders' permissions,[25] and social media sites (such as YouTube) that host allegedly infringing clips from copyrighted audio-visual works.[26]

In the context of trade mark law, the leading modern exemplars of secondary liability claims are actions brought against online auction sites, each essentially alleging that the auction site could have done more to stop the sale of counterfeits or other allegedly infringing items by third parties on its Web site; and claims brought against search engines alleging that the sale of keyword advertising consisting of the trade marks of parties other than the mark owner resulted in infringement (normally, by causing actionable confusion).[27]

In defamation or libel law, web sites (such as the retailer Amazon.co.uk) have been sued where a third party posted an allegedly defamatory book review on the claimant's book product page.[28] And search engines such as Google have been sued for allegedly "publishing" defamatory material that appeared within "snippets" summarising search results for the claimant,[29] or "processing" personal data the publication of which within snippets violated the privacy of individuals to whom the personal data related (even if the data is not removed from the actual publisher's website).[30]

This illustrates the wide variety of online intermediaries pursued as liable for enabling wrongs perpetrated by others, but core internet service providers, such as companies providing access to the internet or web hosting services, are potential defendants in any of these scenarios.[31] And as right-holders—and potentially policymakers—adopt "follow the money" or "least cost avoider" strategies to identify defendants of first resort, the list of relevant online intermediaries may grow further.

---

[24] *See, e.g.,* Sony Corp. of Am. v. Universal City Studios, Inc., 464 U.S. 417 (1984).

[25] *See, e.g.,* Metro-Goldwyn-Mayer Studios Inc. v. Grokster, Ltd., 545 U.S. 913 (2005); A&M Records, Inc. v. Napster, Inc., 239 F.3d 1004 (9th Cir. 2001); *see generally* Jane C. Ginsburg, *Separating The Sony Sheep from the Grokster Goats: Reckoning The Future Business Plans of Copyright-Dependent Technology Entrepreneurs,* 50 ARIZ. L. REV. 577 (2008).

[26] *See* Viacom International, Inc. v. YouTube Inc., 676 F.3d 19 (2d Cir. 2013), *on remand*, 940 F. Supp.2d 110 (S.D.N.Y. 2013).

[27] *See generally* Graeme B. Dinwoodie, *Secondary Liability for Online Trademark Infringement: The International Landscape,* 36 COLUMBIA. J. L. & ARTS 463–501 (2014); *see also* Free Kick Master LLC v. Apple Inc., 2016 WL 77916 (N.D. Cal. Feb 29, 2016) (app store sued for trade mark infringement with respect to apps sold by third parties in its app store).

[28] *See* McGrath v Dawkins, [2012] EWHC B3 (QB) (discussed in National Report of the United Kingdom, at 15).

[29] *See, e.g.,* Metropolitan Schools v DesignTechnica, [2009] EWHC 1765 (QB) (UK); A v Google New Zealand Limited [2012] NZHC 2352 (NZ).

[30] *See* Case C-131/12, Google Spain SL, Google Inc. v. Agencia Española de Protección de Datos (AEPD), Mario Costeja González, 2014 ECLI:EU:C:2014:317 [hereinafter *Google Spain*].

[31] *See, e.g.,* Louis Vuitton Malletier, S.A. v. Akanoc Solutions, Inc., 658 F.3d 936 (9th Cir. 2011).

Thus, companies who process credit card payments have also been sued for facilitating unlawful transactions,[32] and companies merely providing customers access to the internet have been required to block websites in other countries where allegedly infringing content resides.[33]

B. "Secondary" Liability

The term "secondary liability" is an umbrella term encompassing a number of different types of claims. But it causes some terminological difficulties for comparative analysis. In many countries, the term "secondary liability" has a meaning in discrete statutory regimes that covers conduct that would be thought in other countries to be quite clearly a species of primary infringement.[34] I do not regard these statutory labels as sufficient in and of themselves to bring these claims within the scope of the Chapter; often, conceptually, the causes of action relate to what are in substance well-understood as primary acts of infringement. The concept of copyright "authorisation" (in UK, New Zealand and Australian law) inverts this difficulty by establishing an act of *nominally* primary liability that clearly maps in substance to conventional forms of secondary or joint tortfeasor liability.[35] I regard those claims as encompassed by the substantive concept of "secondary liability."[36]

---

[32] *See, e.g.*, Gucci Am., Inc. v. Frontline Processing Corp., 721 F. Supp. 2d 228 (S.D.N.Y. 2010) (allowing secondary infringement claim to proceed against credit card processing companies who provided services to online merchant allegedly selling counterfeit goods); *but cf.* Perfect 10, Inc. v. Visa Int'l Serv., Ass'n, 494 F.3d 788 (9th Cir. 2007) (affirming dismissal of actions against credit card companies).

[33] *See, e.g.*, Cartier Int'l AG v. British Sky Broadcasting Limited [2016] EWCA Civ 658.

[34] *See* National Report of the United Kingdom, at 5 ("Many statutory wrongs define further so-called 'secondary' torts—such as dealing commercially with articles that infringe primary rights … "); *see also* Copyright, Designs and Patents Act 1988, §§ 22–26 (UK). The term "secondary infringement" has a different (narrow) meaning in U.K. trademark law. *See* LIONEL BENTLY & BRAD SHERMAN, INTELLECTUAL PROPERTY LAW 1044 (4th ed. 2014).

[35] *See* Copyright, Designs and Patents Act 1988, § 16(2) (UK); National Report of the United Kingdom, at 10 ("the definition of primary liability includes a party who authorises another to engage in acts restricted by copyright. Such an authorising party is primarily liable, even though their liability is secondary in the sense that it cannot exist until there has actually been a primary infringement of copyright carried out by the third party whom they have authorised to act."); Copyright Act 1968, §§13(2), 36(1), 101(1) (Aus.); National Report of New Zealand, at 2. The statutory concept of authorisation has perhaps been most fully fleshed out—both statutorily and in decisional law—in Australia. *See* Roadshow Films Pty v iiNet Ltd., [2012] HCA 16 (finding Internet service provider not liable for authorising copyright infringement when subscribers infringed copyright by using file sharing software). For discussion of authorisation in UK law, see Richard Arnold and Paul S Davies, *Accessory Liability for Intellectual Property Infringement: The Case of Authorisation*, 133 LAW Q. REV. ___ (2017) (forthcoming).

[36] In the United States, the insertion into the 1976 Copyright Act of the language rendering "authorisation" of an infringing act itself actionable has been explicitly explained as simply offering a statutory hook for secondary liability. *See* Subafilms, Ltd. v. MGM-Pathe Communications Co., 24 F.3d 1088 (9th Cir. 1994) (en banc). In the United Kingdom, there is some—but very

Moreover, causes of action that are the subject of this Chapter are not always *denominated* as actions for "secondary liability"; other common terms used for this type of liability include "accessory liability" or "indirect liability."[37] In this Chapter, I will for convenience use the term "secondary liability" as an umbrella term for these different forms of liability.[38]

Secondary liability can be participant-based or relationship-based. Participant-based liability occurs by virtue of the secondary defendant inducing or contributing to or facilitating the harmful conduct of the primary wrongdoer. This type of claim tends to revolve around the level of knowledge of the defendant concerning the wrongful conduct, sometimes constructively imputed through proxies, and the extent to which the defendant has actively contributed to cause the harm regarded as actionable by the applicable primary law. "Contributory infringement" under U.S. copyright law is an example of this type of liability, holding that "one who, with knowledge of the infringing activity, induces, causes, or materially contributes ... may be held liable as a contributory infringer."[39]

Alternatively, secondary liability may arise where the defendant benefits from the harm and is sufficiently close in relationship to the primary wrongdoer that the law will treat them as one and the same (often because the relationship is sufficiently close that the intermediary could have prevented the primary unlawful acts from occurring).[40] This relationship-based liability reflects the principle of respondeat superior and informs, for example, U.S. law on vicarious trademark or copyright infringement.[41] In U.S. copyright law, vicarious liability arises when the defendant "has the right and ability to supervise the infringing activity and also has a direct financial interest in such activities."[42]

---

little—daylight between the concept of authorisation and joint tortfeasorship (or accessory liability). *See* Arnold and Davies, *supra* note 35, at [21]–[22] (discussing differences). But it is clearly understood as part of the picture of accessory liability.

[37] For use of the term "accessory" liability, see National Report of the United Kingdom, at 2; National Report of Switzerland, at 5. *See generally* Paul Davies, *Accessory Liability: Protecting Intellectual Property Rights* [2011] INTELLECTUAL PROPERTY QUARTERLY 390; DAVIES, *supra* note 7. The term is also found in criminal law. *See* National Report of the United Kingdom, at 4.

[38] Service providers may also be criminally liable as accessories where they participate in criminal wrongdoing. In some jurisdictions, accessorial *criminal* liability has been set out in detail by statute. *See* National Report of the United Kingdom, at 4–5 (discussing Accessories and Abettors Act 1861 and Serious Crime Act 2007, and noting judicial interpretation). This Chapter focuses on civil liability. But there are certain circumstances where criminal accessory liability might be broader than civil liability. *See* National Report of the United Kingdom, at 4.

[39] *See* Gershwin Publishing Corp. v. Columbia Artists Mgmt., Inc., 443 F.2d 1159, 1162 (2d Cir. 1971).

[40] *Cf.* RIORDAN, *supra* note 3, at § 5.13, at 116 (suggesting that one feature distinguishing participatory from relational liability is the lack of causative analysis).

[41] *See* Hard Rock Café Licensing Corp. v. Concession Servs., Inc., 955 F.2d 1143, 1150 (7th Cir. 1992) (noting that vicarious liability for trademark law requires "a finding that the defendant and the infringer have an apparent or actual partnership, have authority to bind one another in transactions with third parties or exercise joint ownership or control over the infringing product").

[42] *See* Shapiro, Bernstein and Co. v. H. L. Green Co., 316 F.2d 304 (2d Cir. 1963); Fonovisa, Inc. v. Cherry Auction, Inc., 76 F.3d 259 (9th Cir. 1996).

Most countries have these latter extensions of liability, and they are largely uncontroversial, albeit with local differences as to the type of relationships (e.g., employees, agents, suppliers, landlords, etc.) that are regarded as sufficiently close to impose liability.[43] These relationship-based forms of liability will tend to be more common across different primary causes of action because the extension of liability derives from the (constant) nature of the connection between the parties rather than the (varying) contribution to the conduct that offends one or more laws.

The essence of secondary liability, as reflected in both these concepts, is that the defendant is held responsible for harm caused by the wrongful conduct of a third party.[44] That is to say, liability is derivative. This feature of secondary liability is seen mostly clearly in common law countries.[45] But some civil law countries have also adopted formulations that emphasise the indirect or derivative nature of liability. This might be because formally such liability often arises from code provisions that are expressed in term of "joint liability," which seems to tie the "secondary" defendant's liability closely to the primary act.[46] And in practice those provisions are frequently applied in ways that emphasise the same elements that have been characterised in common law countries as contributory infringement.[47] For example, Article 185(2) of the Taiwanese Civil Code provides that "[i]nstigators and accomplices are deemed to be joint tortfeasors," which appears to parallel the twin bases of inducement and common design (contributory) infringement found in UK law.[48]

---

[43] *See* National Report of Taiwan, at 4 ("[P]aragraph 1, Article 188 of the Civil Code provides employer's liability for their employees' conducts. The concept of employer's liability under the statute is similar to the concept of vicarious infringement"); National Report of the United Kingdom, at 4.

[44] *See* OBG Ltd. v Allan [2008] 1 AC 1, 27 (Lord Hoffmann) (UK) (characterising accessory liability as "principles of liability for the act of another"); Sony Corp. of America v. Universal City Studios, Inc., 464 U.S. 417, 435 (1984) (explaining that "the concept of contributory infringement is merely a species of the broader problem of identifying circumstances in which it is just to hold one individually accountable for the actions of another").

[45] *See* National Report of the United Kingdom, at 2; *see also* Subafilms, Ltd. v. MGM-Pathe Communications Co., 24 F.3d 1088, 1092 (9th Cir. 1994) (en banc) ("Contributory infringement under the 1909 Act developed as a form of third party liability. Accordingly, there could be no liability for contributory infringement unless the authorized or otherwise encouraged activity itself could amount to infringement.").

[46] *See* National Report of Czech Republic, at 2–3 (discussing application of provision creating "joint liability"). The doctrine of accessory liability in U.K. law is also frequently treated under the label of "joint tortfeasorship". *See generally* DAVIES, *supra* note 7, at 177 ("The wide umbrella of 'joint tortfeasance' has engulfed accessory liability in tort and obscured its constituent elements"); Sea Shepherd v. Fish & Fish [2015] UKSC 10.

[47] See National Report of Taiwan, at 3–4 ("Taiwan's legal system does not expressly recognize concepts of contributory or vicarious liability. Nonetheless, this does not necessarily mean that an ISP will never be held liable for its users' conduct. [A]n ISP is likely to be held liable for indirect infringement under the Civil Code … [which] states that] "[i]f several persons jointly conducted a tort with another person, they are jointly liable for the damage arising therefrom").

[48] Taiwan also has a copyright-specific statutory form of secondary liability. *See id.*, at 5 (reporting liability for copyright infringement under Article 87 where providing "to the public computer programs or other technology that can be used to publicly transmit or reproduce works, with the intent to allow the public to infringe on others' economic rights by means of public transmission

However, even in countries insisting on the derivative nature of secondary liability, it is not always necessary to *prove* the primary infringement.[49] Thus, Jaani Riordan notes that although "all secondary liability is in some sense derivative from primary liability ... the picture becomes somewhat more complicated when one considers that secondary liability does not actually require primary liability to be alleged, much less established, against anyone in particular; a service provider could be targeted alone, and frequently is in internet disputes—where the primary wrongdoer is often anonymous, insolvent or beyond jurisdiction."[50] Instead, U.K. courts have often made relatively cursory assessments of the specific details of the primary infringement and moved on.[51] This is more noticeable—and perhaps most understandable—when the relief being sought against the defendant intermediary is not a determination of monetary liability as an accessory but an order requiring assistance in preventing infringement by typically numerous third parties.[52] The regulatory nature of such orders, and the often blatant piracy or counterfeiting

---

or reproduction by means of the Internet of the copyrighted works of another, without the consent of or a license from the economic right owners, and to receive benefits therefrom"). The provision was added in 2007 to adopt the reasoning of the U.S. Supreme Court in *MGM v. Grokster*, and (based upon the meaning of the "intent" element of the claim) in particular the intentional inducement action articulated in *Grokster*. *See id.*, at 6 ("[T]he ISP's intent [as required by the cause of action] is satisfied if the ISP instigates, solicits, incites, or persuades its users to make use of the computer program or other technology the ISP provides for the purpose of infringing copyrights of others by advertising or other active measures"). This cause of action appears to duplicate the "instigation" or "inducement" version of secondary liability under Taiwan's general joint tortfeasorship provision. However, unlike a joint tortfeasorship claim, this extra statutory copyright liability does not depend upon a primary infringement; if the elements of the statutory inducement cause of action exist, the claim is made out. *See id.*, at 8 ("A comparison of the general standard for establishing secondary liability under the Civil Code and that under the Copyright Act reveals differences between them. For an ISP to be held liable for others' conducts under the Civil Code, there must be a direct infringement. However, under the Copyright Act, an ISP is liable when the four elements are satisfied. In other words, no direct infringement of others is required to establish secondary liability of an ISP under the Copyright Act.")

[49] *See* National Report of Taiwan, at 4 (noting that under the Civil Code of Taiwan the joint liability provision applies "even if the infringed person cannot indicate that who in fact did commit the tort"); *cf.* National Report of Switzerland, at 2; National Report of Portugal, at 6.

[50] National Report of the United Kingdom, at 9.

[51] *See, e.g.,* Twentieth Century Fox Film Corporation v Newzbin Ltd. [2010] EWHC 608 (Ch) (*Newzbin* I) at [97] ("I am prepared to proceed on that basis because I am satisfied that the claimants' copyrights have indeed been infringed by the defendant's premium members. The number of active premium members is very substantial, as evidenced by the defendant's turnover, and those members are primarily interested in films, as the Newzbin website makes clear. In the light of these matters, the nature of Newzbin as I have described it and the interaction between the defendant and its members as shown by the sharing forums, I consider it overwhelmingly likely that the defendant's premium members have made use of the facilities to which they have subscribed and that in doing so a number of them have downloaded copies of the claimants' copyright films, including those specifically identified in these proceedings, all of which are popular titles. The claimants are unable to identify which particular films individual premium members have copied only because the defendant has chosen not to record details of the NZB files they have downloaded...").

[52] *See, e.g.,* Cartier Int'l AG v. British Sky Broadcasting Limited [2016] EWCA Civ 658 at [84]–[85].

involved (which makes the issues primarily factual), easily dissuades courts from insisting on a finely-grained analysis of individual primary liability. But it may make sense in secondary liability actions generally insofar as the attractiveness of the secondary relief is often its capacity to restrain multiple primary infringements by focusing on a single choke point.

The derivative nature of the secondary liability claim may structure the assertion and vindication of rights even when proof of primary liability is not required. For example, as a result of recent legislation in the United Kingdom, a court cannot hear a defamation claim against a secondary party unless the court is satisfied that it is not reasonably practicable for the claimant to proceed against the primary party.[53] It is not clear how courts will apply the limit on secondary liability where the primary defendant acted online under conditions of anonymity.[54] And there remains disagreement among national courts in Europe whether web blocking orders must as a matter of proportionality be sought first from host providers before actions are brought against access providers; this is a slightly different exhaustion of remedies rule, differentiating among service providers rather than between primary and secondary defendants.[55] Finally, in any event, proving primary liability could also be relevant to relief.[56]

In many other countries, however, courts will treat a fact pattern handled by common law jurisdictions under the rubric of secondary liability as involving *direct* liability under tort law for failure to conduct business in a particular way or a failure to take certain reasonable precautions.[57] The ultimate result of this conceptually different

---

[53] *See* Defamation Act 2013 (UK).

[54] *See* National Report of the United Kingdom, at 9 n. 35.

[55] Some commentators have suggested that whether the third party's conduct is notionally unlawful is always relevant under English law, even when the relief sought against an intermediary is non-monetary (and not based upon a finding of illegality by the intermediary). *See id.*, at 20 ("English law recognises a number of [non-monetary] remedies against non-wrongdoers. . . Although they are independent of any finding of primary or secondary liability against the service provider, they require some kind of primary wrongdoing to be demonstrated to the relevant standard of proof.").

[56] *See id.*, at 10 ("The availability of relief against a primary wrongdoer is normally relevant only to proportionate liability. In general, where damage is "indivisible" as between a service provider and primary wrongdoer, both joint causes of harm will each be liable to compensate the whole of the damage. However . . . the service provider may have a claim against the primary wrongdoer under the Civil Liability (Contribution) Act 1978"); *see* Sea Shepherd v. Fish & Fish [2015] UKSC 10 at [57] (Lord Neuberger) ("I agree with Lord Sumption that, once the assistance is shown to be more than trivial, the proper way of reflecting the defendant's relatively unimportant contribution to the tort is through the court's power to apportion liability, and then order contribution, as between the defendant and the primary tortfeasor.")

[57] *See e.g.*, Tribunal de commerce [TC][court of trade] Paris, June 30, 2008, No. 200677799, 11–12 (Fr.) (Louis Vuitton Malletier S.A. v. eBay, Inc.); *See, e.g.*, TC Paris, June 30, 2008, No. 200677799, 11–12 (Fr.) (Louis Vuitton Malletier S.A. v. eBay, Inc.) (unpublished); Tribunal de grande instance [TGI] Troyes, civ., June 4, 2008, No. 060264 (Fr.) (Hermés Int'l v. eBay, Inc.) (unpublished). This is explicitly not the approach taken in the United Kingdom. *See, e.g.*, Cartier Int'l AG v. British Sky Broadcasting Limited [2016] EWCA Civ 658, at [54] ("it is clear in light of the decision of the House of Lords in *CBS Songs Ltd. v Amstrad plc* that [the access providers] do not owe a common law duty of care to [trade mark owners] to take reasonable care to ensure that their services are not used by the operators of the offending websites.").

approach to the question may not be hugely different in concrete cases. Liability flowing directly from the failure to conduct business in a particular way or a failure to take certain reasonable precautions might be understood as an isolated culpability determination giving rise to a claim in tort or unfair competition. But such failures can also be relevant to questions of knowledge and level of contribution, which normally inform claims of secondary liability that are expressly framed in derivative terms.[58]

Moreover, the relationship between primary and secondary liability is important in other respects. As a policy matter, the availability of primary claims to address potentially wrongful conduct should affect the need for development of secondary claims. For example, in the United States, the treatment of the sale by search engines of trade marks as triggers for keyword advertising as a potential act of primary infringement has largely forestalled the litigation of the question of secondary liability for such acts.[59] In contrast, in the European Union, the Court of Justice has held that a search engine is not involved in the type of use that can give rise to primary trademark liability under EU trademark law simply by selling keyword advertising that consists of the trademark of another.[60] This forces trade mark

---

[58] *Cf.* Stacey L. Dogan, *Principled Standards vs. Boundless Discretion: A Tale of Two Approaches to Intermediary Trademark Liability Online*, 37 COLUM. J.L. & ARTS 503, 509 (2014).

[59] *See, e.g.,* Gov't Emps. Ins. Co. v. Google, Inc., No. 1:04CV507, 2005 WL 1903128 (E.D. Va. Aug. 8, 2005) (finding for defendant search engine because no likelihood of confusion was proved at trial); *see also* Eric Goldman, Google Defeats Trademark Challenge to Its AdWords Service, Forbes Online (Oct. 22, 2012, 12:35 PM), http://www.forbes.com/sites/ericgoldman/2012/10/22/google-defeats-trademark-challenge-to-its-adwords-service/ ("In a remarkable litigation tour-de-force, Google has never definitively lost any of these cases in court (though it has occasionally lost intermediate rulings)."). There has as a result been less developed analysis by U.S. courts of the potential secondary liability of search engines for infringements committed by advertisers to whom they have sold keywords tied to the trademarks of others. This is notwithstanding the urging by some scholars that secondary liability is the only appropriate vehicle for assessment of such claims. *See* Stacey L. Dogan & Mark A. Lemley, *Trademark and Consumer Search Costs on the Internet*, 41 HOUS. L. REV. 777 (2004). The only substantial resistance to this occurred in the Second Circuit in 1-800 Contacts, Inc. v. WhenU.com, Inc., *see* 1-800 Contacts, Inc. v. WhenU.com, Inc., 414 F.3d 400 (2d Cir. 2005). and that circuit brought keyword advertising cases back within the primary infringement rubric in its 2009 *Rescuecom Corp. v. Google Inc.* decision. *See* Rescuecom Corp. v. Google Inc., 562 F.3d 123 (2d Cir. 2009); *cf.* Kelly-Brown v. Winfrey, 717 F.3d 295 (2d Cir. 2013). Assessing culpability of search engines under primary infringement standards does not appear to be altering the pro-defendant outcomes that one would expect as a matter of secondary infringement under *Tiffany*. On the whole, search engines are prevailing. To be sure, Google could (even under the *Tiffany* standard) be held contributorily liable if an advertiser is found to be primarily infringing and Google, after notice, does not disable the ad. The frequency of that will however depend upon evolving case law on advertiser liability, which also seems still to be largely pro-defendant. *See, e.g.*, J.G. Wentworth, S.S.C. Ltd. P'ship v. Settlement Funding L.L.C., No. 06-0597, 2007 WL 30115 (E.D. Pa. Jan. 4, 2007) (awarding summary judgment to defendant purchaser of keywords on question of confusion); Network Automation, Inc. v. Advanced Sys. Concepts, Inc., 638 F.3d 1137 (9th Cir. 2011); *cf. 1-800 Contacts*, 722 F.3d 1229.

[60] Joined Cases C-236/08–C-238/08, Google France SARL v. Louis Vuitton Malletier SA, 2010 E.C.R. I-2417, ¶ 55 (CJEU 201). Only uses by a search engine "in its own commercial communication" would fall within the proscription of trademark law; the Court found that this was not the case with the sale of keyword advertising. *Id.* ¶¶ 55–56.

owners either to pursue the primary infringers (the purchasers of the advertise-ments) or sue the search engines for secondary liability.

To date, mark owners in the European Union have had marginally more success bringing primary infringement claims based upon keyword purchases against adver-tisers than their counterparts in the United States, but any restriction of that option might shift their focus back to the filing of secondary liability claims against search engines.[61] Of course, insofar as the claim against the search engine in such cases is truly derivative of advertiser liability, restricting the liability of the advertisers might also limit the claims against search engines. The focus would then possibly turn to autonomous search engine conduct that appears to be detached from particular advertiser behaviour (e.g., the manner of presentation of search results generally on a search engine page).[62]

This set of dynamic interactions in keyword advertising litigation shows the importance of the relationship between primary and secondary infringement. And it works in both directions. As a policy matter, a secondary liability standard that is unlikely to be satisfied will cause claimants to push for the expansion of the scope of primary liability. In contrast, the availability of secondary liability claims (or orders mandating intermediary assistance but falling short of monetary liability) might moderate the demand to hold intermediaries primarily liable.[63]

This type of argument arguably lay behind the submission of the European Copyright Society in *Svensson v Retriever Sverige AB* that hyperlinking was not an act of communication to the public, but might in certain circumstances be actionable for facilitating primary acts of infringement.[64] But any effort to regulate intermedi-ary liability as a matter of accessory liability was rejected by the Court of Justice. Instead, rather than assessing liability for linking to a copyright-protected work as a matter of secondary liability, the Court dealt with the issue by asking whether the act involved communication of the work at issue to a "new public."[65] This is an (at best) enigmatic concept that is still being fleshed out by the Court of Justice in several recent and pending references. It may well be that the variables that determine that question reflect (inter alia) policy concerns that might typically inform assessments

---

[61] *See* Interflora v Marks & Spencer, [2013] EWHC 1291 (Ch), *rev'd*, [2014] EWCA Civ 1403; Cosmetic Warriors Ltd. v Amazon.co.uk Ltd., [2014] EWHC 181 (Ch); Birgit Clark, *Fleurop: When Keyword Advertising May Exceptionally Be a Trade Mark Infringement After All*, 9 J. INTELL. PROP. L. & PRAC., 693 (2014) (discussing German Federal Court of Justice (Bundesgerichtshof), Case I ZR 53/12—"Fleurop" 27 June 2013).

[62] *See* Graeme B. Dinwoodie and Mark D. Janis, *Lessons From the Trademark Use Debate*, 92 IOWA L. REV. 1703, 1717 (2007).

[63] *See infra* Part V.

[64] European Copyright Society, Opinion on the Reference to the CJEU in Case C-466/12 Svensson, at 7, available at https://europeancopyrightsociety.org/opinion-on-the-reference-to-the-cjeu-in-case-c-46612-svensson/

[65] *See* Case C-466/12, Svensson v Retriever Sverige AB, [2014] E.C.D.R. 9 (CJEU 2014).

of secondary liability. *GS Media* certainly elevates intent and responding to take down notices as relevant to primary infringement.[66]

In the European Union context, an additional institutional consideration often goes unstated. If these concerns are developed as a matter of primary infringement, then it will fall to the Court of Justice to guide their development (as would not have been the case had the matter been treated under unharmonised national standards of secondary liability).[67] And elements of the primary claim may also be the locus of the debate handled in other jurisdictions in the context of secondary claims as a result of the historical development of doctrine.[68] Thus, evolution of the concept of actionable "publication" in defamation law would appear to have occurred with an eye to the variables that we might think relevant to questions of secondary liability, such as levels of knowledge and the nature of the intermediary's contribution.[69] In the United Kingdom or New Zealand, a "successful [defamation] claim against a secondary publisher (such as a service provider) is usually treated as an example of primary liability for a second and distinct publication, even though it is derivative from another wrong (the original publication)."[70] Thus, search engines such as Google have been formally found not liable for defamatory material that appeared

---

[66] *See* Case C-348/13, BestWater Int'l GmbH v Mebes and Potsch EU:C:2014:2315 (CJEU 2014); Case C-348/13, C More Entertainment AB v Sandberg, ECLI:EU:C:2015:199 (CJEU 2015); Case C-160/15, GS Media BV v. Sanoma Media Netherlands BV, ECLI:EU:C:2016:644 (CJEU 2016).

[67] In copyright, national courts in the E.U. appear to be quick to assess service provider liability as a matter of primary infringement, perhaps following the lead of the Court of Justice. *See, e.g.,* Public Relations Consultants Association Ltd. v The Newspaper Licensing Agency Ltd. [2013] UKSC 18 (UK) (liability of an online news aggregator service turned on whether it engaged in acts which amounted to reproduction of news headlines without coming within the statutory defence of incidental use); *cf.* Case C-610/15, Stichting Brein v. Ziggo BV, EU:C:2017:99 (AG Szpunar Feb 8, 2017) at ¶3.

[68] *See* National Report of the United Kingdom, at 8–9 (noting that "[i]n trade mark cases, as in other areas of law, English courts frequently call upon primary liability rules as means of demarcating the liability of secondary parties).

[69] *See* National Report of New Zealand, at 3 (discussing *Wishart v Murray* [2013] 3 NZLR 246, 262, and noting that the liability of the operator of the Facebook page where allegedly defamatory material was posted was plausibly a publisher of the third party material but the degree of the defendant's responsibility for and control over the content of the Facebook page was considered as a matter of primary liability); National Report of the United Kingdom, at 9 (noting that "in defamation cases the courts apply the 'long established line of authority' that exempts secondary disseminators from responsibility for conveying a third party's material unless they knew or ought reasonably to have known that the material was likely to be defamatory"); *see also id.*, at 19 ("[M] any of the safe harbours mirror devices in substantive liability doctrines which serve to immunise passive and neutral platforms from *prima facie* liability for third parties' activities (though these depend on different considerations depending on the type of wrongdoing involved). Doctrines of innocent dissemination in defamation and authorisation liability in copyright provide the clearest examples of this tendency in English law").

[70] *See* National Report of the United Kingdom, at 9; National Report of New Zealand, at 3 ("in the current state of New Zealand law, the defendant [search engine] would have been liable primarily as a 'publisher' of the defamatory material. Secondary liability principles are therefore not apposite.")

within "snippets" summarising search results for the claimant because they were not a publisher of the material.[71]

Similarly, the reasons proffered by the United States Court of Appeals for the Second Circuit in *Rescuecom v. Google* for distinguishing its prior decisions on trade mark liability for the sale by search engines of keyword advertising reek of the considerations that inform analysis of secondary infringement claims: to what extent did Google's Keyword Suggestion Tool effectively induce the primary infringing conduct, and to what extent did the selling by Google of particular marks affect the level of contribution and causality relevant to Google's culpability?[72] Certainly, these do not seem questions typically relevant to assessing primary trademark infringement. Indeed, this disjunction has caused Stacey Dogan to label the *Rescuecom*-endorsed cause of action as a "curious branch of direct trademark infringement designed to distinguish between the innocent intermediary, and one whose technology and business model deliberately seeks to confuse."[73]

In light of the permeability of the line between primary and secondary liability, it should not be a surprise therefore that service providers are often sued under every conceivable basis, nor that courts are not always careful about the precise basis on which liability is found.[74] Thus, in the United Kingdom "in *Twentieth Century Fox Film Corp v Newzbin Ltd*, the operator of a Usenet binary storage service was liable both for itself communicating and for authorising the communication to the public of copyright works uploaded by others, and also for procuring and engaging in a common design with users to infringe copyright. It made little difference... whether liability was classified as primary or secondary."[75]

II. The Rise of Secondary Liability Claims

Claimants might strategically prefer to bring a secondary liability claim instead of suing the third party wrongdoer for any number of reasons. A secondary infringement action may increase efficiency by allowing the claimant to secure, in a single

---

[71] *See, e.g.,* Metropolitan Schools v DesignTechnica, [2009] EWHC 1765 (QB); A v Google New Zealand Limited [2012] NZHC 2352 (NZ).

[72] *See* Rescuecom Corp. v. Google Inc., 562 F.3d 123, 129 (2d Cir. 2009). On this latter point, the Second Circuit distinguished between the practice of the defendant in *1-800 Contacts*, which had sold advertising keyed to categories of marks (e.g., selling the right to have an ad appear when a user searches for a mark connected to eye care products, but not disclosing to the advertiser the proprietary mapping of marks and categories), and that of Google (which sold advertising tied directly to single marks). *See id.* at 128–29.

[73] Stacey L. Dogan, *"We Know It When We See It": Intermediary Trademark Liability and the Internet,* 2011 STAN. TECH. L. REV. 7 at [19] (2011).

[74] *See* Dogan & Lemley, *supra* note 59 (discussing trademark liability of intermediaries).

[75] National Report of the United Kingdom, at 10 (citing *Newzbin* [2010] EWHC 608 (Ch)).

proceeding, relief against a party whose conduct is simultaneously enabling multiple wrongful acts by a number of primary tortfeasors. The intermediary offers a "choke point." As a result, it may ensure more *effective* enforcement of rights.[76] The advent of the Internet has only enhanced these benefits. The efficiency gains are magnified substantially when the number of wrongs to which the secondary liability defendant contributes are multiplied many times over and the "whack-a-mole" problem becomes even more acute (as it is online with ease of replication and distribution).

The effective consolidation of complaints can occur across borders as well as across defendants. Thus, secondary liability claims might also allow de facto worldwide relief depending upon the extent to which the intermediary's business is globally integrated.[77] As the territorial nature of much law comes under increasing pressure in an era of free trade and cross-border digital communication, this is an attractive feature to claimants (though one that might undermine good reasons for territorial laws, such as the ability of different countries to regulate conduct within their borders according to their different policy preferences).[78]

A secondary liability action might shift some of the costs of enforcement to intermediaries. This occurs whether the claimant directly secures relief from a court that requires an intermediary to undertake certain detection and prevention measures, or because the intermediary adjusts its practices to be more conservative in light of an award made against it (or similarly situated OSPs) under principles of secondary liability. Secondary liability actions may thus enable certain claimants (such as copyright or trade mark owners) to affect the future structure of business models employed by intermediaries and the direction of technological development considered by intermediaries.[79] Thus, the benefit to right holders of efficient enforcement simultaneously creates the specter of intrusive regulation of the business of intermediaries operating in the online environment.[80] And this could hamper the substantial contribution that internet services have made to recent economic

---

[76] *See* Sony Corp. of Am. v. Universal City Studios, Inc., 464 U.S. 417, 442 (1984) (making point in relation to copyright and patent law).

[77] *See* Graeme B. Dinwoodie, Rochelle C. Dreyfuss & Annette Kur, *The Law Applicable to Secondary Liability in Intellectual Property Cases*, 42 N.Y.U. J. INT'L. L. & POL. 201 (2009).

[78] *See* GRAEME B. DINWOODIE & ROCHELLE C. DREYFUSS, A NEOFEDERALIST VISION OF TRIPS: THE RESILIENCE OF THE INTERNATIONAL INTELLECTUAL PROPERTY REGIME (Oxford University Press 2012).

[79] *See* National Report of the United Kingdom, at 1 ("Service providers point out that wider secondary liability could hinder [economic] growth and have harmful effects upon innovation, while also doubtlessly threatening to erode their business models").

[80] As a result, the U.S. Supreme Court has been reluctant to find secondary liability based upon design choices *alone*. But the Court has recognized the relevance of design choice to determinations of inducement liability when combined with other evidence. *See* MGM Studios Inc. v. Grokster, Ltd., 545 U.S. 913, 939 (2005) ("[T]his evidence of unlawful objective is given added significance by MGM's showing that neither company attempted to develop filtering tools or other mechanisms to diminish the infringing activity using their software ...").

growth,[81] and to integration of global markets.[82] This risk has been recognized by legislatures in creating safe harbours,[83] and enacting provisions guaranteeing that OSPs cannot be placed under any general duty to monitor the activities of their subscribers, and (in the European Union) by courts insisting on taking into account the fundamental right of OSPs to conduct business before shifting enforcement costs to them.[84]

Broad secondary liability for intermediaries may also chill speech or stop legal conduct as a result of over-enforcement. This may occur because of a spillover effect on legitimate users (discussed below in the context of "overblocking"). And it might also result from the practical realities of operating notice and takedown systems. Intermediaries will for quite rational, pragmatic reasons tend to take down allegedly infringing postings and err on the side of preserving the immunity that comes from compliance and risk the ire of a few customers (some of whom may indeed be acting wrongfully) and consumer groups. Indeed, in the *eBay v. Tiffany* ligation in the United States, the court noted that eBay never refused to remove a listing reported by Tiffany as an alleged infringement and did so very quickly (within twenty-four hours, and typically within twelve hours).[85]

III. Defining Liability Positively or Negatively

Although the nature of the internet has increased the attractiveness of secondary liability actions against intermediaries, the standards under which an online service provider will be held liable for the conduct of third parties remains unclear.[86] This partially

---

[81] *See* National Report of the United Kingdom, at 1 ("Internet services contribute as much 5.4 per cent of GDP and account for up to 23 per cent of recent British economic growth" (citing McKinsey Global Institute, *Internet Matters: The Net's Sweeping Impact on Growth, Jobs, and Prosperity* (May 2011) 15–16).

[82] National Report of Taiwan, at 1.

[83] *See infra* text accompanying notes 148–193.

[84] *See* 17 U.S.C. § 512 (US) (safe harbour); E-Commerce Directive, arts. 12–14 (EU) (safe harbour); E-Commerce Directive, art 15 ("Member States shall not impose a general obligation on providers, when providing the services covered by Articles 12, 13 and 14, to monitor the information which they transmit or store, nor a general obligation actively to seek facts or circumstances indicating illegal activity."); Case C-314/12, UPC Telekabel Wien GmbH v. Constantin Film Verleih GmbH, EU:C:2014:192, [2014] E.C.D.R. 12 (CJEU 2014) (recognizing role of right to conduct business under Article 16 of the Charter).

[85] Tiffany (NJ) Inc. v. eBay Inc., 600 F.3d 93, 99 (2d Cir. 2010). Not all intermediaries are equally compliant. *See* National Report of the United Kingdom, at 31 (calculating that, according to the Google Transparency Report, Google "refused to comply with over 2.3 million take-down requests made by BPI in 2013. On average, this figure amounted to around 9.4% of the URLs specified in each take-down request. This may suggest a tendency for false positive requests on the part of rights-holders. On the other hand, many of the ignored requests related to duplicate materials, and the remainder tended to relate to obviously infringing materials").

[86] *See* National Report of New Zealand, at 1.

reflects the doctrinal variance internationally in the underlying concept of secondary liability noted in Part I of this Chapter. But this uncertainty is often true even within single jurisdictions or among supposedly harmonized legal systems. And the difficulty in identifying a clear standard is compounded by the fast-changing and diverse nature of online intermediaries. Moreover, in practice, effective online enforcement occurs has depended heavily upon private ordering mechanisms, reducing the level of public guidance or scrutiny that courts have been able to offer. And those decisions that have been handed down, while apparently relevant to the question of intermediary liability, may assume the character of regulatory norms rather than assessments of individual private liability. Thus, rather than delineate in detail the conditions that pertain in a wide array of jurisdictions, I focus in this Part on how these standards come to be established, using specific standards in a range of countries as illustrations.

The development of conditions under which online service providers might effectively be responsible for the wrongful conduct of others has occurred through two distinct, but related, dynamics. First, courts have applied long-standing principles of secondary liability found in national private law—whether denominated as such—to new intermediaries operating in the online environment, whether by analogy to the offline world or by reference to broad policy considerations. In this Chapter, I have referred to this dynamic as a "positive" approach to defining service provider liability.

A second approach, which I have described as a "negative" approach, comes at the endeavour from the other direction and defines the circumstances in which an online provider will be *immune* from liability. Under this approach, courts focus less on whether the intermediary conduct evinces sufficient fault (or on whether the relationship between the intermediary and the primary wrongdoer is sufficiently close) to extend liability, and more on whether the intermediary has complied with a series of legislated conditions for immunity. This latter approach has been more dominant than one might expect in defining online service provider liability, possibly because statutes immunizing online service providers from liability arose on a widespread basis soon after initial commercial exploitation of the internet and are arguably the most common expression of legislative decisions regarding ISPs.

Although the standards for OSP liability can thus be developed positively or negatively, there has been far greater legislative activity flowing from the negative approach of defining zones of immunity. This might affect the juridical character of the standard. Legislative conditions for immunity (especially if quite detailed) operate in effect as a form of business regulation for online service providers, a conceptualization which might in turn shape the way in which courts frame questions of secondary liability and thus articulate relevant standards.[87]

---

[87] The level of detail in immunity legislation varies substantially. The process of notice and takedown is laid out in immense detail, including precise time periods, in the US Digital Millennium Copyright Act ("DMCA"). *See* 17 U.S.C. § 512. The parallel EU provision talks of "expeditious" action by an OSP, which has been fleshed out (and was intended to be fleshed out even more than it has been) by private ordering and negotiated statements of best practices. *See* Dinwoodie, *supra* note 1, at 170–172 (discussing efforts to fund and encourage industry agreements in the European Union).

A. The Positive Approach: Standards Establishing Secondary Liability

In common law countries, the standard for holding intermediaries liable for copyright or trade mark infringement based on conduct and knowledge has proved hard to satisfy. It is not impossible to satisfy, but the cases where it has successfully been asserted have tended to be blatant cases of intermediary wrongdoing.[88]

In particular, where a product or service can be used for both lawful and unlawful purposes, its use by third parties for unlawful purposes does not typically render the service provider secondarily liable.[89] Most countries reject a standard that would expect intermediaries to monitor the activities of its users and thus insure against the non-infringement of rights. Such a principle (which rejects the alternative approach advanced in the copyright context in the White Paper issued by the Clinton Administration in 1995)[90] is enshrined in the legislation of a number of countries, but is also clearly given weight by courts.

Thus, mere knowledge on the part of the supplier of equipment or a service that it *might* be used to engage in unlawful conduct does not make out the claimant's case, not even when combined with proof that it did in fact facilitate infringement. The strict reading by common law courts of the state of mind necessary to establish secondary liability has largely prevented right owners from shifting enforcement costs entirely to intermediaries.

For example, in the United Kingdom, the applicable standard for liability across a range of causes of action in tort is drawn from the general law of "joint tortfeasorship," which premises so-called accessorial liability on (1) procurement of an infringement by inducement, incitement or persuasion, or (2) a common design.[91] These concepts are not defined in any statute, but have instead evolved from case law dealing with offline and online conduct.[92] A summary by Arnold J. in his *eBay v. L'Oreal* opinion captures the essence:

---

[88] *See, e.g.,* Metro-Goldwyn-Mayer Studios Inc. v. Grokster, Ltd., 545 U.S. 913, 936 (2005); Twentieth Century Fox Film Corporation v Newzbin Ltd. [2010] EWHC 608 (Ch).

[89] That principle is also sometimes endorsed via provisions or decisions granting immunity from certain claims. *See, e.g.,* National Report of New Zealand, at 4 (discussing Section 92B(2) of the New Zealand Copyright Act); *cf.* Sony Corp. of Am. v. Universal City Studios, 464 U.S. 417, 442 (1984) (protection for technologies having substantial non-infringing uses).

[90] *See* WORKING GROUP ON INTELLECTUAL PROP. RIGHTS, INFO. INFRASTRUCTURE TASK FORCE, INTELLECTUAL PROPERTY AND THE NATIONAL INFORMATION INFRASTRUCTURE (1995), *available at* http://www.uspto.gov/web/offices/com/doc/ipnii/ipnii.pdf [hereinafter White Paper] at 114–24. The White Paper analyzed ISP liability based on temporary copies made in random access memory of computers as direct infringements of copyright, although it also discussed contributory and vicarious liability.

[91] *See* National Report of the United Kingdom, at 3. In addition, particular regimes augment the general joint tortfeasorship standard with statutory forms of liability (such as authorisation of copyright infringement) that appear to possess a derivative character but be labelled a form of primary liability. *See supra* notes 35–36.

[92] *See* National Report of the United Kingdom, at 3.

> Mere assistance, *even knowing* assistance, does not suffice to make the "secondary" party liable as a joint tortfeasor with the primary party. What he does must go further. He must have *conspired* with the primary party or *procured* or *induced* his commission of the tort … ; or he must have joined in the common design pursuant to which the tort was committed …[93]

This should not be taken to rule out liability based upon assistance, but only *mere* assistance without more. Recent Supreme Court opinions (in *Sea Shepherd v. Fish & Fish*), stress that assistance-based liability is possible provided there is some other circumstance at play.[94] In that case, Lord Sumption focused on the nature of intent as that additional factor: "What the authorities, taken as a whole, demonstrate is that the additional element which is required to establish liability, over and above mere knowledge that an otherwise lawful act will assist the tort, is a shared intention that it should do so."[95] The focus on intent is particularly clear from some of the variants of liability. For example, the 'inducement' basis for liability requires both causation and willful conduct, the latter meaning "that the procurer must intend the acts constituting the wrongful conduct to occur in a particular way."[96]

A common design claim may be based upon an explicit or implicit agreement. But mere unilateral invitation, instruction or silence is insufficient. The common design test is a hard standard to satisfy, and it seems increasingly relevant to framing all variants of joint tortfeasorship (including liability grounded on assistance).[97]

---

[93] L'Oréal S.A. v. eBay Int'l AG, [2009] EWHC 1094 at [350] (*quoting* Credit Lyonnais Bank Nederland NV v. Export Credit Guarantee Dep't, [1997] EWCA (Civ) 2165, [1998] 1 Lloyd's Rep 19, [48] (Eng.)) (emphasis supplied).

[94] *See* Sea Shepherd v. Fish & Fish [2015] UKSC 10 at [41] (Lord Sumption, dissenting) ("I do not think that … Lord Templeman [in CBS] was seeking to limit liability as a joint tortfeasor to cases of inducement or procurement, as opposed to assistance. When read with his general statement of the elements of liability as a joint tortfeasor, it is clear that he was intending to limit it to cases of common intent. Inducing or procuring a tort necessarily involves common intent if the tort is then committed. Mere assistance may or may not do so, depending on the circumstances. The mere supply of equipment which is known to be capable of being used to commit a tort does not suggest intent. Other circumstances may do so."); *id.* at [57] (Lord Neuberger) ("the defendant should not escape liability simply because his assistance was (i) relatively minor in terms of its contribution to, or influence over, the tortious act when compared with the actions of the primary tortfeasor, or (ii) indirect so far as any consequential damage to the claimant is concerned."). Although Lord Sumption dissented from the result in *Sea Shepherd*, it is not clear that the judges were divided on the applicable legal principles. *See id.* at [61] (Lord Neuberger) ("I do not detect any significant difference between this analysis of the law and the rather fuller analyses advanced in the judgments of Lord Sumption and Lord Toulson"); *id.* at [91] (Lord Mance dissenting) ("At the end of the day, the difference of opinion in the court about the outcome of this appeal derives from a difference not about the legal principles which it involves, but about their application to the facts").

[95] *See id.* at [44]; see also id at [54] (Lord Neuberger) ("the assistance must have been pursuant to a common design on the part of the defendant and the primary tortfeasor that the act be committed").

[96] *See* National Report of the United Kingdom, at 3–4.

[97] It is most commonly satisfied by facts that are different from the facilitation scenario, and look not far removed from vicarious liability claims. *See, e.g.*, Cosmetic Warriors Ltd. v. Amazon.co.uk, [2014] EWHC 181 (Ch) (Eng.) (holding a U.K. and a Luxembourg entity jointly liable for trade mark infringement since they "joined together and agreed to work together in the furtherance of a common plan").

There must be "'concerted action to a common end', rather than independent but cumulative or coinciding acts."[98] Under this test of joint tortfeasorship, for example, the UK courts suggested (without deciding) that eBay was unlikely to be liable for the infringing uses of marks posted on its auction site.[99]

The precise relationship between the different variants of this tort—conspiracy, procurement, inducement or common design—has never been entirely clear.[100] Lord Hoffmann has crystallised the essence of all the claims of joint tortfeasorship as "whether the acts were done pursuant to a common design so that the secondary party has made the act his own."[101] Jaani Riordan notes that these categories of liability correspond broadly to the forms of liability described as "inducement" and "contributory" in other countries (such as the United States).[102]

And indeed, eBay likewise avoided secondary trademark liability in the United States for use of its online auction platform by sellers of fake Tiffany products under similar standards.[103] Again, the level of knowledge demanded by the U.S. court in that case before it would contemplate secondary liability was substantial. Under the applicable (*Inwood*) standard,[104] Tiffany was required to show either that eBay "intentionally induces another to infringe a trademark" or continued to offer its services "to one whom it knows or has reason to know is engaging in trademark

---

[98] National Report of the United Kingdom, at 3.

[99] Although the case settled before the issue was decided, Arnold J. suggested in his judgment referring question to the European Court of Justice that he would come down on eBay's side on the question. See *L'Oréal*, [2009] EWHC 1094, at [359]–[372].

[100] *Cf.* Sea Shepherd v. Fish & Fish [2015] UKSC 10 at [37] (Lord Sumption, dissenting) ("The legal elements of liability as a joint tortfeasor must necessarily be formulated in general terms because it is based on concepts whose exact ambit is sensitive to the facts.")

[101] SABAF Spa v MFI Furniture Centres Ltd. [2004] UKHL 45 at [40]; *see also* Sea Shepherd v. Fish & Fish [2015] UKSC 10 at [38] (Lord Sumption, dissenting) ("It is now well established that if these requirements are satisfied the accessory's liability is not for the assistance. He is liable for the tortious act of the primary actor, because by reason of the assistance the law treats him as party to it"); *see also id.* at [19]–[20] (Lord Toulson) (setting out variants of liability); *cf. id.* at [59] (Lord Neuberger) ("I have some concerns about the notion that the defendant has to "[make the tortious act] his own" ... While it can be said that it rightly emphasises the requirement for a common design, this formulation is ultimately circular and risks being interpreted as putting a potentially dangerous gloss on the need for a common design.").

[102] *See* National Report of the United Kingdom, at 3.

[103] *See* Tiffany (NJ) Inc. v. eBay Inc., 600 F.3d 93, 99 (2d Cir. 2010).

[104] Both the district court and the Second Circuit applied the *Inwood* test, notwithstanding that the case involved the continued provision of services. Following Ninth Circuit precedent in *Lockheed Martin Corp. v. Network Solutions, Inc.*, 194 F.3d 980, 984 (9th Cir. 1999), the district court held in *Tiffany* that "*Inwood* applies to a service provider who exercises sufficient control over the means of the infringing conduct," and eBay satisfied this test because of the control it retained over the transactions and listings on its Web site. *Tiffany*, 576 F. Supp. 2d at 505–07. Although amici challenged whether *Inwood* applied to services, eBay did not include this argument in its appeal and thus the Second Circuit did not reach that question. *See Tiffany*, 600 F.3d at 106 ("On appeal, eBay no longer maintains that it is not subject to *Inwood*."); *see also id.* at 105 n.10 (noting amici arguments). Other cases have operated on the same assumption. *See, e.g.*, Rosetta Stone Ltd. v. Google, Inc., 676 F.3d 144 (4th Cir. 2012).

infringement." Tiffany did not suggest that eBay had induced the sale of counterfeit goods, and thus the case turned on knowing continued supply. eBay did not continue to supply its services to third-party sellers when Tiffany put it on actual notice of particular listings of counterfeit goods.[105] Indeed, eBay never refused to remove a reported Tiffany listing, and did so very quickly (within 24 hours, and typically within 12 hours).[106]

Thus, Tiffany's claim rested on the allegation that eBay had continued the provision of service as regards other listings which, by virtue of its generalized knowledge of infringement on its Web site, it had "reason to know" were infringing.[107] A "significant portion" of the Tiffany jewelry listed on the eBay Web site was counterfeit and eBay knew that some portion of the Tiffany goods sold on its Web site might be counterfeit.[108] Essentially, the argument was that the widespread nature of infringement amounted to constructive knowledge. However, the Second Circuit held that:

> For contributory trademark infringement liability to lie, a service provider must have more than a general knowledge or reason to know that its service is being used to sell counterfeit goods. Some contemporary knowledge of which particular listings are infringing or will infringe in the future is necessary.[109]

The *Tiffany* court rejected efforts by Tiffany and several amici to establish a secondary liability standard (derived from the *Restatement (Third) of Unfair Competition*) that would have required an intermediary to take reasonable precautions to prevent infringement by the direct infringers when it could reasonably anticipate that infringements would occur.[110] Such an approach would allow an obligation to act to be triggered by generalized knowledge, but the extent of that

---

[105] *Tiffany*, 600 F.3d at 106.

[106] *Id.* at 99.

[107] *See id.* at 103.

[108] *Id.* at 98.

[109] *Id.* at 107. The U.S. court did leave one opening for mark owners who had not given notice regarding individual listings, suggesting that willful blindness may satisfy the knowledge standard. *Tiffany*, 600 F.3d at 109–10 ("A service provider is not, we think, permitted willful blindness. When it has reason to suspect that users of its service are infringing a protected mark, it may not shield itself from learning of the particular infringing transactions by looking the other way... [W]illful blindness is equivalent to actual knowledge for purposes of the Lanham Act." (internal quotation marks omitted). The court did not find willful blindness in *Tiffany* merely because eBay knew generally that infringements were occurring on its site and eBay did not ignore the information it was given. *Id.* at 110 ("eBay appears to concede that it knew as a general matter that counterfeit Tiffany products were listed and sold through its website. Without more, however, this knowledge is insufficient to trigger liability under *Inwood*... eBay did not ignore the information it was given about counterfeit sales on its website."). But this opening may still leave later courts with some room to reach a different result when a defendant with a less legitimate business model makes less reasonable efforts at preventing infringement.

[110] Tiffany (NJ) Inc. v. eBay Inc., 600 F.3d 93, 109-10 (2d Cir. 2010); *see* RESTATEMENT (THIRD) OF UNFAIR COMPETITION § 27 (1995) ("[Contributory liability attaches when] the actor fails to take reasonable precautions against the occurrence of the third person's infringing conduct in circumstances in which the infringing conduct can be reasonably anticipated.").

obligation would then be determined by considerations of reasonableness. Adoption of that position might have required the Second Circuit to align itself with arguments that the Supreme Court had appeared to reject in *Inwood*.[111] But, as the *Tiffany* court noted, the Supreme Court precedent was hardly constraining on the knowledge question.[112] And the generalized knowledge standard does seem to comport with the approach adopted by many courts in the offline environment.[113]

Some scholars have suggested that, despite the narrowness of the doctrinal language in *Tiffany*, the Second Circuit may have effectively decided the case based upon an assessment of the reasonableness of the measures adopted by eBay.[114] Certainly, the court's recitation of eBay's efforts to detect and limit infringement

---

[111] Reading *Inwood* is complicated by the fact that Justice White's concurring opinion ascribes to Justice O'Connor's opinion a treatment of the lower court's opinion that she herself disclaimed. In particular, Justice White expressed concern that the majority was endorsing a knowledge standard based on whether the defendant "could reasonably have anticipated" the third parties' infringing conduct; Justice O'Connor explicitly rejected this charge. *Compare* Inwood Labs., Inc. v. Ives Labs., Inc., 456 U.S. 844, 854 n.13 (1982) (O'Connor, J.) (rejecting Justice White's concern that the majority opinion endorsed any change to the contributory infringement test), *with id.* at 859–60 (White, J., concurring) ("The mere fact that a generic drug company can anticipate that some illegal substitution will occur to some unspecified extent, and by some unknown pharmacists, should not by itself be a predicate for contributory liability. I thus am inclined to believe that the Court silently acquiesces in a significant change in the test for contributory infringement.").

[112] It is worth noting the extent to which the Second Circuit felt that *Inwood* spoke to the precise question before the *Tiffany* court. The court opened its opinion by noting the "paucity of case law to guide us" and the fact that the few cases decided in the Second Circuit contained little detail, leaving the "law of contributory trademark infringement ill-defined." *Tiffany*, 600 F.3d at 103, 105. And the Supreme Court in *Inwood* only applied the inducement prong of the test, so one could even regard what *Inwood* said about the continued supply prong to be dicta: indeed, the Second Circuit acknowledged that *Inwood* did not "establish[] the contours of the 'knows or has reason to know' prong." *Id.* at 108. Yet the Court's conclusion turned to some extent on a very careful parsing of the language of the *Inwood* test, which referenced supply to "one" whom the defendant knew was engaging in infringement, and on dicta in *Sony* that ventured a guess as to how that case would have come out under *Inwood* as opposed to the copyright standard. *See id.* at 108–09. Certainly, the court's protestations about judicial circumspection—that "we are interpreting the law and applying it to the facts of this case [and] could not, even if we thought it wise, revise the existing law in order to better serve one party's interests at the expense of the other's"—do seem a little strained given the room for maneuver that appeared to exist. *Id.* at 109.

[113] *See, e.g.*, Fonovisa, Inc. v. Cherry Auction, Inc., 76 F.3d 259, 264 (9th Cir. 1996). Indeed, this approach in offline cases does not seem to have been altered by the *Tiffany* opinion. *See* Coach, Inc. v. Goodfellow, 717 F.3d 498 (6th Cir. 2013).

[114] *See* Dogan, *supra* note 58, at 510 n. 38 (noting that judicial assessment of whether the intermediary is in a position where it can stop infringement and can reasonably be required to do so is "often implicit" and channeled "through doctrinal standards of 'knowledge' and 'substantial assistance'"); *cf.* Barton Beebe, *Tiffany* and *Rosetta Stone*: Intermediary Liability in U.S. Trademark Law, Paper Delivered at UCL Institute of Brand and Innovation Law 4 (Feb. 15, 2012), *available at* http://www.ucl.ac.uk/laws/ibil/docs/2012_beebe_paper.pdf ("[W]ho… doubts that if eBay had done nothing to minimize counterfeit sales, the Second Circuit would have found contributory liability, as federal courts routinely have for flea market operators, whether these operators had general knowledge, actual knowledge, constructive knowledge, or something else? One is tempted to say that the issue was ultimately not so much one of negligence as of who would be the lowest-cost enforcer of the right.").

hints at an approach more sensitive to individual context than the doctrinal holding suggests.[115] And this sensitivity would be valuable, because the legitimacy of the behavior of intermediaries occupies a spectrum that requires greater flexibility (and room for more subtle calibration) than formal secondary liability doctrine might seem to allow.[116] Indeed, such flexibility is necessary also to account for the different demands that might be imposed on smaller entities without the capital or sophistication of eBay.[117]

Openly assessing the reasonableness of intermediary behavior might at some level appear to be a more genuine engagement with a tort-grounded notion of secondary liability. It might also provide greater comfort to mark owners, who see preventative and prospective measures as more useful in fighting infringement than the specific notice and takedown regime that *Tiffany* appears to establish as the almost exclusive device for affording assistance.[118]

The *Tiffany* approach has largely been followed by other U.S. courts. In *Rosetta Stone Ltd. v. Google, Inc.*, a trademark infringement case based on keyword advertising, the Fourth Circuit articulated a reading of the *Inwood* standard regarding knowledge that is very much in line with the Second Circuit's standard in *Tiffany*: "It is not enough to have general knowledge that some percentage of the purchasers of a product or service is using it to engage in infringing activities; rather, the defendant must supply its product or service to 'identified individuals' that it knows or has reason to know are engaging in trademark infringement.[119] To meet this standard, the plaintiff argued that Google had allowed "known infringers and counterfeiters" to bid on the Rosetta Stone marks as keywords after Rosetta Stone had notified Google of two hundred instances of sponsored links that led to

---

[115] *See* Dogan, *supra* note 58, at 516 ("The [*Tiffany*] opinion suggested—without holding—that eBay's generalized knowledge of widespread counterfeiting required it to do *something* to facilitate the detection and removal of counterfeit goods. The problem with Tiffany's argument was that eBay *had* done something; indeed, it had done quite a lot.").

[116] *See generally* Jane C. Ginsburg, *Separating the* Sony *Sheep from the* Grokster *Goats: Reckoning the Future Business Plans of Copyright-Dependent Technology Entrepreneurs*, 50 ARIZ. L. REV. 577 (2008) (noting the need for copyright law to have tools by which to recognize the different culpability of a range of intermediaries even though all operate dual purpose technologies).

[117] *Cf.* Frederick W. Mostert and Martin B. Schwimmer, *Notice and Takedown for Trademarks*, 101 TRADEMARK REP. 248, 264 (2011) (speculating as to some of the misjudgments that might have prompted the lack of responsiveness of the defendant in Louis Vuitton Malletier, S.A. v. Akanoc Solutions, Inc., 658 F.3d 936 (9th Cir. 2011)). In *Louis Vuitton Malletier, S.A. v. Akanoc Solutions, Inc.*, the Ninth Circuit affirmed a jury verdict of over $10 million against a Web hosting business that leased, inter alia, server space to customers trafficking in counterfeit goods. In that case, the defendant (a small company) failed to respond expeditiously to takedown requests, and thus fell afoul of *Tiffany* because it had received actual notice. *See* Louis Vuitton Malletier, S.A. v. Akanoc Solutions, Inc., 658 F.3d 936, 941 (9th Cir. 2011).

[118] *See* Report from the Commission to the European Parliament and the Council on the Functioning of the Memorandum of Understanding on the Sale of Counterfeit Goods via the Internet, at 6, 11, COM (2013) 0209 final 6, 11 (Apr. 18, 2013) [hereinafter Report from the Commission], available at http://eurlex.europa.eu/LexUriServ/LexUriServ.do?uri=COM:2013:0209:FIN:EN:PDF.

[119] Rosetta Stone Ltd. v. Google, Inc., 676 F.3d 144, 163 (4th Cir. 2012).

Web sites offering counterfeits. The Fourth Circuit reversed the grant of summary judgment to Google, but only because the lower court's reliance on the comparative level of knowledge in *Tiffany* failed to account for the different procedural posture of the *Rosetta Stone* case. According to the Fourth Circuit, there was at least a factual question as to whether Google "continued to supply its services to known infringers." Although there is perhaps some ambiguity in the court's opinion,[120] on balance, the Fourth Circuit suggests that generalized knowledge will be of no relevance to the question of secondary liability.

Although the Fourth Circuit in *Rosetta Stone* appeared to mimic the specific knowledge requirement of *Tiffany*, the fact that the Fourth Circuit contemplated possible liability based upon allowing "known infringers and counterfeiters" to bid again on the Rosetta Stone marks suggested a slight loosening of the specificity requirement beyond an item-by-item (or listing-by-listing) takedown system.[121] A subsequent decision of the Tenth Circuit may best reflect this evolution of the *Tiffany* court's approach. In *1-800 Contacts, Inc. v. Lens.com, Inc.*, the plaintiff mark owner brought an action, *inter alia* for contributory infringement, against a competitor whose affiliate had engaged in the alleged acts of primary infringement, namely, the purchase of keywords consisting of the marks of the plaintiff.[122] The plaintiff filed a complaint in federal court against the defendant that included screen shots of an allegedly offending ad, but the plaintiff did not identify which of the defendant's ten thousand affiliates had purchased the ad.[123] It took the defendant 3 months to take corrective action against the affiliate in question.[124] The Tenth Circuit reversed a grant of summary judgment against the plaintiff on its claim of contributory infringement because "a rational juror could find that [the defendant] knew that at least one of its affiliates was [engaging in infringement] yet did not make reasonable efforts to halt the affiliate's practice."[125] According to the court, the defendant had an effective tool to stop infringement (sending out a notice to all affiliates instructing them not to purchase the keyword in question) but did not use it.[126] This duty, according to the

---

[120] *See id.* at 163 ("The district court recognized that Rosetta Stone had come forward with evidence relevant to its contributory infringement claim. The most significant evidence in this regard reflected Google's purported allowance of known infringers and counterfeiters to bid on the Rosetta Stone marks as keywords … .").

[121] *See also* Dogan, *supra* note 58, at 517 (discussing Tiffany (NJ) LLC v. Dong, No. 11 Civ. 2183(GBD)(FM), 2013 WL 4046380 (S.D.N.Y. Aug. 9, 2013)); *cf.* Plaintiffs' Motion for Partial Summary Judgment at 9, Chloé SAS v. Sawabeh Info. Servs. Co., No. CV-11-4147-GAF (MANx), 2012 WL 7679386 (C.D. Cal. Oct. 8, 2013) (copy on file with author at 14 (issuing order against business-to-consumer Web sites found to be contributorily liable requiring them to "monitor their websites on an ongoing basis for compliance and must upon written notice remove or disable access to any listing that plaintiffs identify as infringing or that otherwise comes to defendants' attention as infringing plaintiffs' marks").

[122] *See* 1-800 Contacts, Inc. v. Lens.com, Inc., 722 F.3d 1229, 1234 (10th Cir. 2013).

[123] *Id.* at 1252.

[124] *Id.*

[125] *Id.*

[126] *Id.*

court, arose even through the defendant did not know the specific identity of the alleged infringer. A jury could find that this would have been "reasonable action" to take: the court would not insist on a "rigid line requiring knowledge of [an infringer's specific] identity, so long as the remedy does not interfere with lawful conduct."[127]

However, this evolution remains something short of the "reasonable anticipation" standard that Tiffany had urged on the Second Circuit, or a full-blown lowest cost avoider analysis consistent with a tort based approach to secondary liability. Those alternative approaches might impose on intermediaries a greater obligation to engage in affirmative steps to prevent future infringement (depending upon assessment of costs and benefits). As Stacey Dogan astutely notes, the implicit reasonableness analysis that can be discerned in U.S. case law to date does not go that far: the cases are converging around a principle of "reasonableness in receiving and responding to specific notices of infringement."[128] That is to say:

> The "reasonableness" principle seems—albeit implicitly—to be pursuing a modified "best-cost-avoider" strategy: it places responsibility for infringement detection and elimination with the party best positioned to accomplish each task. Detection falls on the intellectual property owner, who is best suited to recognize unauthorized versions of its work or trademark. Responsibility for terminating the infringement, in turn, rests on the intermediary, assuming that it has control over the means used to infringe. If the structure of a defendant's system and its relationship to infringers gives it the power to stop them without threatening lawful use, then the law requires it to exercise that power.[129]

Dogan uses the phrase "modified 'best-cost-avoider' strategy" to describe this position because "a pure best-cost-avoider approach would take system design into account and require intermediaries to take reasonable steps to head off infringement before it occurs."[130] It might also be called "lowest-cost-terminator," insofar as it retains an essentially backwards-looking posture.

Like the UK courts, the US courts have long endorsed standards that permit liability based on either inducement or contribution. But the formulations in U.S. courts much more clearly treat these as alternatives. For example, the *Inwood* test noted above explicitly offers a claimant two avenues by which to establish contributory secondary trademark liability: (1) intentional inducement; and (2) continued supply with actual or constructive knowledge of infringement.[131] More recently U.S. copyright doctrine has also distinguished between the features of inducement and contributory liability, allowing claims under the latter (but not the former) to be defended by the assertion that the intermediary's product or facility could be used for substantial non-infringing purposes.[132]

---

[127] *Id.* at 1254.

[128] Dogan, *supra* note 58, at 509–510.

[129] *Id.* at 509–510.

[130] *Id.* at 509 n.35.

[131] Inwood Labs., Inc. v. Ives Labs., Inc., 456 U.S. 844, 854 (1982).

[132] *See* MGM Studios, Inc. v. Grokster, Ltd., 545 U.S. 913 (2005) (noting that the doctrine of substantial non-infringing use would not immunize a copyright defendant secondarily liable under intentional inducement standard).

In addition to these participant-based grounds for secondary liability, most systems have permitted liability based upon the closeness of the relationship between the primary actor and in the intermediary. Typically, these forms of liability require a close nexus (often requiring an ability stop the primary infringement), and some financial benefit accruing to the intermediary. For example, in U.S. copyright law, vicarious liability arises when the defendant "has the right and ability to supervise the infringing activity and also has a direct financial interest in such activities."[133] The control element can be challenging; a landlord cannot be liable if it is unaware of infringing activity and has no capacity to stop it.

Vicarious liability has received far less prominence as a separate basis of liability in the literature and the case law. To some extent, this might be supervising given the elements of vicarious claims. Thus, in the United Kingdom, this form of liability most closely approximates to that part of the joint tortfeasorship test that looks to common design. But, although common design overlaps with vicarious liability (and indeed with the procurement variant of joint tortfeasorship),[134] it is not entirely congruent. Claims based upon vicarious liability may succeed when an allegation of common design does not.[135] Likewise, in the United States, although the safe harbours introduced into copyright law by the Digital Millennium Copyright Act ("DMCA") appeared to immunize an intermediary from damages claims for direct or contributory infringement, one of the conditions for securing immunity might arguably have taken vicarious liability out of the scope of immunity.[136] However, these opportunistic features do not appear to have been exploited by those seeking to hold online intermediaries liable.

The practical upshot of this is that in common law countries, claimants seeking to hold online intermediaries liable for the conduct of third parties have often been forced to make more tenuous claims of primary liability. And they have argued for evolutions of the so-called notice and takedown system to secure assistance with day-to-day enforcement that some hoped would be incentivized by more viable secondary liability claims.

---

[133] See Shapiro, Bernstein and Co. v. H. L. Green Co., 316 F.2d 304 (2d Cir. 1963); Fonovisa, Inc. v. Cherry Auction, Inc., 76 F.3d 259 (9th Cir. 1996); see also Hard Rock Cafe Licensing Corp. v. Concession Services, Inc., 955 F.2d 1143 (7th Cir. 1992) (same concept in trademark law).

[134] Twentieth Century Fox v. Newzbin Ltd. (*Newzbin I*), [2010] EWHC 608 (Ch) (Eng.) at [103] (Kitchin J) ("it will seldom if ever be possible to establish joint tortfeasance on the basis of procurement where there is no common design. Procurement is probably a species of common design. Procurement in its various forms—inducement, incitement and persuasion—is sometimes going to be the clearest way to show that there was a common design"); CBS Songs Ltd. v Amstrad Consumer Electronics plc [1988] AC 1013 (HL); cf. Unilever plc v Gillette (UK) Ltd. [1989] RPC 20 (Mustill LJ) (suggesting that there were the two distinct routes to joint tortfeasance).

[135] See, e.g., Vertical Leisure Limited v. Poleplus Limited [2015] EWHC 841 (IPEC) (rejecting claim of common design but finding vicarious liability for cybersquatting variant of passing off based upon vicarious liability as employee or agent).

[136] Digital Millennium Copyright Act, Pub. L. No. 105–304, 112 Stat. 2860 (1998); see also Mark A. Lemley, *Rationalizing Internet Safe Harbors*, 6 J. TELECOMM'S & HIGH TECH. L. 101, 104 n. 23 (2007) (critiquing that view).

The notice and takedown procedures adopted by most intermediaries grew in the United States out of judicial interpretation of the secondary liability standard in copyright and trademark law.[137] Thus, in *Religious Tech. Ctr. v. Netcom On-Line Communication Servs., Inc.*, a federal district court suggested that the failure of an online intermediary (an internet access provider) to investigate a notice of alleged copyright infringement and take down the infringing material could amount to a substantial contribution to user infringement and hence potential contributory infringement liability under the standard noted above.[138] In response, intermediaries implemented systems to receive notices from copyright owners and take down allegedly infringing material, and thus avoid the possible secondary liability to which the *Netcom* court alluded.[139]

In the trademark context, the leading action that consolidated the use of notice and takedown was brought against the eBay auction site by Tiffany many years after *Netcom* (and the arguable codification of *Netcom* in the DMCA).[140] eBay had by then implemented a number of measures to ensure the authenticity of products advertised on its site. For example, eBay: (1) developed software—a so-called "fraud engine"—to ferret out illegal listings, including of counterfeit goods; (2) offered mark owners space on its site (an "About Me" page) to warn users about suspected fakes; and (3) suspended hundreds of thousands of sellers each year whom it suspected of having engaged in infringing conduct.[141] Most importantly, however, eBay maintained and administered the Verified Rights Owner (VeRO) Program, a notice and takedown system that allowed trademark owners to submit a Notice of Claimed Infringement (NOCI) to eBay identifying listings offering infringing items, so that eBay could remove such reported listings. eBay never refused to remove a reported Tiffany listing, and did so very quickly (within 24 hours, and typically within 12 hours).[142] Thus, eBay's conduct was lawful under *Inwood* because it did not continue to supply its services to third-party sellers when Tiffany put it on actual notice of particular listings of counterfeit goods.[143] The notice and takedown system obviated any potential secondary liability.

In countries where the liability of intermediaries has been resolved under the rubric of direct liability—albeit in circumstances treated by this Chapter as in substance involving secondary liability—claimants appear to have had more success. For example, prior to the decision of the Court of Justice of the European Union in *eBay v. L'Oreal*, claims had been brought against eBay (and other auction sites) for

---

[137] *See* Tiffany v. eBay 600 F.3d 93 (2d Cir. 2010).

[138] *See* Religious Tech. Ctr. v. Netcom On-Line Commc'n Servs., Inc., 907 F. Supp. 1361 (N.D. Cal. 1995).

[139] The copyright notice and takedown mechanism has since that decision been further embedded in the operations of online intermediaries in the United States as much through its inclusion as one of the conditions of safe harbor immunities, as discussed below in Part IV.

[140] Tiffany (NJ) Inc. v. eBay Inc., 600 F.3d 93 (2d Cir. 2010).

[141] *See id.* at 98–100.

[142] *Id.* at 99.

[143] *Id.* at 106.

almost identical conduct in a number of European countries. And in some (civil law) jurisdictions where the secondary infringement claims were denominated or even conceptualized differently, eBay had a more mixed record than in the UK and US. For example, in France, the claim against eBay was advanced as a direct viola-tion of the general fault provision in the Civil Code. Under this provision, the French courts could hold intermediaries liable based upon a finding that they had disre-garded a duty to act reasonably.[144] But this too could encourage the development of a notice and takedown system insofar as it represented one way of demonstrating that the intermediary was conducting its business in a reasonable fashion. And inso-far as the intermediaries were transnational actors, the more direct pressure to create such a mechanism in light of the evolving standard in common law countries surely made notice and takedown a convenient way of showing compliance with a duty of care.

The position was not, however, consistent among civilian countries, even with almost identical fact patterns. The German courts appeared less willing than the French or Belgian courts to impose full liability upon auction sites based upon merely facilitating infringement.[145] In this, the German courts appeared to endorse some of the same propositions as the UK or US courts. But from an early stage, German courts adopted a middle ground, and were willing to offer the claimants limited injunctive relief—essentially mandating takedown, but potentially requiring some pro-active filtering for infringements—based upon a German remedial doc-trine discussed below, but resisting full liability claims absent specific and concrete knowledge of the allegedly infringing acts.[146]

Thus, even if a notice and takedown system did not flow inevitably in every country from the conditions positively establishing secondary liability, it did so in a sufficient number of important countries that—when also seen as one means of complying with the slightly different standards in others—it became commercially attractive in minimizing risk and pursuing uniform procedures across borders.[147] This pressure emanating from case law positively establishing standards for the

---

[144] See e.g., Tribunal de commerce [TC][court of trade] Paris, June 30, 2008, No. 200677799, 11–12 (Fr.) (Louis Vuitton Malletier S.A. v. eBay, Inc.); See, e.g., TC Paris, June 30, 2008, No. 200677799, 11–12 (Fr.) (Louis Vuitton Malletier S.A. v. eBay, Inc.) (unpublished); Tribunal de grande instance [TGI] Troyes, civ., June 4, 2008, No. 060264 (Fr.) (Hermés Int'l v. eBay, Inc.) (unpublished).

[145] See Matthias Leistner, Structural Aspects of Secondary (Provider) Liability in Europe, 9 J. INTELL. PROP. L. & PRAC. 75 (2014).

[146] The clarity of the German position as regards liability for trademark infringement is somewhat muddied by a line of case law proceeding under the general clause (Section 3) of the German Unfair Competition law. See Annette Kur, Secondary Liability for Trademark Infringement on the Internet: The Situation in Germany and Throughout the EU, 37 COLUM. J.L. & ARTS 525 (2014). Insofar as such a claim conceptualizes the liability of the intermediary as an autonomous tort, it comes closer to the French approach. See Case Comment, Trademark Law—Infringement Liability—European Court of Justice Holds that Search Engines Do Not Infringe Trademarks: Joined Cases C-236/08, C-237/08 & C-238/08, Google France SARL v. Louis Vuitton Malletier SA, 2010 ECJ EUR-Lex LEXIS 119 (Mar. 23, 2010), 124 HARV. L. REV. 648 (2010).

[147] See Dinwoodie, supra note 1, at 173.

secondary liability of OSPs was buttressed by another development, namely, provisions granting OSPs immunity from liability.

B. Immunity Provisions Precluding Liability

While there has been scant legislative development of a standard for when online intermediaries will be liable for secondary infringement, in many countries statutory safe harbours have been adopted that dictate when they will be *immune* from liability.[148] These provisions have been introduced to provide a climate of certainty in which to develop e-commerce. The immunity typically prevents monetary liability, but in most countries following this model some form of injunctive relief remains a possibility.[149]

Like the standard for secondary liability, provisions conferring immunity on intermediaries may in different countries be either subject-specific or horizontal.[150] In some countries, immunity provisions are restricted to particular types of claims.[151] For example, in New Zealand in the principal areas covered in this Chapter, there are four different states of possible immunity: "The only context in which immunities from secondary liability exist is the copyright law ... No safe harbours have been created in the trademark context, either by legislation or in case law. And ... in the defamation context any safe harbours that might develop are likely to concern primary liability for 'publication' of the defamatory statements. No safe harbours have developed in the privacy law context."[152]

In the United States, there is an even more complex matrix of immunity.[153] Section 230 of the Communications Decency Act provides the strongest and most unconditional form of immunity from liability for providers and users of an "interactive computer service" who publish information provided by others.[154] However,

---

[148] These safe harbours have arisen as a result of legislative provisions, judicial creativity, and some degree of private ordering. *See* National Report of the United Kingdom, at 2 ("[N]ew limitations are emerging. These are both formal (specific statutory defences) and informal (codes of practice, privately-administered takedown procedures and other alternatives to litigation)").

[149] *See, e.g.,* National Report of New Zealand, at 5. New Zealand also appears, despite its safe harbour, to leave open the possibility of secondary liability for a primary copyright infringement that consists of communicating a work to the public. *See id.,* at 5 (concluding that "if a service provider could be secondarily liable for a primary infringer's act of transmitting copyright-protected material to the public—or liable for "authorising" such infringement—the safe harbour provisions would appear not to apply").

[150] The scope of the immunity provision in a particular country will not necessarily map to the scope of the secondary liability standard. Nor logically must it do so.

[151] *See* National Report of Taiwan, at 3, 11 (noting copyright specific statutory immunities mirrored on those available under the DMCA); National Report of New Zealand, at 4 (safe harbour in copyright, paralleling with some differences the DMCA).

[152] National Report of New Zealand, at 4.

[153] *See* Lemley, *supra* note 136 (calling for rationalization of the matrix).

[154] *See* Communications Decency Act of 1996, 47 U.S.C.A. § 230(c)(1).

that provision does not apply to federal intellectual property claims, which represent a large portion of the disputes covered in this Chapter. Instead, the DMCA introduced a detailed set of provisions into Section 512 of the Copyright Act establishing a series of immunities from damages for copyright infringement for intermediaries engaged in certain types of behaviour: providing internet access as conduits, hosting web sites, or offering "information location tools".[155] It also provided a safe harbor for caching, from which many intermediaries benefit. The copyright-specific DMCA has been replicated in a number of countries.[156]

To avail oneself of the different immunities under the DMCA, the statute created two *general* conditions. Thus, OSPs must "adopt and reasonably implement, and inform its subscribers and account holders, of a policy for termination of repeat infringers"; and they must "accommodate and not interfere with standard technical measures used by copyright owners to identify or protect copyrighted works".[157] In Taiwan, the conditions were more precise and required the adoption of policies of terminating services for repeat infringers after three instances of infringement. A similar provision is found in New Zealand. These provisions incentivize private entities developing policies that achieve some of what has been sought to implement through graduated response systems. However, this type of provision attracted controversy in some countries. The legislative provision in question never entered into force in New Zealand and was repealed in 2011.

In addition, each safe harbour (e.g., access versus hosting versus caching) has its own specific requirements that vary slightly. But the crucial hosting immunity in Section 512(c) of the Copyright Act is based upon inter alia the operation of a notice and takedown system by the intermediary. Section 512(c) is an elaborate provision that (like each safe harbor in the statute) imposes a series of specific conditions—in addition to the general conditions noted above—that are quite detailed. Thus, "the (hosting) provider must not have actual knowledge that the material or an activity using the material is infringing, is not aware of facts or circumstances from which infringing activity is apparent, and upon obtaining such knowledge or awareness, it acts expeditiously to remove or disable access to the material."[158] In addition, where the provider has the right and ability to control such activity, it must not receive a financial benefit directly attributable to the infringing activity.[159]

---

[155] 17 U.S.C. § 512 (2000). The Second Circuit has read the DMCA safe harbor expansively to limit claims not only under the federal copyright statute but also under state copyright law, which is the means by which sound recordings were protected in the United States prior to 1972. *See* Capitol Records, LLC et al. v. Vimeo LLC, 826 F.3d 78 (2d Cir. 2016).

[156] *See, e.g.*, National Report of Taiwan, at 11–17 (discussing 2009 amendments to the Taiwanese Copyright Act).

[157] *See* 17 U.S.C. § 512 (i)–(j). For recent applications, see BMG Rights Management v. Cox Communications 199 F. Supp.3d 158 (E.D. Va. 2016); EMI Christian Music, Inc. v. MP3tunes, LLC, 844 F.3d 79 (2d Cir. 2016), 2016 WL 7235371.

[158] In addition, the provider must have publicly designated an agent to receive notifications of claimed infringement. The Copyright Office has recently revised this provision.

[159] *See* 17 U.S.C. § 512 (c).

Most importantly, where an OSP receives a takedown notice alleging infringement that complies with the detailed provisions of the statute, it must respond expeditiously to remove or disable access to the material alleged to be infringing. And the statute then orchestrates in precise terms the actions that the OSP must take to preserve its immunity, including how it must respond to any counter notice that it receives from any person in response to its good faith disabling of access to, or removal of, material.[160] The counter-notice procedure of the DMCA is intended to preserve some balance between the rights of the customers of the OSP, who might have valid grounds for believing that their conduct is not infringing, and those of the copyright owner (with whose notice the OSP is likely to comply in order to maintain immunity). These provisions take the form of conditions which an OSP must satisfy to protect itself from secondary liability for hosting infringing content or direct liability to a customer for removing what turns out to be lawful material. But the benefits of compliance are so strong that the conditions in the statute function as a form of business regulation. These immunity provisions have as result proved of immense importance in framing the effective standard of secondary liability applicable to OSPs.

Trademark claims are encompassed within neither US safe harbour regime[161]; liability of intermediaries for trade mark infringement thus is determined by the positive standard articulated in *Inwood* and applied in *eBay v. Tiffany*.[162] But as noted above, that has ultimately has led to the development of similar notice and takedown practices, though without a detailed statutory footing.[163] And in any event,

---

[160] See 17 U.S.C. § 512 (g).

[161] Some have argued for a DMCA-like system of notice and takedown, tied to immunity, to be extended to trademarks. *See* Susan D. Rector, *An Idea Whose Time Has Come: Use of Takedown Notices for Trademark Infringement*, 106 Trademark Rep. 699 (2016); Mostert and Schwimmer, *supra* note 117, at 265. The OSP community is split. *See* Comments of Etsy, Foursquare, Kickstarter, Meetup, Shapeways, In the Matter of Joint Strategic Plan on Intellectual Property Enforcement (Oct. 16, 2015) (asking for consideration of trade mark safe harbour).

[162] In the United States, the Lanham Act does contain provisions that immunize certain intermediary conduct from liability. *See, e.g.,* 15 U.S.C. §1114(2)(B)-(C) (protecting innocent publisher or distributor of newspaper, magazine or electronic communication from liability for trademark infringement based upon use in paid advertising matter against liability for damages); *see also* 15 U.S.C. §1125(c)(3) (defense against dilution liability based upon facilitating fair use, although it the use facilitated was fair it is hard to see much need for this provision). Some scholars have seen Section 32(2) as possessing the potential for operating as a general trade mark immunity. *See, e.g.,* Lemley, *supra* note 136, at 105–106. Indeed, Mark Lemley has argued that that immunity provision should provide the model for a uniform safe harbor for intermediaries across all causes of action, in preference to that found in the Communications Decency Act or the DMCA. *See id.* at 115–118. However, the courts have rarely found reason to apply Section 32(2). *Cf.* Hendrickson v. eBay, Inc., 165 F. Supp. 2d 1082 (C.D. Cal. 2001).

[163] Domain name registrars also benefit from provisions drafted in ways that confer limited immunity from liability for trade mark infringement under U.S. law. *See* 15 U.S.C. § 1114(2)(D)(iii) (West 2014) ("A domain name registrar, a domain name registry, or other domain name registration authority shall not be liable for damages under this section for the registration or maintenance of a domain name for another absent a showing of bad faith intent to profit from such registration or maintenance of the domain name."); *see also* Lockheed Martin Corp. v. Network Solutions, Inc., 141 F. Supp. 2d 648 (N.D. Tex. 2001).

the overlap of copyright and trademark claims—some real and some contrived—is such that the detailed copyright system would inevitably inform the development of the trademark system, just as the more explicit common law standards inevitably informed the way that transnational actors complied with their duty to act reasonably in civil law jurisdictions.[164]

In the EU, reliance on the immunity provisions as the effective source of a standard of secondary liability is even more true, perhaps because there has been no harmonisation of national secondary liability standards, leaving a gap for immunity provisions to shape the space. The E-Commerce Directive has filled that gap, creating immunity for service providers: (1) who are mere neutral and transient conduits for tortious or unlawful material authored and initiated by others (Article 12); (2) for caching local copies of third parties' tortious or unlawful data (Article 13)[165]; and (3) who store third parties' tortious or unlawful material while having neither actual knowledge of the "unlawful activity or information" nor an awareness of facts or circumstances from which that "would have been apparent (Article 14).[166]

The types of activity protected by the safe harbours varies slightly among jurisdiction. For example, in New Zealand there are storage and caching safe harbours, along with a provision (which arguably serves as immunity for a mere conduit) affirming that service provider liability cannot be based on mere use of its facilities by a primary infringer. The EU legislation picks up three of the four safe harbours found in the DMCA.

Unlike the U.S. safe harbors, however, those required by the E-Commerce Directive apply horizontally to tort claims under national law as well as trade mark or unfair competition law claims. Indeed, they apply horizontally across all forms of civil and criminal wrongdoing, subject to a number of exclusions set out in Article 3.[167] Thus, although approaches to secondary liability are not harmonized, the conditions under which online intermediaries will be *immune* from liability under any law *are* harmonized at the European level. Transcending the terminological and conceptual differences among secondary liability would have been politically

---

[164] In some countries where the DMCA has been copied, the suggestion has been advanced that the system should be extended to trade mark law. *See* National Report of Taiwan, at 21–22.

[165] To qualify for protection, caching must be 'automatic, intermediate and temporary', and for the sole purpose of making onward transmission more efficient. Service providers will lose protection if they do not act expeditiously to remove cached information upon obtaining actual knowledge that the original copy has been removed or its removal ordered by a competent authority. *See* E-Commerce Directive, art. 13.

[166] *Id.* arts. 12–14; *see also* Case C-484/14, Tobias McFadden v Sony Music, EU:C:2016:689 (CJEU 2016).

[167] The safe harbours limit monetary liability whether classified as primary or secondary. This is important. While in common law systems, they will operate most often in limiting secondary liability, *see* National Report of the United Kingdom, at 18 ("[These safe harbours are] not restricted to secondary liability, though in practice the nature of the conditions mean that primary wrongdoers will rarely qualify (since they would tend to have authority or control over the author of data, or have knowledge of the relevant unlawful activity)"), many countries treat these fact patterns as a matter of primary liability. *See supra* text accompanying notes 46–48. And even common law countries might treat liability under particular regimes (e.g., defamation) as a matter of primary liability.

fraught; enacting anew a common set of online immunities was a more achievable task. By adopting a forward-looking orientation—rather than an historical debate about the fundamentals of tort liability—the European Union was able to conclude a directive on the subject, even if variation remains in terms of its national implementation.[168]

Article 14 of the E-Commerce Directive, which provides a hosting safe harbor not unlike that found in Section 512(c) of the U.S. Copyright Act, has received the greatest attention in litigation and is similar to immunity provisions in many jurisdictions.[169] Article 14 provides that:

1. Where an information society service is provided that consists of the storage of information provided by a recipient of the service, Member States shall ensure that the service provider is not liable for the information stored at the request of a recipient of the service, on condition that:

    (a) the provider does not have actual knowledge of illegal activity or information and, as regards claims for damages, is not aware of facts or circumstances from which the illegal activity or information is apparent; or
    (b) the provider, upon obtaining such knowledge or awareness, acts expeditiously to remove or to disable access to the information.

2. Paragraph 1 shall not apply when the recipient of the service is acting under the authority or the control of the provider.[170]

The Court of Justice has interpreted this provision in two leading trademark cases, *Google France* and *eBay v. L'Oreal*.[171] The Court held that whether the intermediaries in question—Google or eBay—came within the safe harbor would depend upon two considerations.

---

[168] *Contra* National Report of the United Kingdom, at 12 (suggesting that uncertainty as to the scope of safe harbours "in large part [derives] from the fact that they define negatively what conduct is incapable of giving rise to liability, rather than stating positively what a service provider must do to avoid liability. This unsatisfactory position has hindered efforts at harmonisation."). Given the apparent historical contingency as to why courts have shaped service provider liability through an affirmative standard of secondary liability, the development of a zone of immunity, or the evolution of elements of primary liability to cater for intermediary concerns, it should be no surprise that (despite the professed generality of certain principles), once can sometimes find reference to different considerations in different contexts. *See id.* at 19 ("[M]any of the safe harbours mirror devices in substantive liability doctrines which serve to immunise passive and neutral platforms from prima facie liability for third parties' activities (though these depend on different considerations depending on the type of wrongdoing involved). Doctrines of innocent dissemination in defamation and authorisation liability in copyright provide the clearest examples of this tendency in English law").

[169] *See id.* at 15 ("Of the three safe harbours, the storage safe harbour is most frequently relied upon by service providers, most frequently in the context of defamation claims.").

[170] E-Commerce Directive, art. 14.

[171] *See* Joined Cases C-236/08–C-238/08, Google France SARL v. Louis Vuitton Malletier SA, 2010 E.C.R. I-2417, ¶¶ 106–20; Case C-324/09, L'Oréal SA v. eBay Int'l AG, 2011 E.C.R. I-6011, ¶ 113.

First, as a threshold matter, the intermediary could not have been "active" in the allegedly illegal activity; the safe harbor protects conduct of a "mere technical, automatic and passive nature."[172] To take advantage of the safe harbor, an intermediary must be a "neutral" actor.[173] Thus, Google's liability would depend upon the role it played in the selection of keywords.[174] Google's Keyword Suggestion Tool might render its activities "non-neutral" and make it vulnerable to loss of immunity under Article 14. Likewise, in *eBay* L'Oréal argued that Article 14(1) could not apply because the activities of eBay went "far beyond the mere passive storage of information provided by third parties."[175] On the contrary, L'Oréal alleged, eBay "actively organized and participated in the processing and use of the information to effect the advertising, offering for sale, exposing for sale and sale of goods (including infringing goods)."[176] The Court of Justice found first that eBay was potentially entitled to the benefit of Article 14 as an information society service provider. Whether eBay was such a provider within the protection afforded by the safe harbor would as a threshold matter depend upon how active it was in the allegedly illegal activity.[177] If it had been involved in "optimising the presentation of the offers for sale in question or promoting those offers, it [would] be considered not to have taken a neutral position … [and thus unable] to rely, in the case of those [offers]," on the Article 14 exemption.[178] Determination of that question was, however, left to national courts.[179]

Second, by the terms of Article 14, even if an intermediary was insufficiently inactive to qualify as an information services provider it would lose immunity if it was put on knowledge of a wrong and did not act expeditiously. Thus, for example, if eBay "was aware of facts or circumstances on the basis of which a diligent

---

[172] Case C-324/09, L'Oréal SA v. eBay Int'l AG, 2011 E.C.R. I-6011, ¶ 113; *see id.* ¶ 115 ("[T]he mere fact that the operator of an online marketplace stores offers for sale on its server, sets the terms of its service, is remunerated for that service and provides general information to its customers cannot have the effect of denying it the exemptions from liability provided for by [the E-Commerce Directive].")

[173] The Advocate-General in *eBay* had questioned the application of the neutrality condition to immunity under Article 14; the concept is referenced in a recital addressing another provision. *See eBay*, ¶¶ AG 139–46. But the Court of Justice adopted the requirement in both *Google France* and *eBay*. *See Google France*, at ¶¶ 113–16.

[174] *See* L'Oréal ¶ 119.

[175] *L'Oréal*, [2009] EWHC 1094, [437].

[176] *Id.*

[177] Case C-324/09, L'Oréal SA v. eBay Int'l AG, 2011 E.C.R. I-6011, ¶ 113; *see id.* ¶ 115. The fact that certain services are automated should not always mean that the provider of those services is of itself not "active." Algorithmic development is surely relevant behavior in assessing degrees of activity. *Cf.* Lenz v. Universal Music Corp., 815 F. 3d 1145 (9th Cir. 2016) (amending opinion to reflect algorithmic review.) *But see* Kur, *supra* note 146, at 532 (noting the majority position in German case law that has treated suggestions proffered by operation of algorithms as within the protection of Article 14).

[178] *L'Oréal*, 2011 E.C.R. I-6011, ¶ 116.

[179] *Id.* ¶ 117.

economic operator should have realised that the offers for sale in question were unlawful and, in the event of it being so aware, failed to act expeditiously."[180] Google could invoke Article 14 only if it disabled ads upon receiving notice.[181] As Jaani Riordan astutely notes, the Court of Justice's reading of the second condition imposes "a hybrid standard that assesses what the defendant actually knew according to the standards of a reasonable service provider in the defendant's position."[182]

Although the standards for liability here are derived from the *immunity* provision, one focus again has been whether the knowledge or awareness that triggers a responsibility on intermediaries to act could be general in nature. In *L'Oréal*, the Court of Justice indicated that the knowledge or awareness requiring removal of a listing could arise either from eBay's own investigation or notification from a third party (most likely the mark owner).[183] But it did not tackle the central question about the nature of the requisite knowledge. Arnold J has read the requirement to be one of "actual" knowledge, but that does not fully dispose of the question.[184]

The Court of Justice has not finally decided on the availability of the EU safe harbor for either search engines or auction sites (or other intermediaries), leaving that to be fought out among the lower national courts. In decisions handed down on similar facts to *eBay v. L'Oreal*, the French Supreme Court held that eBay could not avail itself of the protection under Article 14 because it had played too active a role by assisting sellers in the promotion and fostering of sales.[185] In contrast, the Madrid Court of Appeals has held that Google (as the owner of YouTube) was acting as a

---

[180] *Id.* ¶ 124 (quoting Article 14 of the E-Commerce Directive).

[181] *See id.* ¶ 120.

[182] National Report of the United Kingdom, at 15.

[183] *See* L'Oréal ¶ 122.

[184] The knowledge condition that might take a service provider out of the hosting safe harbor varies slightly from country to country. For example, the New Zealand safe harbor is dependent upon the service provider no knowing or have reason to know that the material infringes copyright in the work. But it is clear that receipt of notice from a copyright owner does not of itself constitute knowledge. *See* National Report of New Zealand, at 6–7. That would of course put the service provider on notice of a claim of infringement. (And as New Zealand decided not to penalise the service of false or misleading notices, *see id.* at 7, unlike for example section 512(f) of the US Copyright Act, it is possible that incorrect notices might be served). But that is not the statutory knowledge standard, the establishment of which might require a court to look at the quality of the notice and the information it provides. For example, does it contain sufficient information to enable a service provider to determine whether the allegedly infringing conduct is justified by a defense? See *id.* at 7.

[185] *See* Cass. com., May 3, 2012, Bull. civ. IV, No. 89 (Fr.) (eBay Inc. v. LVMH, Parfums Christian Dior); *see also* Beatrice Martinet Farano, French Supreme Court Denies eBay Hosting Protection, Transatlantic Antitrust & IPR Developments, Issue No. 3/2012 (June 21, 2012), available at http://ttlfnews.wordpress.com/2012/07/02/transatlantic-antitrust-and-ipr-developments-issue-no-32012-june-21-2012/ (discussing the French Supreme Court's decision, in which the court specifically relied on the "active role" standard from the Court of Justice's L'Oréal and Google France decisions).

host under Article 14 in the context of a copyright infringement case.[186] The Italian courts have reached similar outcomes regarding Youtube.[187] Variations clearly remain at the national level within Europe.[188]

The safe harbor in the E-Commerce Directive has thus strongly encouraged the development of a notice and takedown procedure in Europe, not unlike that which has arisen in some countries by virtue of the affirmative secondary liability standard. However, there is an important difference to note when the practice evolves from the immunity standard rather than the liability standard. Under EU law, failing to comply with notice and takedown (or any other element of the safe harbor) doesn't mean liability; it just means no automatic immunity.[189] In Europe, if an intermediary falls outside safe harbor because it did not take down on notice, liability will only arise if the standard for secondary liability is also met under the applicable national law.[190] As noted above, these standards vary. But in the United Kingdom, it is entirely possible that one could fail to come within the safe harbor but still not be held secondarily liable as a joint tortfeasor. This is true also of the copyright safe harbours in the United States. However, under *Tiffany* in the United States, an auction site that continues to supply its services after a notice will trigger trademark liability.

Case law in some countries suggests a closer convergence.[191] And certainly the conduct required of an intermediary to avail itself of (some of) the safe harbours is to some extent that which will assist avoiding liability under the affirmative test for secondary liability. The UK National Report highlights this point (both as regards

---

[186] *See* Audiencia Provincial Sentencia [A.P.S.] [Provincial Appellate Court Sentence] Madrid, Jan. 14, 2014 (No. 11/2014) (Spain) (Gestevision Telecinco, S.A. v. YouTube, LLC), *translated in* Decision No. 11/2014 on YouTube v Telecinco, Hogan Lovells Global Media & Comm. Watch (Feb. 14, 2014), http://www.hlmediacomms.com/files/2014/02/Telecinco-v-Youtube_EN.pdf.

[187] *See* Martini Manna & Elena Martini, The Court of Turin on the Liability of Internet Service Providers (June 10, 2014), http://www.lexology.com/library/detail.aspx?g=ec9e1298-7367-4f0a-a263-0cbaff076e29 (reporting on similar outcomes in Italy on similar grounds in Delta TV Programs s.r.l. v. Google Inc., Tribunale Ordinare [Trib.] [Ordinary Court of First Instance] Turin, 23 giugno 2014, Docket No. 38113/2013 (It.), available at http://www.ipinitalia.com/Ordinanza%20Youtube%20Torino.pdf).

[188] *See* Kur, *supra* note 146, at 531–32 (noting different approaches in Germany and France).

[189] This would also have been the approach adopted in New Zealand had the recommendations of the New Zealand Law Commission been accepted. In 1999, the Commission proposed what looked like a horizontal standard for secondary liability of online service providers. *See* New Zealand Law Commission, *Electronic Commerce: Part Two* 137, ¶ 333 (Wellington, New Zealand 1999). But strictly what it proposed would have created immunity where the provider had no knowledge of the existence of the unlawful information or acted to remove such information. That is to say, even if knowledge were established, the claimant would still have to show the other elements of an applicable cause of action. *See* National Report of New Zealand, at 1–2 n.3.

[190] *See, e.g.*, L'Oréal S.A. v. eBay Int'l AG, [2009] EWHC 1094, [344] (Ch) (Eng.) (noting that the question of secondary liability is a matter for national law).

[191] *See, e.g.*, National Report of the Czech Republic, at 10.

the E-Commerce Directive and more generally).[192] And the historical development of the safe harbours in the DMCA (now Section 512 of the US Copyright Act) also suggests this affinity. Those standards largely codified (in greater detail) an approach that had been developed by the district court in the *Netcom* case, where the judge had sought to apply the contributory infringement standard under US copyright law to a service provider.

The conceptual relationship between liability standards and safe harbors can plausibly follow a number of forms. For example, where statutory safe harbors have been enacted, they could represent the line between the area of immunity conferred by defenses and the zone of liability. Alternatively, safe harbors may, like many defenses, immunize conduct that would otherwise be infringing because the benefits that flow from the online services provided by the intermediary outweigh the costs of the infringement that they facilitate. Alternatively, the safe harbor may simply create reassurance about activities that may or may not have been infringing.[193] The benefit of such a scheme is that, despite its apparent redundancy, the safe harbor provides certainty, which is important because even valuable non-infringing activity can be chilled by the threat of liability. Conduct that falls outside the safe harbor, but does not clearly attract secondary liability, remains conduct for which an intermediary might expect to be challenged by claimants trying to test the margins of secondary liability. Finally, such a scheme may also, by virtue of the conditions on which it offers immunity, nudge service providers towards modes of conduct that we deem desirable.

IV. Notice and Takedown

The notice and takedown mechanism has been mentioned several times above. Clearly, it is becoming the norm by which intermediaries in a number of countries are enlisted in the task of separating lawful from unlawful conduct.[194] The nature of

---

[192] *See* National Report of the United Kingdom, at 13–14 ("In essence, [the mere conduit] safe harbour confirms that mere assistance by passively transmitting will not be sufficient for liability, which largely mirrors the position of English law in relation to secondary liability for defamation, copyright and trade mark infringement" (citing Bunt v Tilley [2006] EWHC 407 (QB)); *id.* at 14 ("[L]ike the mere conduit safe harbour, the service provider [seeking to benefit from the storage safe harbour] must not have had authority or control over the person who provided the data. This excludes cases of common design, procurement, and vicarious liability."); *id.* at 19 ("[M]any of the safe harbours mirror devices in substantive liability doctrines which serve to immunise passive and neutral platforms from *prima facie* liability for third parties' activities (though these depend on different considerations depending on the type of wrongdoing involved). Doctrines of innocent dissemination in defamation and authorisation liability in copyright provide the clearest examples of this tendency in English law").

[193] *See id.* at 15 ("Safe harbours are often invoked by service providers to achieve summary disposal of claims against them, particularly in the context of defamation").

[194] *See* National Report of Taiwan, at 13 (in copyright law).

the notice and takedown system thus may serve in practice to define the liability of intermediaries for the acts of third parties.

As noted above, some notice and takedown systems have grown out of the judicially-developed standards for secondary liability of intermediaries; cases such as *Tiffany* encouraged the development of such as system in the United States as a means of avoiding liability for trademark infringement. Other notice and takedown systems have been introduced statutorily as a condition of an intermediary availing itself of immunity under safe harbour. These statutory regimes may be skeletal (as in the EU E-Commerce Directive) or they may be quite detailed (as in the US DMCA). Finally, some of these systems are growing up organically, either as a policy adopted by the relevant intermediary (e.g., the policies adopted unilaterally by search engines as to when they will disable ads triggered by or containing the marks of others)[195] or under an agreement negotiated by intermediaries and interested e-commerce players.[196]

In this last category, the most prominent example is a 2011 memorandum of understanding between thirty brand owners and Internet platforms in Europe regarding their respective roles in tackling counterfeiting online (the "EU MOU"). These are also MOUs (including some applying outside Europe) between specific traders and platforms, such as that between Louis Vuitton and Taobao.com (a Chinese e-commerce site owned by Alibaba that is similar to Amazon or eBay).[197] And in July 2013, the U.S. Office of the Intellectual Property Enforcement Coordinator announced that a number of companies were adopting Best Practices Guidelines designed to reduce the flow of ad revenue to infringing Web sites.[198]

---

[195] *See* e.g., AdWords Trademark Policy, Google, https://support.google.com/adwordspolicy/answer/6118?hl=en. The fact that a service provider adopts a corporate or industry policy (for example, of notice and takedown) should in theory not be dispositive of secondary liability. *Cf.* Interflora, Inc. v. Marks & Spencer Plc & Anor (C-323/09), [2012] E.T.M.R. 1 (CJEU 2011) (questions referred). But according to the UK National Report, "[a]lthough not decisive on their own, they contribute to the courts' assessment of service providers' passivity and neutrality, whether those providers authorise copyright infringements by their users, whether they engage in communications to the public of the relevant copyright works, and whether they intervene in a way that creates tortious secondary liability." National Report of the United Kingdom, at 27.

[196] Memorandum of Understanding on the Sale of Counterfeit Goods via the Internet, May 4, 2011 [hereinafter Memorandum of Understanding], *available at* http://ec.europa.eu/internal_market/iprenforcement/docs/memorandum_04052011_en.pdf.

[197] *See, e.g.*, Meng Jing, Taobao Steps up Fight Against Fake LV, China Daily USA (Nov. 11, 2013, 4:13 PM), http://usa.chinadaily.com.cn/business/2013-10/11/content_17025360.htm.

[198] *See* OFFICE OF THE U.S. INTELLECTUAL PROP. ENFORCEMENT COORDINATOR, INTELLECTUAL PROPERTY SPOTLIGHT 2 (July/August 2013), *available at* http://www.whitehouse.gov/sites/default/files/omb/IPEC/spotlight/ipec_spotlight_july_aug_2013.pdf. The signatories to the MOU have been discussing its possible extension to payment service providers. *See* European Comm'n, *Memorandum of Understanding on the Sale of Counterfeit Goods via the Internet–Autumn Meeting–13 November 2013–Summary*, at 4, E.C. Doc. MARKT D3/FT/cm-Ares(2013)3821568 [hereinafter November 2013 Meeting Report], *available at* http://ec.europa.eu/internal_market/iprenforcement/docs/summary-of-the-autumn-meeting-20131113_en.pdf.

The EU MOU is limited to the provision of services in the European Economic Area,[199] and addresses only counterfeit goods, rather than disputes about parallel imports or selective distribution systems.[200] Moreover, the agreement does not address all intermediaries, only "providers whose service is used by third parties to initiate online the trading of physical goods."[201] Notice and takedown procedures are central to the MOU, which is not surprising: the platforms all already operate them.[202] The parties commit to continue operating such systems, but also agree to details that differ from the type of system that arguably now flows from hard law secondary liability rules. In particular, in addition to item-based notice and take-down, the agreement allows trademark owners to notify the platforms of sellers who are generally engaged in the sale of counterfeits, and the platforms will "take this information into consideration as part of their pro-active and preventive measures."[203] This is clearly an effort to move away from the specificity of notice that some case law may require to confer knowledge sufficient to establish secondary liability.[204]

The notice and takedown system has proved an efficient and largely accepted way of reconciling the obligation of intermediaries to assist in preventing their services being used to perpetrate unlawful conduct with a concern about over-regulation that would stifle e-commerce. Yet, the mechanism does warrant close consideration. The system established by the DMCA in the United States is thought to have resulted in over-compliance and thus the prohibition of lawful conduct[205]; the speed and rate of compliance by eBay reported in the *Tiffany* case raises similar concerns in non-statutory systems.

Liberal enforcement regimes can cause undue restraint of legal activities. The development of secondary liability rules has historically occurred with an eye to such concerns. In the online context, these concerns are aimed at (1) unwarranted demands by claimants to take down lawful material (2) the over-compliance by service providers with claimant demands, and (3) the imposition of remedial measures that not only restrain illegal activity but necessarily impede lawful activity or interfere with the rights of persons other than the wrongdoers. Tackling these concerns requires attention not only to the substantive rules of liability, but also to embedded incentives for particular behavior (often caused by litigation costs rules or assymetric commercial interests) and the tailoring of remedies.

---

[199] Memorandum of Understanding, *supra* note 196, at 1.

[200] *Id.* at 1.

[201] *Id.* ¶ 1(1).

[202] *Id.* ¶¶ 11–12; *see generally* Mostert & Schwimmer, *supra* note 117.

[203] Memorandum of Understanding, *supra* note 196, at ¶ 12.

[204] *See supra* text accompanying notes 103–113.

[205] For an early assessment, see Jennifer M. Urban & Laura Quilter, *Efficient Process or "Chilling Effects"? Takedown Notices Under Section 512 of the Digital Millennium Copyright Act*, 22 SANTA CLARA COMPUTER & HIGH TECH. L.J. 621 (2006); *see also* JENNIFER M. URBAN, JOE KARAGANIS AND BRIANNA L. SCHOFIELD, NOTICE AND TAKEDOWN IN EVERYDAY PRACTICE (2016).

As to unsubstantiated claims of wrongdoing, an aggrieved party may have a number of claims depending upon the particular wrongdoing that is alleged.[206] Most countries have a general tort of abuse of process that may in certain circumstances offer relief.[207] But the uncertain nature of these mechanisms and the (lack of) incentive to pursue them in light of litigation costs means that in most countries a retaliatory action for an unwarranted assertion made to a service provider is unlikely in practice to prevail.[208]

In the United States, the DMCA requires that any notice of claimed infringement was served in good faith.[209] Moreover, Section 512(f) of the DMCA creates a cause of action for a service provider or an end-user (for actual damages, as well as costs and attorney's fees) against "[a]ny person who knowingly materially misrepresents under this section (1) that material or activity is infringing, or (2) that material or activity was removed or disabled by mistake or misidentification".[210] The DMCA also contains put-back and counter-notification provisions[211] that allow the primary actor easily to dispute the unlawfulness of its conduct and the service provider thus to maintain the hosting of the copyright material in question.[212] (Some variants on this, such as in New Zealand, require the service provider to notify the user that material has been deleted or access to it prevented, but does not statutorily create a counter-notice system that would require the service provider to put back the material.[213] The notice requirement arguably might give rise to informal counter-notices, but immunity for service provider putting back might make such a development more likely.[214]) The burden is then on the right holder to initiate proceedings and establish infringement. Although this mechanism is surely under-used, it does rebalance the incentives to contest disputed conduct in ways that can minimize abu-

---

[206] See National Report of the United Kingdom, at 30 (noting inter alia actions for unjustified threats in respect of several intellectual property rights).

[207] See id. at 30 (discussing Media CAT Ltd. v Adams [No 2] [2011] FSR 297; Golden Eye (International) Ltd. v Telefónica UK Ltd. [2012] EWHC 723 (Ch), rev'd in part [2012] EWCA Civ 1740).

[208] See National Report of the United Kingdom, at 30 ("it seems unlikely that retaliatory action would be practicable in response to a groundless takedown request in all but the most exceptional circumstances").

[209] See 17 U.S.C. § 512(c)(3).

[210] See Rossi v. Motion Picture Ass'n of America, 391 F.3d 1000, 1004 (9th Cir. 2004); Lenz v. Universal Music Corp., 572 F.Supp.2d 1150, 1154–55 (N.D. Cal. 2008).

[211] See 17 U.S.C. § 512(g)(1).

[212] This mechanism also exists in countries that have copied the detail of the DMCA. See, e.g., National Report of Taiwan, at 15.

[213] See National Report of New Zealand, at 6.

[214] This approach in New Zealand is, however, consistent with its decision also not to immunise from other causes of action service providers who act in good faith on a notice and take down material. See id. at 6; compare 17 USC § 512. This omission in theory might make more balanced the strategic choices faced by service providers receiving notice, but the immunity from exposure to copyright liability coupled with the lack of any clear cause of action through which to pursue wrongful takedowns means that this difference is likely to be theoretical only.

sive claims being acted on. It is thus regrettable that not all statutory notice and takedown systems incorporate this feature; the E-Commerce Directive, for example, has no such provision. Of course, European service providers could adopt their own counter-notice system as a matter of policy, but it is possible that absent statutory authorisation, they might take themselves outside the protection of the Directive's safe harbours.

National courts might also devise mechanisms to monitor for over-reaching in order to vindicate the fundamental rights of the primary defendants (as the English courts have done in issuing cooperation orders). But the latitude for authorizing service providers to do so more generally may depend upon creatively reading the strict language of the E-Commerce Directive (perhaps through the lens of competing fundamental rights). Developments in the English courts in the context of defamation might offer a generalisable model. Thus, one court has held that a service provider could resist a takedown notice alleging defamatory material without placing itself outside the hosting safe harbour status if it lacks information that would allow it to assess the strength of available defences.[215] Moreover, even privately developed notice and take down systems can incorporate these types of balancing mechanisms. Thus, the EU MOU also tackles abuse of the notice and takedown system, an issue addressed in some statutory notice and takedown systems.[216] Under the MOU, if a trademark owner makes notifications to an intermediary without exercising appropriate care, the owner may be denied future access to the system and must pay the platform any fees lost due to such notifications.[217] And "sellers should be informed where an offer has been taken down, including the underlying reason, and be provided with the means to respond including the notifying party's contact details."[218] These provisions are designed to ensure some balance in the system.

Despite the complaints tendered about Section 512 of the US Copyright Act, a statutory regime of that detail has some advantages. It provides certainty about the form of notices and expected responses, and that certainty allows online actors to structure behavior accordingly[219]; in this regard, it might offer advantages over the skeletal nature of the EU legislation.[220] Thus, under the E-Commerce Directive, any

---

[215] *See* National Report of the United Kingdom, at 31–2.

[216] *Cf.* 17 U.S.C. § 512(f) (West 2014) (imposing liability on "[a]ny person who knowingly materially represents … that material or activity is infringing" or "that material or activity was removed or disabled by mistake or misidentification").

[217] *See* Memorandum of Understanding, *supra* note 196, at ¶ 16.

[218] *Id.* ¶ 19.

[219] *Cf.* Mostert & Schwimmer, *supra* note 117, at 265 (noting that at present intermediaries would be advised to respond even to notices that were "written in crayon"); *see also* National Report of the United Kingdom, at 15 (discussing adequacy of the notice given in *Tamiz v. Google, Inc.* [2012] EWHC 449).

[220] Annette Kur has observed that "the German Federal Supreme Court has given certain guidelines as to when a notification is to be considered as serious and plausible enough to give rise to removal and prevention claims, but those guidelines inevitably remain fairly vague and general." Kur, *supra* note 146, at 534–35; *see also* BGH Aug. 17, 2011, 191 BGHZ 19, (Ger.) (*Perfume Stick*), *translated in* 44 INT'L REV. INTELL. PROP. & COMPETITION L. 123, 128, ¶ 28 (2013) ("This requires that

communications sent by the claimant to the service provider in order to trigger the
second condition of Article 14 must be sufficiently specific to give notice of unlawful conduct. But what does that mean? In another context, in *Tamiz v Google Inc.*,
the trial judge held that "notices alleging defamation—without giving any details of
the statements' falsity or the inapplicability of obvious defences—were neither sufficiently precise nor substantiated that a diligent host should have identified the
statements' unlawfulness."[221] In this regard, although mundane, statutory clarity is
helpful. Where such regulatory specificity is lacking, OSPs are likely (and well
advised) to treat even vague notices as requiring some reaction. The detailed delin-
eation of the US copyright system also allowed Congress to incorporate mechanisms, such as counter-notification and actions for bad faith notice, which would
address to some extent the possible chilling effect of over-enforcement that notice
and takedown might create.

Of course, this can be done in regimes other than the detailed statutory one found
in the DMCA. Thus, in the orders granted under Section 97A of the CDPA and in
Cartier, Arnold J included provision for affected third parties to intervene before the
court to seek to vary his orders. Indeed, his orders went so far as to detail the notices
that must appear when users sought to access a blocked web site, so as to alert them
to the reasons for that result and to enable such interventions. These intervention
options might not be as easy or cheap for users to exercise as the counter-notification
mechanism introduced by the US statute. The allocation of burdens in the DMCA—a
quid pro quo for the fact that the system established will in all likelihood routinely
implement the wishes of the notifying party in almost all cases—arguably makes for
better balance. But there is no reason that judges cannot vary the procedural mechanism they construct in light of experience with prior orders. This is evident in the
discussion of the content of prior orders by Arnold J in *Cartier.*[222] Indeed, the ability
of the judiciary to adapt these procedures over time is perhaps the strongest reason
to resist excessive statutory detail. And in doing so they will have to take account of
the interests of absent parties, as the Court of Justice in *Telekabel* required national
courts to devise a mechanism by which to allow the vindication of the fundamental
right of users.[223]

---

the notice be worded so concretely as to allow the addressee of the notice to detect the violation
easily—meaning without a thorough legal or factual examination. The amount of effort to be
expected of an operator of an internet trading platform in performing any such examination
depends upon the circumstances of the individual case, especially upon the gravity of the reported
infringements on the one hand and the operator's opportunities to learn of them on the other
hand.").

[221] National Report of the United Kingdom, at 15 (discussing Tamiz v. Google, Inc. [2012]
EWHC 449).

[222] Cartier Int'l v. British Sky Broadcasting, [2014] EWHC 3354 (Ch), [2016] EWCA Civ 658.

[223] Some scholars have argued that the focus on human rights has impoverished the analysis of the
Court of Justice. *See* Martin Husovec, *Holey Cap! CJEU Drills (Yet) Another Hole In The E-
Commerce Directive's Safe Harbors*, 12 J. INTELL. PROP L. & PRAC. 115 (2016) (discussing deci-
sion in *Tobias McFadden v Sony Music*). It certainly runs the risk of fossilising legal norms.

In some countries, most notably in Canada with respect to copyright infringement, the concerns about the over-enforcement resulted in a slightly modified system, called notice and notice.[224] Under such a system, when an OSP receives
notification from a rights holder that one of its subscribers is allegedly hosting
infringing material, the ISP undertakes to forward the notice to the subscriber, and
to monitor that subscriber's activities for a period of time. The OSP will not takedown the material, which may be lawful, absent court order. It thus seeks to prevent
a system that embeds an incentive to over-enforce. To be sure, the counter-
notification system in the US DMCA is intend similarly to ensure balance, but it is
rarely used. Moreover, under the Canadian regime, the OSP does not reveal the
subscriber's personal information absent a court order.

On the other hand, although implementing and operating the notice and takedown system was not without some costs for OSPs—all in, eBay's measures to stop
infringement involved an expenditure of $20 million per year—it also required vigi-
lance on the parts of right owners, and much litigation about Notice and Takedown.
Thus, in other countries, such as Germany, the "whack-a-mole" problem persuaded
courts to impose measures on intermediaries that began to approximate a system
that rights owners would prefer to see replicated legislatively (or though interpretation of existing law), namely, "notice and staydown".[225] Within the European Union,
some national courts have taken the view that an obligation for providers of hosting
services to prevent the reappearance of contents they have already taken down
("notice and stay down") might amount to a general monitoring obligation prohibited
by Article 15 of the E-Commerce Directive.[226] These initiatives have largely been
developed under the rubric of additional obligations to assist, and are thus discussed
in greater detail below in Part V.

The EU MOU also addresses what are called "[p]ro-active and [p]reventive measures," thus addressing the prevention of future infringement by ensuring that offers
of counterfeit goods do not appear online.[227] This commitment by the Internet platforms extends to "appropriate, commercially reasonable and technically feasible
measures, taking into consideration [their] respective business models."[228] Examples
revealed by reports on the MOU include detection technologies and pre-emptive

---

[224] *See* Copyright Modernisation Act (Bill C-11) (effective January 2, 2015).

[225] *Cf.* Kur, *supra* note 146, at 536 (noting that German courts applying the Störerhaftung doctrine
to Internet auction sites "did not confine [filtering] duties to future identical offers made by the
same person, but extended them to infringements that were essentially of the same character").

[226] *See* Cour de cassation [Cass.] [supreme court for judicial matters] 1e civ., July 12, 2012, Bull.
civ. I, No. 167 (Fr.) (Google Inc. v. Bac Films); Cass. 1e civ., July 12, 2012, Bull. civ. I, No. 166
(Fr.) (Google Inc. v. Bac Films); Cass. 1e civ., July 12, 2012, Bull. civ. I, No. 162 (Fr.) (consoli-
dated appeals, Google Inc. & Aufeminin.com v. André Rau, H & K); *see also* Christine Gateau &
Pauline Faron, *Take Down, Stay Down: Paris Court of Appeal Confirms Hosting Providers Have
No General Monitoring Obligation*, 20 COMPUTER & TELECOMM. L. REV. 12 (2014) (U.K.) (dis-
cussing Cour d'appel [CA] [regional court of appeal] Paris, 2e civ., June 21, 2013 (Fr.) (Societe des
Producteurs de Phonogrammes en France v. Google) (unpublished)).

[227] Memorandum of Understanding, *supra* note 194, ¶ 25.

[228] *Id.* ¶ 27.

takedowns, or the vetting of some sellers prior to allowing them to sell online.[229] This may be closer to hard European law, which in Article 11 of the Enforcement Directive contemplates future-looking measures. But, if extended to the United States, this would go beyond the prevailing view of current hard law obligations.

And the reaction in the United States to this type of approach in SOPA and PIPA highlights that it is by no means an accepted way of tackling secondary liability. The acceptability of such proactive measures in the EU at present is surely bolstered by the fact they are imposed under the rubric of what I call "obligations without liability" (discussed below in Part V) without suggesting any exposure to monetary liability. If, instead, they were imposed as a condition of immunity or as essential to negate an element of secondary liability, this would be more troubling.

Critics of SOPA and PIPA feared the former, and the recent proposal of the European Commission for a new copyright instrument has raised similar concerns.[230] Proposed Article 13 would require OSPs to

> take measures to ensure the functioning of agreements concluded with rightholders for the use of their works or other subject-matter or to prevent the availability on their services of works or other subject-matter identified by rightholders through the cooperation with the service providers. Those measures, such as the use of effective content recognition technologies, shall be appropriate and proportionate. The service providers shall provide rightholders with adequate information on the functioning and the deployment of the measures, as well as, when relevant, adequate reporting on the recognition and use of the works and other subject-matter.[231]

The system of notice and takedown, whether through legislation or private practice, carries with it the typical costs and benefits of private ordering generally. To the extent that it precludes frequent and more concrete judicial development of the legal obligations under which intermediaries operate, it impedes a process that will over time inculcate greater certainty. As a result, it is incumbent upon parties to (or superintending) private arrangements to continue to disseminate information about the practices developing thereunder, as it has done to date. This will allow critical scrutiny, but also give to those practices some of the public-regarding value that attaches to judicial opinions.

As demonstrated by the list of parties to the MOU—mostly multinational companies—[232] there is a risk that legal norms applicable to all actors grow up around

---

[229] Report from the Commission, *supra* note 118.

[230] *See* Lemley et al., *supra* note 6.

[231] *See* European Commission, Proposal for a Directive of the European Parliament and of the Council on copyright in the Digital Single Market – COM(2016)593 (Sept. 14, 2016), art 13. The UK Goverment has been facilitating talks among Google, movie studies and the recording studios intended to negotiate a voluntary code of conduct that would reduce prominence of links to infringing contents.

[232] The existing signatories have begun to consider procedures for accession of new brand owners and platforms to the MOU. *See* European Comm'n, *Memorandum of Understanding on the Sale of Counterfeit Goods via the Internet – Spring Meeting – 1 April 2014 – Summary*, at 1, E.C. Doc. MARKT D3/FT/cm-Ares(2013)1215882, *available at* http://ec.europa.eu/internal_market/iprenforcement/docs/mou_meeting_summary_20140401_en.pdf.

the capacity and sophistication of large economic players—though this can also happen if litigation is the vehicle for development of the principles. Again, public scrutiny is essential.[233]

On the other hand, the MOU reduces costs by limiting litigation expenses and effectuating the lowest cost avoider principle to which many countries are (with varying degrees of transparency) committed. The EU MOU commits parties (which include Internet stalwarts such as eBay and Amazon, as well as brand owners) beyond obligations that might presently flow from hard law secondary liability standards. The memorandum makes explicit that it does not replace or interpret the existing legal framework, and it cannot be used in evidence in legal proceedings.[234] But it clearly has an eye to litigation. The exchange of information contemplated by the MOU is not to constitute actual or constructive notice.[235] And the parties to the MOU agreed to a 1-year moratorium on lawsuits against each other regarding matters within the scope of the agreement.[236]

And, as the reports on the progress of the arrangement suggest, the private ordering process may offer a far greater chance of developing an international norm than negotiation of a multilateral instrument of a public international law character. Finally, it would do so while retaining flexibility to adapt procedures to reflect both new forms of trading and changing technological capacity. Thus, some have argued for further evolution to allow for so-called Notice and Trackdown (though it is not clear precisely what this would entail).[237]

V. Obligations Without "Liability"

Some countries have developed mechanisms under which intermediaries can be required to assist in preventing wrongdoing regardless of fault.[238] At the EU level, these obligations exist with respect to infringement of intellectual property rights. But in some countries, the doctrine on which the intellectual property-related measures have been ordered is grounded in general law.[239] These mechanisms might be

---

[233] See November 2013 Meeting Report, *supra* note 197, at 2–3 (discussing public availability of key performance indicators).

[234] See Memorandum of Understanding, *supra* note 196, at 1.

[235] *Id.* ¶¶ 22, 25.

[236] *Id.* ¶ 10.

[237] See UK Intellectual Property Office, Protecting Creativity, Supporting Innovation: IP Enforcement 2020 (May 2016).

[238] See National Report of the United Kingdom, at 1 ("service providers have lately been subjected to a range of new regulatory schemes and litigation intended to deputise their services for various public and private purposes, including law enforcement, upholding intellectual property rights, and detecting and prosecuting wrongdoing").

[239] See *id.* at 20–23; National Report of Germany, at 2.

thought to function to some extent as forms of quasi-secondary liability.[240] Certainly, they are an important variable to take into account in setting out the scope of online service provider liability.[241]

A. Explicit Regulatory Responses

These provisions also operate to create a quasi-regulatory network of obligations without imposition of full monetary liability. Thus, as a preliminary matter it is worth noting that several countries have quite explicitly introduced regulatory systems designed to structure service provider behavior in order to reduce wrongful conduct of others, particularly in the field of copyright law. The leading examples of regulatory mechanisms are Italy (AGCOM), France (HADOPI), and New Zealand[242]; the French system was repealed in 2013, and in fact had resulted in very few suspensions of internet users (as contemplated by the system).

The United Kingdom also enacted legislation (the Digital Economy Act 2010) establishing a regulatory framework for instituting a graduated response system with respect to copyright infringement only. The difficulties with its implementation perhaps illustrate some of the challenges. The UK system would impose both an initial set of informational obligations (so-called "notice-notice" provisions requiring passing on to their subscribers any notifications of infringement received from copyright owners, and the maintenance of a list of repeat infringers that would be accessible to copyright owners), and a collection of technical obligations (such as the suspension of a subscriber's access) that would only come into effect if the initial obligations failed to reduce copyright infringement by a specified amount (70–80%) over 12 months.[243] In addition, Sections 17–18 authorised the government to make provision for the imposition of blocking injunctions. But those provisions were repealed by the Deregulation Act in light of the success of Section 97A of the Copyright Act[244] and the regulations requiring ISPs to disconnect repeat infringers is unlikely to come into effect.[245] (Indeed, even the initial informational obligations are not yet in place.) The inclusion of informational and blocking injunctions in the government's conceptions of this administrative apparatus serves only

---

[240] *See* National Report of the United Kingdom, at 2 ("[service providers] may be liable in the very loose sense of being contractually required to act under an applicable industry code or self-regulatory regime, or of being liable to a declaration under such a regime.").

[241] However, the need to comply with such obligations does not as strict matter render the service providers "liable." *See* R (British Telecommunications plc) v Secretary of State [2011] EWHC 1021 (Admin), *aff'd* [2012] EWCA Civ 232.

[242] *See* National Report of New Zealand, at 2–3. For a critical discussion of several graduated response systems, see Rebecca Giblin, *Evaluating Graduated Response*, 37 COLUM. J. L. & ARTS 147 (2014).

[243] *See* National Report of the United Kingdom, at 28; *see* Digital Economy Act 2010, § 9.

[244] *See* Deregulation Act 2015 c. 20 § 56 (May 26, 2015).

[245] *See* National Report of the United Kingdom, at 28 ("it appears unlikely at this stage that they will ever be implemented").

to highlight the regulatory function of these new mechanisms, and they should perhaps be viewed in that light as well as part of the fabric of service provider "liability".

The system in New Zealand has been functioning for a couple of years. It is aimed at a subset of copyright problems, namely infringement by "file-sharing," although that is broadly defined.[246] The New Zealand system is operated in the first instance by the New Zealand Copyright Tribunal, but appeals lie to the High Court (although there have been no High Court decisions to date).[247] The principal sanction that the Tribunal will impose after three notices is a monetary award[248]; the statute provides for a court requiring a service provider to suspend the subscriber's internet account for 6 months, but that provision has not yet been activated.[249] Unlike the copyright safe harbor mechanism, the New Zealand graduated response system incorporates a statutory counter-notice available to users receiving a notice, which triggers an obligation on the right-holder to provide reasons for the infringement allegation.[250] Like the UK system, the service provider is obliged to maintain information about the subscribers who receive infringement notices, but (unlike the UK system) the service provider is precluded from giving the subscriber name and contact details to the claimant absent an order from the Tribunal.[251]

Two additional features of the New Zealand system are worth noting. First, the system is aimed at account holders rather than infringers.[252] Second, actions before the Tribunal can be brought by agents of the copyright owners, which helps to overcome coordination problems.[253]

B. Cooperative Mechanisms: Assistance in Preventing Unlawful Conduct

As Jaani Riordan notes, "claimants are increasingly seeking remedies to remove, prevent, block access to, de-index, and identify the authors of internet materials said to be unlawful."[254] Although the principal remedies that have been granted are the disclosure of information about the primary wrongdoer, and blocking of access to unlawful materials uploaded by the primary wrongdoer, the principles that have

---

[246] See National Report of New Zealand, at 3 & 8; see also Copyright Act 1994, § 122A ("file sharing is where—(a) material is uploaded via, or downloaded from, the Internet using an application or network that enables the simultaneous sharing of material between multiple users; and (b) uploading and downloading may, but need not, occur at the same time").

[247] See National Report of New Zealand, at 8.

[248] The award can rise to as much as NZ $ 15,000. See id., at 8. However, although some awards have approached NZ $ 1000, most have been lower. See id. at 12.

[249] See id., at 8.

[250] See id., at 9.

[251] See id., at 12.

[252] See id., at 13.

[253] See id., at 9.

[254] National Report of the United Kingdom, at 1–2.

given rise to and shaped these mechanisms may well justify different measures in future technological and social conditions.[255]

Although two pieces of parallel EU legislation now provide the basis for these types of measures in intellectual property litigation, courts in several European countries had begun to offer these forms of relief under prior national mechanisms. Indeed, to the extent that the EU legislation in question takes the form of directives, national courts may formally be extending assistance under these pre-existing doctrines as adapted to take account of the principles of EU law announced by the legislation in question and interpreted by the Court of Justice.

First, there is the question of disclosure of information. Article 8 of the EU's Enforcement Directive contains a provision specifically authorizing this possibility and the copyright decision of the Court of Justice in *Promusicae v. Telefonica* confirmed it would be permissible.[256] The Court of Justice has recently acknowledged that measures regarding provision of information should take into account privacy concerns—a point that had also been raised by the Court of Justice in *Scarlet*—though banking secrecy laws cannot be employed fully to negate these kinds of orders.[257]

In the United Kingdom, so-called *Norwich Pharmacal* orders have for some time been available as a discretionary equitable remedy to provide right-holders with information about the primary wrongdoer that will enable the right holders to pursue that person.[258] "The claimant must establish: (1) an arguable case of primary wrongdoing by someone; (2) facilitation of that wrongdoing by the defendant service provider; (3) that disclosure is necessary to enable some legitimate action against the wrongdoer; and (4) that disclosure would, in the circumstances, be proportionate."[259]

---

[255] *See id.* at 23 ("it is theoretically possible for English courts to issue de-indexing and payment freezing orders by analogy with *Norwich Pharmacal* disclosure under the equitable protective jurisdiction. However, no such orders have yet been made").

[256] Directive 2004/48/EC of the European Parliament and of the Council of 29 April 2004 on the Enforcement of Intellectual Property Rights, 2004 O.J. (L 195) 22 [hereinafter Enforcement Directive]; Case C-275/06, Promusicae v. Telefonica, 2008 E.C.R. I-271. *See also* Case C-461/10, Bonnier Audio AB v. Perfect Commc'n Sweden AB, 2012 E.C.R. I-00000, 2 C.M.L.R. 42 (2012) (order for information regarding subscriber to whom ISP gave IP address which was allegedly used to commit infringement); Golden Eye v. Telefonica, [2012] EWHC 723 (Ch) (Eng.) (order to similar effect).

[257] *See* Case C-580/13, Coty Germany GmbH v. Stadtsparkasse Magdeburg, EU:C:2015:485 (CJEU 2015) (Article 8(3)(e) over-rides a national provision which allows, "in an unlimited and unconditional manner," a banking institution to invoke banking secrecy in order to refuse to provide information concerning the name and address of an account holder").

[258] *See* Norwich Pharmacal Co v Customs and Excise Commissioners [1974] AC 133. "Two cases suggest that the English courts are alive to the risk that blanket disclosure orders against ISPs may be abused by claimants for the purpose of extracting monetary settlements from subscribers accused of copyright infringement." National Report of the United Kingdom, at 21 (citing Media CAT Ltd. v Adams [No 2] [2011] FSR 29, 724–5; Golden Eye (International) Ltd. v Telefónica UK Ltd. [2012] EWHC 723 (Ch), [146]; *rev'd in part* [2012] EWCA Civ 1740.)

[259] National Report of the United Kingdom, at 20.

Consistent with principles later endorsed by the Court of Justice, the English courts will take into account confidentiality or privacy concerns. Proportionality appears increasingly to be the dominant concern, and that likewise requires attention to countervailing fundamental rights such as privacy.[260] Under the proportionality rubric, courts can also consider a number of other factors, including (as they do with blocking orders) the cost and inconvenience of giving disclosure and whether there are any realistic alternatives to the relief sought. In the disclosure context, this latter concern means the possibility of trap purchases or spot checks.[261] And they are also alert to fears of over-enforcement, insofar as disclosure might affect the personal information of those other than arguable wrongdoers.[262]

Second, courts have issued a series of more intrusive injunctive orders against intermediaries, especially under EU legislation. Article 11 of the Enforcement Directive (with respect to intellectual property generally) requires member states to ensure that "right holders are in a position to apply for an injunction against intermediaries whose services are used by a third party to infringe an intellectual property right."[263] Article 8(3) of the Information Society Directive provides the same with respect to copyright.

These provisions may be applicable to provide relief where an intermediary is between the zone of nationally determined liability and EU-guaranteed safe space (under Article 14 of the E-Commerce Directive). Indeed, they might offer the possibility of injunctive relief even when the intermediary is *within* the safe harbor and thus immune from monetary liability.[264]

---

[260] *Id.*, at 20.

[261] *See* The Rugby Football Union v. Viagogo Limited [2011] EWCA Civ. 1585 [27]–[29] (Longmore LJ).

[262] *See* The Rugby Football Union v. Consolidated Info. Servs. (Viagogo), [2012] UKSC 55, [43]–[46] (Lord Kerr JSC).

[263] Enforcement Directive art. 11. *See also* Directive 2001/29/EC of the European Parliament and of the Council of 22 May 2001 on the Harmonisation of Certain Aspects of Copyright and Related Rights in the Information Society, 2001 O.J. (L 167) 10 [hereinafter Information Society Directive]. Although the claimant is not required to prove the primary wrongdoing, courts do have regard to the nature, seriousness and strength of the primary claim. *See* National Report of the United Kingdom, at 20–21. As an equitable matter, the service provider's conduct might also be relevant. *See* G v Wikimedia Foundation Inc. [2010] EMLR 14, 364.

[264] *See* E-Commerce Directive, recital 45; *id.*, arts. 12(3), 13(3), 14(3). *Cf.* National Report of the United Kingdom, at 16 (noting that an injunction "under the Equality Act may not impose on such a service provider any liability which would 'contravene' articles 12–14 of the E-Commerce Directive, or amount to a general obligation within Article 15. This goes further than the safe harbours, since it suggests that no non-monetary liability can attach in respect of activities protected by a safe harbour."). Riordan argues that whether the third party's conduct is notionally unlawful is always relevant, even when the relief sought against an intermediary is non-monetary (and not based upon a finding of illegality by the intermediary). *See* National Report of the United Kingdom, at 20 ("English law recognises a number of [non-monetary] remedies against non-wrongdoers… Although they are independent of any finding of primary or secondary liability against the service provider, they require some kind of primary wrongdoing to be demonstrated to the relevant standard of proof.").

In its first discussion of Article 11, in *eBay v. L'Oreal,* the Court of Justice held that, in light of the objective of the Enforcement Directive to ensure effective protection of intellectual property, injunctions to prevent *future* infringement were permissible.[265] But what measures might these be? The court recognized that national courts would devise precise measures using the procedural devices available to them,[266] but sketched the general parameters, offered some non-exhaustive possibilities and stressed some limits on this form of relief.[267] As a general matter, the court held that measures must be "effective and dissuasive," but also proportionate.[268] Measures "need to strike a fair balance between the various rights and interests."[269] English courts have read proportionality to require consideration whether "alternative measures are available which are less onerous."[270]

The *eBay* court offered minimal guidance to national courts or online actors. As concrete illustrations, the court mentioned only the possibility of an order "to suspend the perpetrator of the infringement … in order to prevent further infringements of that kind by the same seller in respect of the same trade marks," and orders to take measures to make it easier to identify the primary infringers.[271] Quite what this means in practice is not yet entirely clear. The case was sent back to Mr. Justice Arnold to work out how to craft the appropriate order. But the case settled, so the U.K. court did not have an opportunity to devise precise measures for the *L'Oréal* parties in accordance with the Court of Justice's guidance.[272]

Given the conceptual similarities between Article 11 and the so-called *Störerhaftung* doctrine under German law, German case law in particular might offer hints about additional contours. The obligation imposed on a defendant under the *Störerhaftung* doctrine related to the prevention of *future* infringements.[273] But this was not unlimited. Although the defendant's obligations were not limited to "future identical offers made by the same person, [they extended only to] infringements that

---

[265] *L'Oréal* ¶ 131. The applicability of Article 11 would not render the intermediary liable for damages.

[266] *See* Enforcement Directive, recital 23; *see also* Information Society Directive, recital 59.

[267] *See L'Oréal*, 2011 E.C.R. I-6011, ¶¶ 136–43.

[268] *See id.* ¶¶ 136, 141.

[269] *See id.* ¶ 143.

[270] Cartier Int'l v. British Sky Broadcasting, [2014] EWHC 3354 (Ch), at [162].

[271] *See L'Oréal*, ¶¶ 141–42.

[272] *See* William Horobin & Greg Bensinger, *L'Oréal, eBay Settle Dispute over Counterfeit Goods: French Company Was Dissatisfied with Internet Retailer's Response to Concerns*, WALL ST. J. ONLINE (Jan. 15, 2014, 3:18 PM), http://online.wsj.com/news/articles/SB10001424052702304 41910457932287080181859O. *But see infra* notes 306–351 and accompanying text (discussing Mr. Justice Arnold's later copyright rulings, where he relied on the Court of Justice's *L'Oréal* opinion in interpreting Article 8(3) of the Information Society Directive).

[273] *See* Kur, *supra* note 146, at 535–39 (discussing the relevant line of German cases).

were *essentially of the same character*."[274] Otherwise, this might amount to a general obligation to monitor. Under the German doctrine, courts have had regard to whether there exists a causal link between the defendant's acts (or omissions) and the infringement, and whether the defendant is able factually and legally to prevent or terminate the infringement.[275] Moreover, intermediaries have not been obliged to act as regards infringements whose character was doubtful or which could not be detected by reasonable means.[276]

Applying these principles, courts have ordered the takedown of infringing listings by online auction sites and imposed obligations on intermediaries to engage in a measure of technological checking (such as limited filtering) to minimise future infringement.[277] At some point, there is a risk that any measures (especially if sought by numerous mark owners) could effectively amount to a general obligation on intermediaries to monitor the activity of their customers, violating the principles that are both enshrined in Article 15 of the E-Commerce Directive and behind the reluctance (in the United States and Europe) to create broad secondary liability.[278]

---

[274] *See id.* at 9; *see also* Leistner, *supra* note 145, at 79 ("Typically, the injunctions also require the interferer ... to take reasonable measures, such as filtering, to prevent comparable infringements in the future. However, such preventive measures must not be unreasonably burdensome in the sense that the provider is required to take steps which would jeopardise its entire business model. Instead, only reasonable and technically possible measures in order to identify comparable infringements, *i.e.*, offers of the same trader or comparable counterfeit goods, should be imposed."). Kur does express a concern that the second *Rapidshare* judgment may move too far beyond this important limit and flirt with a violation of Article 15 of the E-Commerce Directive. *See* Kur, *supra* note 146, at 539–40 (citing the second *Rapidshare* judgment, BGH Aug. 15, 2013, GRUR 1030, 2013 (Ger.) (*File-Hosting-Dienst*)).

[275] *See* Kur, *supra* note 46, at 532; *cf.* Dogan, *supra* note 58, at 516 n. 85 ("'Reasonableness,' of course, does not require the intermediary to do the impossible, and many courts have made clear that contributory infringement defendants must have 'sufficient control over the infringing activity to merit liability'." (citing Gucci Am., Inc. v. Frontline Processing Corp., 721 F. Supp. 2d 228, 248 (S.D.N.Y. 2010))).

[276] *See* Leistner, *supra* note 145 at 78–82; *see, e.g.,* BGH Mar. 11, 2004, 158 BGHZ 236 (Ger.) (*Internet Auction I*), *translated in* 2005 E.T.M.R. 25, ¶ 41 ("In the event of an order on injunctive relief being issued, then the defendant would only be held liable for breaches if it committed a culpable act... For trade mark infringements which cannot be recognised in the filtering procedure applied in advance (because, for example, a counterfeit Rolex watch is on offer at a price appropriate for an original, without reference to its counterfeit character), no fault would be attached").

[277] *See generally* Leistner, *supra* note 145 at 79; *see, e.g.*, BGH Mar. 11, 2004, 158 BGHZ 236 (*Internet Auction I*), *translated in* 2005 E.T.M.R. 25, ¶ 41 ("[E]very time that the defendant is notified of a clear breach of the law, it must not only block the concrete offer without delay ... but also apply preventative measures to ensure as far as possible that no further trade mark infringements of this kind occur. In the case in dispute, ... a number of clearly recognisable trade mark infringements have occurred. The defendant must see these events as a reason to subject offers for Rolex watches to specific checking... It may be possible for the defendant to make use of a software program in this connection to reveal the corresponding suspicious cases, with the relevant suspect points being perhaps the low price and also any reference to imitation.").

[278] The E-Commerce Directive recognizes that preventing recurring infringement may require some specific monitoring, and that if properly tailored this will not violate Article 15. *See* E-Commerce Directive, recital 48.

Most German courts have been alert to this concern in applying the *Störerhaftung* doctrine, as was the Court of Justice in *L'Oréal*.[279]

The *L'Oréal* court made clear that measures under Article 11 could not impose an obligation actively to monitor all of the "data of each of its customers in order to prevent any future infringement of intellectual property rights via that provider's website."[280] According to the court, this would not be fair and proportionate.[281] Indeed, the court also instructed national courts that measures could not create barriers to legitimate trade by effecting "a general and permanent prohibition on the selling, on that marketplace, of goods bearing those trademarks."[282]

The operative principles found in German law, which are consistent with the Court of Justice's *L'Oréal* ruling, are similar to those considerations that might inform a lowest cost avoider or reasonableness analysis under U.S. tort law.[283] Indeed, the mechanism introduced by the Enforcement Directive is clearly motivated by that kind of normative principle.[284] Similar considerations may be silently informing the application by U.S. courts of the doctrinally narrow *Inwood/Tiffany* standard.[285] But to the extent that there are parallels in substance, it is important to note that Article 11 will not (unlike secondary liability) expose intermediaries to damages liability. This is not insignificant.

The Court of Justice has not had an opportunity to comment further on the application of Article 11 (other than in the offline context). And this might raise one further concern about the potential costs of this mechanism: legitimate and innovative business models might be chilled as much by the uncertainty of legal standards as by risk of monetary liability. The experience of the German courts applying its *Störerhaftung* doctrine may provide only limited comfort.

However, Article 11 extended to all intellectual property the mechanism previously introduced into copyright law by Article 8(3) of the Information Society

---

[279] Kur, *supra* note 146, at 534–538; Leistner, *supra* note 145, at 79; *see* BGH Mar. 11, 2004, 158 BGHZ 236 (*Internet Auction I*), *translated in* 2005 E.T.M.R. 25, ¶ 40 ("A company which ... operates a platform for third-party auctions on the internet cannot reasonably be expected to investigate every offer prior to publication on the internet to identify any possible breach of the law. Any such obligation would place in question the entire business model...").

[280] Case C-324/09, L'Oréal SA v. eBay Int'l AG, 2011 E.C.R. I-6011, ¶ 139.

[281] *Id.*

[282] *Id.* ¶ 140.

[283] *See* Leistner, *supra* note 145, at 88 (identifying as considerations relevant to the imposition of relief, based upon an analysis of European law: "[T]he degree of (objective) risk caused by the secondary infringer as well as the degree of control the secondary infringer has in relation to the acts of direct infringement ... Moreover, the (objective) design of a business model of an intermediary might establish tortious liability where the business model is specifically designed to profit from direct acts of infringement ..."); *cf.* MGM Studios, Inc. v. Grokster, Ltd., 545 U.S. 913, 939 (2005) (relevance of design); Sony Corp. of Am. v. Universal City Studios, Inc., 464 U.S. 417, 442 (1984) (extent of non-infringing use); Dogan, *supra* note 58, at 509–10 (discussing control).

[284] *See* Information Society Directive, recital 59 ("In many cases, intermediaries are best placed to bring ... infringing activities [of third parties using their services] to an end").

[285] *See* Dinwoodie, *supra* note 27, at 489–93.

Directive.[286] The developing jurisprudence under that parallel provision should perhaps give some further reassurance to those concerned that the mechanism is simply a back-door way of placing the burden of policing on intermediaries.

For example, the Court of Justice decided two cases brought by a Belgian collecting society (SABAM) seeking an order for an Internet service provider (Scarlet) and a social media network (Netlog) to install a filter system to help prevent future copyright infringement facilitated by their respective services.[287] In both cases, the court found that the immunities of the E-Commerce Directive were applicable to prevent full liability. And in both cases, the court found that Article 8(3) of the Information Society Directive was potentially applicable. But in both cases, the court found that the imposition of the filtering obligation that had been sought was not justified.

The court concluded that the question of the appropriate relief had to be considered in light of other European instruments, as well as general principles of European Union law and fundamental rights enshrined in the Charter of Fundamental Rights of the European Union. In particular, the court also looked to: (1) the E-Commerce Directive, which in Article 15 prohibits a general obligation on Internet service providers actively to monitor; (2) the general principle of EU law that any relief be proportionate; and (3) the fundamental right to have "freedom to conduct a business" as guaranteed by Article 16 of the Charter.[288] The court found that the requested obligation to filter all data from all customers for any future infringement of intellectual property for an unlimited time violated each of these principles.[289]

---

[286] Information Society Directive, art. 8(3).

[287] Case C-360/10, (SABAM) v. Netlog NV, 2 C.M.L.R. 18 (2012); Case C-70/10, Scarlet Extended SA v. SABAM, 2011 E.C.R. I-11959.

[288] Similar considerations have informed German law applying the *Störerhaftung* doctrine. *See* Kur, *supra* note 146, at 533, 536–37.

[289] The Court of Justice may have been more solicitous of the fundamental rights of the Internet service provider than was the European Court of Human Rights in its first case addressing the liability of ISPs. *See* Delfi v. Estonia, App. No. 64569/09, Eur. Ct. H.R. (2013). In that case, the Court of Human Rights held that it was not necessarily incompatible with the free expression guarantee of Article 10 of the European Convention on Human Rights for an Estonian news portal to be held monetarily liable as a result of anonymous (allegedly defamatory) comments posted on its Web site by users in response to a news story on the site. Indeed, the *failure* to contemplate such liability might implicate the rights of the defamed party under Article 8 of the Convention. *See id.* ¶ 91. The Court of Human Rights did not, however, need to have regard to the prohibition against general monitoring found in secondary EU law, and it likewise did not have to weigh in the balance the freedom to conduct business under Article 16 of the Charter of Fundamental Rights. Thus, although the Convention on Human Rights has been more clearly incorporated into EU law post-Lisbon Treaty, see Treaty on European Union, art. 6(3), 2010 O.J. (C 83–13) ("Fundamental rights, as guaranteed by the European Convention for the Protection of Human Rights and Fundamental Freedoms and as they result from the constitutional traditions common to the Member States, shall constitute general principles of the Union's law."), the variety of norms at play in a case before the Court of Justice suggests that a different outcome might ensue under EU law. *See id.* art. 6(1) ("The Union recognises the rights, freedoms and principles set out in the Charter of Fundamental Rights of the European Union ..., which shall have the same legal value as the Treaties."). This implicates the difficult relationship between the European Court of Human Rights and the Court of Justice of

The *SABAM* cases thus establish an extremely balanced framework for courts thinking through the competing values at stake in these types of cases.[290] And the Court's subsequent Article 8(3) decision in *UPC Telekabel Wien GmbH v. Constantin Film Verleih GmbH* likewise exemplifies this nuance and balance.[291] There, owners of the copyright in cinematographic works applied to the Austrian courts for an order requiring an Austrian Internet service provider (ISP) to block the access of its customers to Web sites on which allegedly infringing copies of the claimant's works were available. The ISP against whom the orders were sought did not host the Web sites where the alleged infringements were available. Rather, the ISP provided Internet access to its customers, who were thus able to access the Web site where the infringements were occurring.[292] The order was sought pursuant to an Austrian procedure implementing Article 8(3) of the Information Society Directive. The Court of Justice was required to consider whether the order was of the type that member states were required by Article 8(3) to make available to copyright owners, and whether the particular Austrian procedure—which resulted in a general order to the ISP to achieve a particular outcome without detailing specific measures by which to do so—was compatible with the fundamental rights of the ISP and its customers.[293]

The Court of Justice concluded that the particular Austrian procedure could, under certain conditions, fulfill Austria's obligations under Article 8(3) and be compatible with fundamental rights as enshrined in the Charter of Fundamental Rights of the European Union.[294] The court deduced these conditions from a balancing of

---

the European Union, which is beyond the scope of this Chapter. *See generally* Sionaidh Douglas-Scott, *The European Union and Human Rights After the Treaty of Lisbon*, 11 HUMAN RIGHTS L. REV. 645 (2011); Charter of Fundamental Rights of The European Union 2000 OJ (C 346/3) (Dec 18, 2001) art. 52(3); *see also* Agyar Tartalomszolgáltatók EgyesüLete and Index.Hu Zrt v. Hungary, Application No. 22947/13 (May 2, 2016) (ECHR).

[290] *See.* Leistner, *supra* note 145, at 79 (suggesting that the solution found by the German courts "strikes a fair balance between IPRs and the interest of genuinely 'neutral' host providers").

[291] *See* Case C-314/12, UPC Telekabel Wien GmbH v. Constantin Film Verleih GmbH, EU:C:2014:192 (CJEU 2014).

[292] Article 8(3) did not require a specific relationship between the alleged infringer and the intermediary against whom an injunction was sought. *See Telekabel*, ¶¶ 35, 38. The Court did not address the suggestion by the Advocate-General that the principle of proportionality might, however, warrant the infringers or the host ISP being pursued first. *See Telekabel*, ¶ AG 107 ("[I]t should be noted that the ISP is not in a contractual relationship with the operator of the copyright-infringing website. As a consequence …, a claim against the ISP is, admittedly, not completely out of the question, but the originator must, as a matter of priority, so far as is possible, claim directly against the operators of the illegal website or their ISP."); *see also* Disturber Liability of an Access Provider 2016 I.I.C 481 at ¶ 83 (BGH).

[293] The modalities of injunction practice under Article 8(3)—like Article 11 of the Enforcement Directive—are a matter of national law. So variation among member states is allowed subject to the constraints of EU law. *See Telekabel*, ¶¶ 43–44. That is to say, the Court of Justice was considering whether the Austrian approach was permissible, and not whether that procedure was required in all member states. *See infra* text accompanying note 336 (comparing nature of blocking orders in the United Kingdom).

[294] *See Telekabel*, ¶ 64.

the three fundamental Charter interests implicated by this type of order, none of which were absolute: the interests of intellectual property owner under Article 17; the freedom of the ISP to conduct business under Article 16; and the freedom of expression of Internet users under Article 11.[295] In the course of its judgment, the court fleshed out the content of the three Charter rights at issue in ways that suggest continuing sensitivity to the competing concerns expressed in the *SABAM* cases and earlier in this Chapter. And these considerations are surely transferable both to remedial mechanisms employed in other member states and with respect to orders involving other intellectual property rights under Article 11 of the Enforcement Directive.[296]

In short, the Court of Justice required that procedural protections be in place for ISP and user interests. The ISP's freedom to conduct business under Article 16 was not absolute; but it is obviously affected by the type of order at issue in the case.[297] And this may be exacerbated by the lack of specific measures in the outcome-based order imposed by the Austrian court, which may impose high uncertainty costs on the ISP.[298] However, the Court of Justice found the outcome-based order acceptable because, by leaving to the ISP the power to determine the measures to be taken to achieve the result desired, the ISP could "choose to put in place measures which are best adapted to the resources and abilities available to him and which are compatible with the other obligations and challenges which he will encounter in the exercise of his activity."[299]

This reasoning, informed by the principle of proportionality—which overlays all enforcement questions under EU law—implicitly expresses some support for the sentiment that the measures appropriately imposed on one intermediary might differ from those to be implemented by another of quite different size and sophistication.[300] The proportionality of the Austrian procedure was also supported by the fact that it allowed the ISP to show that it had "taken all reasonable measures," ensuring

---

[295] *See id.* ¶ 47.

[296] Because different member states might implement their obligations under Article 8(3) or Article 11 in different ways, the *precise* conditions that the court imposed on the Austrian courts might not apply to orders of a slightly different nature issued by other courts. For example, some of the detailed procedural protections identified in *Telekabel* flow from the outcome-based order issued by the Austrian court, and may not apply to orders detailing specific measures that have to be implemented. *See* Martin Husovec, *CJEU Allowed Website-Blocking Injunctions with Some Reservations*, 9 J. INTELL. PROP. L. & PRAC. 631, 633 (2014) (commenting upon applicability of *Telekabel* to orders that have been issued by the U.K. courts). But the broad principles announced by the Court of Justice will apply to slightly different orders issued by non-Austrian courts.

[297] *See Telekabel*, ¶¶ 50–51.

[298] In other countries (such as the United Kingdom) orders issued under the relevant provisions have been far more specific, and have been framed in terms of the measures to be implemented rather than the general outcome to be achieved. *See infra* notes 306 & 311 (listing U.K. cases). In those countries, ongoing mechanisms devised by the court to accord legal certainty to ISPs will be less crucial to ensuring compliance with the fundamental rights of the ISP.

[299] *See Telekabel*, ¶ 52.

[300] *See supra* text accompanying notes 116–117.

that it would not be "required to make unbearable sacrifices,"[301] which would of course *have* interfered with the essence of its Article 16 right to conduct business. Moreover, lest uncertainty serve to disrupt this balance, the court emphasized that it had to be possible for the ISP "to maintain before the court, once the implementing measures which he has taken are known and before any decision imposing a penalty on him is adopted, that the measures taken were indeed those which could be expected of him in order to prevent the proscribed result."[302]

Perhaps more intriguingly, the court gave substantial weight to the free expression interests of Internet users (the customers of the ISP).[303] Indeed, it required the *ISP* to take account of those interests when it chose the measures to be adopted in order to comply with the injunction, requiring that such measures be "strictly targeted, in the sense that they must serve to bring an end to a third party's infringement of copyright or of a related right but without thereby affecting internet users who are using the provider's services in order to lawfully access information."[304] Moreover, it insisted that national courts must be able to check on compliance with that obligation, requiring "national procedural rules [that make it possible] for internet users to assert their rights before the court once the [ISP's measures] are known."[305]

Telekabel thus suggests that the Court of Justice will give a reading to Article 11 of the Enforcement Directive that is properly attentive to the range of interests involved in cases alleging (what some would call) secondary liability. And developments in the U.K. courts applying Article 8(3) and Article 11 are largely consistent

---

[301] *See Telekabel*, ¶ 53.

[302] *Id.* ¶ 54.

[303] Typically, in copyright cases involving the direct interpretation of copyright legislation, the Court of Justice views copyright as a "right" of authors and treats limitations thereon designed to further user freedoms as interests or exceptions. *But see* Case C-201/13, Deckymyn v. Vandersteen EU:C:2014:2132 at ¶ 26 (CJEU 2014) (exceptions designed to achieve fair balance). This is a debate about the juridical character of user interests that can be seen worldwide. But when filtered through the balancing mechanisms of a fundamental rights analysis, the user interests are put on a par with those of intellectual property owners, and with those of intermediaries. The effects of injecting fundamental rights discourse into intellectual property adjudication on a routine basis remain to be seen. But one possible effect might be the readjustment of the conceptual relationship between author and user interests.

[304] *See Telekabel*, ¶¶ 55–56. This requirement appears to suggest that measures resulting in over-enforcement may be problematic, although it is not clear how much over-enforcement would violate fundamental user rights. The Court of Justice seemed less concerned with under-enforcement; attention to under-enforcement has prompted some national courts to deny relief on the basis that futile measures are not proportionate. The court's conclusion is to some extent the result of recognizing that intellectual property rights are themselves not absolute. *See id.* ¶¶ 58–60. According to the court, effective protection of the Article 17(2) right requires that measures "must have the effect of preventing unauthorised access … or, at least, of making it difficult to achieve and seriously discouraging internet users who are using the services [from accessing the infringing work made available in violation of Art 17(2)], thus largely rejecting the futility argument. *Id.* ¶ 63. Combined with concerns about over-enforcement, arguments highlighting under-enforcement sought to hold right-holders to an impossible "goldilocks" standard. Courts have rightfully resisted such a rule.

[305] *Telekabel*, ¶ 57; *cf.* 17 USC § 512(g) (West 2014) (counter notice possibility for those on whom DMCA takedown notice served).

with this prediction. Although Mr. Justice Arnold was not required to hand down any opinion in *L'Oréal* on remand because the case settled, he has tackled the question in the copyright context, first in *Twentieth Century Fox Film Corp. v. British Telecommunications PLC (Newzbin II)*.[306] As *Newzbin II* was decided just a few weeks after *L'Oréal* was handed down by the Court of Justice, Mr. Justice Arnold was able to rely on *L'Oréal* in interpreting Article 8(3).[307] In *Newzbin II*, he held that an order to an ISP to block access to pirate Internet sites was permissible, provided that the order was clear and targeted—that is to say, proportionate.[308] It was extremely important to him that the cost of implementation to the ISP would be modest and proportionate.[309]

Since *Newzbin II*, Arnold J has issued a series of orders and judgments regarding the application of these provisions.[310] Most of these have involved applications by copyright owners, although a recent judgment (*Cartier*) has applied broadly similar principles in a case brought by mark owners against internet service providers. Through those decisions, Arnold J has distilled the essential principles applicable when a claimant seeks an order against a service provider requiring blocking of web sites.[311]

Most importantly, perhaps, these orders can be issued against a service provider lacking the level of knowledge necessary for full secondary liability. Thus, although

---

[306] Twentieth Century Fox Film Corp. v. Britich Telecommc'ns PLC (*Newzbin II*), [2011] EWHC 1981 (Ch) (Eng.).

[307] "Section 97A [which implements Article 8(3)] must be interpreted and applied consistently with the Court of Justice's guidance in *L'Oréal v eBay*. In my judgment the court's reasoning demonstrates that the jurisdiction is not confined to the prevention of the continuation, or even repetition, of infringements of which the service provider has actual knowledge. On the contrary, an injunction may be granted requiring the service provider 'to take measures which contribute to … preventing further infringements of that kind.' Although such measures may consist of an order suspending the subscriber's account or an order for disclosure of the subscriber's identity, the Court of Justice makes it clear that these examples are not exhaustive, and that other kinds of measures may also be ordered." *Newzbin II*, [2011] EWHC 1981, [156].

[308] *See id.* at [177].

[309] *Id.* at [200] ("The order is a narrow and targeted one, and it contains safeguards in the event of any change of circumstances. The cost of implementation to BT would be modest and proportionate.").

[310] Similar measures have also been sought in other countries, but have not given rise to the extensive reasoning provided by Arnold J. The most significant judicial discussion of these issues outside the United Kingdom has occurred in the Netherlands and Germany. *Cf.* Stichting Brein v. Ziggo BV., EU:C:2017:99 (AG Szpunar Feb 8, 2017).

[311] *See* Paramount Home Entertainment International Ltd. v British Sky Broadcasting Ltd. [2013] EWHC 3479 (Ch) (Arnold J); The Football Association Premier League Ltd. v British Sky Broadcasting Ltd. [2013] EWHC 2058 (Ch) (Arnold J); EMI Records Ltd. v British Sky Broadcasting Ltd. [2013] EWHC 379 (Ch); Twentieth Century Fox Film Corporation v British Telecommunications plc [2011] EWHC 1981 (Ch) (Arnold J); [2011] EWHC 2714 (Ch) (Arnold J) (*Newzbin II*); Dramatico Entertainment Ltd. v British Sky Broadcasting Ltd. [2012] EWHC 268 (Ch); [2012] EWHC 1152 (Ch) (Arnold J); Golden Eye v. Telefonica, [2012] EWHC 723 (Ch); Football Ass'n Premier League v. British Telecomm. Plc [2017] EWHC 480 (Ch) (order directed at streaming services).

Arnold J has held that the service provider must have actual knowledge of the infringing use of its service "[t]his has been interpreted broadly to include constructive knowledge, and to require knowledge only of at least one instance of infringement. Unlike other forms of secondary liability, Section 97A [the UK implementation of Article 8(3)] does not require knowledge of a specific infringement of a specific work by a specific individual."[312]

The UK courts have also been alert to the policy concerns of over-compliance that the structure of secondary liability puts in play. In particular, Arnold J has taken account of the risks of over-blocking both in the different technical measures ordered in different factual contexts[313] and in devising mechanisms designed to allow the competing interests of internet users and service providers to be asserted.[314] (The latter innovation is one that the Court of Justice in *Telekabel* may have required in order properly to comply with the Charter of Fundamental Rights). Moreover, recognizing the "whack-a-mole" problem that creates pressure to loosen standards of secondary liability, he has built dynamic features in to the orders to allow right-owners to amend and augment blocking requests without having to initiate new, separate proceedings.[315]

In *Cartier*, the first trademark case, Arnold J's approach was endorsed by the Court of Appeal. There, the claimant sought orders requiring the five main retail internet (access) service providers in the UK (with a market share of 95% of UK broadband users) to block access by their subscribers to six websites which advertise and sell counterfeit goods. There was no suggestion that the ISPs themselves liable for trade mark infringement. There has been no specific implementation of Article 11 of the Enforcement Directive in the UK (unlike Section 97A of the CDPA, implementing Article 8(3) of the Information Society Directive). However, the court applied the CJEU ruling in *L'Oreal v. eBay* that Article 11 could be used to stop future infringements (in that case, order against host provider). At the time

---

[312] National Report of the United Kingdom, at 21.

[313] For example, in assessing the particular technical response required of the service providers, Arnold J has taken into account whether the defendant shared an IP address with other web sites. *See* Dramatico Entertainment Ltd. v British Sky Broadcasting Ltd. [2012] EWHC 1152 (Ch), [13] (Arnold J) ("IP address blocking is generally only appropriate where the relevant website's IP address is not shared with anyone else. If it is shared, the result is likely to be overblocking. In the present case, however, [the web site's] IP address is not shared. Thus IP address blocking is appropriate.") (citation omitted); The Football Association Premier League Ltd. v British Sky Broadcasting Ltd. [2013] EWHC 2058, [56] (Ch) (Arnold J) ("[T]he orders require IP address blocking of the IP address for FirstRow's domain name firstrow1.eu. FAPL's evidence is that this will not result in over-blocking since that IP address is not shared. The orders also require IP address re-routing and URL blocking for URLs at any shared IP addresses.")

[314] *See* The Football Association Premier League Ltd. v British Sky Broadcasting Ltd. [2013] EWHC 2058 (Ch), [57]–[58] (Arnold J) (endorsing permission for any affected service provider to apply to set aside the order).

[315] National Report of the United Kingdom, at 22 ("To guard against circumvention, the injunction [in *Newzbin II*] provided for a mechanism by which additional IP addresses or URLs could be added to the blocking measures, provided they had the sole or predominant purpose of enabling or facilitating access to the blocked website.").

of the application, there had been no prior application against an access provider in national court in Europe in a trade mark case (except possibly one Danish application). Arnold J relied on general injunctive powers under Section 37(1) of the Senior Courts Act 1981 to ensure compliance with EU law, and his order was upheld by the Court of Appeal.[316] The Court affirmed that relief depended upon satisfying four threshold conditions: (1) defendant is a service provider; (2) users and/or the operator of the website infringe the claimant's rights; (3) users and/or the operator of the website use the defendant's services to do that; and (4) defendant has actual knowledge of this.

But the decision turned heavily on interpreting provisions of the Charter and—apparently even more so—principles of proportionality. The Court held that "in considering the proportionality of the orders sought, the following considerations are particularly important: The comparative importance of the rights that are engaged and the justifications for interfering with those rights; the availability of alternative measures which are less onerous; the efficacy of the measures which the orders require to be adopted by the ISPs, and in particular whether they will seriously discourage the ISPs' subscribers from accessing the Target Websites; the costs associated with those measures, and in particular the costs of implementing the measures; the dissuasiveness of those measures; and the impact of those measures on lawful users of the internet".[317] Applying these considerations, Arnold J found relief warranted given the "clear infringement of trade mark owner rights", and the lack of any need for ISPs to develop new technology and the minimal costs to ISPs. There was, the judge stressed, no right of users to engage in counterfeiting.[318]

The question of alternative measures raised the question of to whom those measures must be less burdensome. The website operators were outside UK and so beyond the jurisdiction of the Court. A notice and takedown action might have been possible against the host site but Arnold J regarded this a likely to be less effective for right-owners (due to ease of movement to a new host); the judge also considered de-indexing, customs enforcement, and police action as alternatives but found these to be not as effective.[319] He also considered that the empirical evidence regarding the effect of orders under Section 97A on copyright infringement in the United Kingdom spoke well to the efficacy of these types of measures.[320]

Perhaps the most important question in *Cartier* was one of costs, including costs of implementation. Under the rules that Arnold J had developed in the copyright context, rightholders bear the costs of an unopposed application, while ISPs bear the costs of implementation (regarded by the courts as costs of doing business).

---

[316] Cartier Int'l v. British Sky Broadcasting [2014] EWHC 3354 (Ch), *aff'd*, [2016] EWCA Civ 658.

[317] *Id.* at [189].

[318] *Id.* at [196].

[319] *Id.* at [197]–[217].

[320] *Id.* at [236] There is a vibrant debate ongoing about the effect of such measures. *See, e.g.,* Brett Danaher, Michael D. Smith and Rahul Telang, Website Blocking Revisited: The Effect of The UK November 2014 Blocks On Consumer Behavior (April 2016).

As Arnold J noted, in these cases, the costs of implementation were modest compared to the scale of ISP business (and the costs could be passed on to customers). Arnold J was conscious, consistent with *Telekabel*, to bear in mind the impact of those measures on lawful users of the internet: this is most directly reflected in the option to engage in DNS blocking if the website shares an IP address with a legitimate website.

Arnold J also considered safeguards against abuse as required by Article 3(2) of the Enforcement Directive.[321] Thus, he established an option for ISPs or website operates or affected subscribers to apply to the court for variance; he insisted that a message be generated when access was denied, giving information on how to apply to a court; and, he made the order subject a 2 year sunset clause.[322]

In the Court of Appeal, Kitchin LJ upheld Arnold J's order as well as his reasoning.[323] Kitchin LJ also accepted that Arnold J was entitled to order that ISPs bear the costs of implementation. However, on this point, Briggs LJ dissented, and would have imposed upon the applicant for a trademark blocking order the "specific cost incurred by the respondent ISP in complying with that order, but not the cost of designing and installing the software with which to do so whenever ordered." Despite the modest costs, he thought the judge was wrong in principle in concluding that the ISP ought usually to pay the costs of implementation.[324] And the question of costs is one of the central points of debate as seen also in the recent first decision of an Australian court ordering the award of injunctive web blocking measures.[325] In a follow-up opinion, *Cartier II*, the defendants did not appear, and adopted a neutral stance, reflecting practice that had grown up in case of Section 97A orders.[326] There, HHJ Hacon also noted that notices of infringement had previously been sent to website operators before turning to access providers (an issue being debated in national courts in the copyright context).

The system that is being constructed through these applications has a regulatory (quasi-public law) character, as is perhaps inevitable when the underlying EU legislation on which it is built is aimed at preventing and deterring (a broad unidentified group of)[327] future infringements. This character is cemented by a number of aspects of the litigation practice that has developed, all of which further univeralise the particularities of a single injunction. Thus, although the first order issued by the English courts was issued against a single service provider, equivalent orders were soon obtained against the other principal UK service providers once its legality was

---

[321] *Cartier*, at [264].

[322] *Id.* at [265].

[323] [2016] EWCA Civ 658 at [35].

[324] *Id.* at [211].

[325] *See* Roadshow Films Pty Ltd. v Telstra Corporation [2016] FCA 1503 (Dec 15, 2016); *see also* § 115A of the Copyright Act 1968 (Cth).

[326] Cartier Int'l v. British Telecomm [2016] EWHC 339 (Ch) [*Cartier II*].

[327] Injunctions typically are aimed at particular conduct of particular defendants. And the notice and takedown cases in Germany have struggled with how far they can depart from this principle under the storerhaftung doctrine. *See supra* note 274.

established.[328] Indeed in later cases, the claimants sought the relevant order simultaneously against all major service providers.[329] And the position has now been reached (for costs reasons discussed below) where service providers rarely contest the orders.[330] This does not mean that the service providers are not involved in the process. Indeed, they frequently negotiate the terms of the orders in private with the claimants, and then do not publicly contest the application made by the claimants.

These developments suggest that perhaps there is a smaller gap than might be thought not only between this private law device and public law regulation, but also between this developing system and more obvious instances of private ordering. Like those instances (such as the Memorandum of Understanding concluded in the EU between Internet platforms and large brand owners, or the notice and takedown policies established by individual service providers and their collective organisations) a large swathe of online conduct is being regulated by private agreement that is more or less transparent.[331] Although several of those private initiatives in include mechanisms to enhance transparency[332] or address over-enforcement,[333] the system devised through successive judicial interventions does appear potentially to provide the most comprehensive suite of opportunities for public scrutiny, protections against abuse and over-compliance, and vindication of competing interests.[334] Moreover, the continual refinement of the terms of the orders issued is testament to the capacity of the judicial system to adapt and improve mechanisms in light of experience and technological evolution.[335]

Different national courts have adopted different approaches to the detail to be set out in the orders. The Austrian orders considered by the Court of Justice in *Telekabel*

---

[328] *See* National Report of the United Kingdom, at 22.

[329] *See id.* at 22; *see also* EMI Records Ltd. v British Sky Broadcasting Ltd. [2013] EWHC 379 (Ch); Dramatico Entertainment Ltd. v British Sky Broadcasting Ltd. [2012] EWHC 268 (Ch); [2012] EWHC 1152 (Ch).

[330] *See* National Report of the United Kingdom, at 31.

[331] For a critique in the copyright context, see Matthew Sag, *Internet Safe Harbors and the Transformation of Copyright Law*, 83 NOTRE DAME L. REV . – (2017) (forthcoming). Private corporate policies, or Best Practice statements have grown up in a number of areas. Some are adopted by particular organisation; others are industry standards. *See* National Reporter of the United Kingdom, at 24–26 (discussing a number of policies voluntarily adopted by service providers as regards blocking of indecent images of children, deletion of links to unlawful conduct from search indices, demoting sites that contain a high percentage of infringing material, and abusive registration of domain names).

[332] *See* Google Inc., *Transparency Report—Requests to Remove Content* (December 31, 2013) http://www.google.com/transparencyreport/ .

[333] *See supra* text accompanying notes 216–218 (MOU abuse protections).

[334] Regulatory systems such as the so-called graduated response often also make provision for appeals by subscribers who are affected by service provider conduct. *See* National Report of the United Kingdom at 29.

[335] *See* National Report of the United Kingdom, at 31 (suggesting that the lack of any appearance in practice of either the service provider or website operator "has the potential for orders to be made on what is effectively an ex parte basis, as a result of which unintended over-blocking (or other downstream complications) are inevitably likely to occur").

were open-ended, allowing the service provider to craft measures reasonable in light of its capacities. However, courts in the United Kingdom have issued very specific orders.[336] This recognizes that procedural matters largely remain a matter of national competence within the EU, to be devised within the jurisprudential tradition of the different member states. And this diversity might also contribute to the learning process, treating the member states as laboratories in which to test the ideal approach—if such exists—to these questions before imposing a European norm.

One of the explanations for the rise of the secondary liability has been an effort to shift the costs of deterring wrongful conduct to intermediaries. And the question of costs (both litigation and compliance) can also shape intermediary behavior within the systems of shared enforcement responsibility that have arisen in recent years—whether under notice and takedown regimes, or under a system of disclosure and blocking.[337]

The costs regime that has arisen in the United Kingdom departs somewhat from the general rule in that jurisdiction. Thus, a service provider that reasonably resists disclosure of customer information will have its litigation costs paid by the claimant who secures a disclosure order.[338] But in blocking cases, a service provider who unsuccessfully resists an application will have to pay a share of the claimant's litigation costs.[339] As a result, service providers tend not to appear to oppose applications for blocking orders.[340]

It is clear that service providers can voluntarily contribute to efforts to restrain unlawful conduct, or to make access to unlawful material more difficult.[341] It is a separate question, of course, whether they should be forced or encouraged to make those contributions and who should pay for them.[342] There is no general Good

---

[336] *See id.* at 23 ("English courts have directed very specific measures (including the URLs to be blocked, the blocking technology to be adopted, the methods for adding mirror websites to the block-list, and permitted maintenance periods)").

[337] It is perhaps unsurprising that the draft costs-sharing order published by the UK regulator (Ofcom) to govern the UK's graduated response system was challenged by service providers. *See* R (British Telecommunications plc) v Secretary of State, 2011] EWHC 1021 (Admin), *aff'd* [2012] EWCA Civ 232. Much of the fight is over the cost of enforcement and responsibility therefor.

[338] *See* National Report of the United Kingdom, at 23.

[339] *See id.* at 23.

[340] *See id.* at 23; *id.* at 31 ("The emerging practice under Section 97A is problematic. ISP respondents tend not to appear or make submissions in relation to website blocking orders that are sought against them, largely on account of the risk of an adverse costs order. The website operators themselves are very unlikely to appear. This has the potential for orders to be made on what is effectively an ex parte basis, as a result of which unintended over-blocking (or other downstream complications) are inevitably likely to occur.").

[341] *See id.* at 25 ("Since August 2012, Google has taken account of the number of valid copyright complaints made against a website when determining its search ranking, demoting sites which contain a high percentage of infringing material. Keywords related to copyright infringement are no longer suggested to users by Google's search auto-completion tools. The same approach is taken to prevent suggesting terms related to pornography, violence, and hate speech. Upon notification, Google also excludes the display of paid keyword advertising on websites it considers to be associated with copyright infringement.") (citations omitted).

[342] *See* L'Oréal S.A. v. eBay Int'l AG, [2009] EWHC 1094, [370]–[375] (Arnold J).

Samaritan policy embedded in most laws.[343] The path being forged by Arnold J and other courts in Europe under these mechanisms is an attempt to reconcile that capacity to help (and thus effectuate legal rights) with the many dangers of commandeering service providers to police online behavior just because they can. In this respect, the tension is no different than plays out in assessing the wisdom of secondary liability proper. But the mechanism may offer a less binary option than that presented by the secondary liability debate.

One of the other attractions of pursuing a secondary claim against an online service provider is the possibility of effectively restraining unlawful conduct in a number of jurisdictions, which might otherwise require serial actions against primary wrongdoers in a number of countries. This strategic benefit appears most likely to be realized where the mechanism for relief against the service provider is one devised through private ordering or voluntary cooperation. Thus, the many efforts voluntarily undertaken by Google to police third party wrongdoing apply across borders.[344]

This should not be a surprise; the nation-state has historically been slower to disregard political borders than commercial actors.[345] However, disregard of territorial boundaries has in some contexts prompted objections,[346] and courts do pay regard to territorial concerns in deciding cases of service provider secondary liability.[347] Remedies do tend to be territorially defined, even as regards service providers.[348]

The approach that courts have adopted to make assessments of proportionality has been articulated and approved in a number of decision of the English courts.[349] The UK National Report notes that it has proved to be a limited practical constraint on English courts granting injunctive measures,[350] although the Court of Justice has clearly outlawed particular measures as both falling afoul of the general principle of proportionality and violating the fundamental rights of the internet user.

---

[343] Such rhetorical statements are often (sometimes helpfully) used to frame the debate, but resolution requires some compromise of competing claims. *See, e.g.,* National Report of the United Kingdom, at 28 (describing the cost sharing order proposed as part of the UK graduated response system as "a compromise [for] both service providers and right holders, who continue to fight between the 'polluter pays' principle (arguing that because ISPs enjoy benefits from infringement in the form of higher subscription revenues, they should contribute to right holders' enforcement costs) and the 'private property' principle (arguing that because copyright is simply private property, its enforcement should not be subsidised by third parties).").

[344] *See id.* at 25. It is not clear whether there is a universal corporate standard that continues to be applied nationally, or whether a single Google action will apply to national versions of its search engine. Obviously, the latter most fully ameliorates the costs of territoriality.

[345] *See* Graeme B. Dinwoodie, *A New Copyright Order: Why National Courts Should Create Global Norms*, 149 U. PENN. L. REV. 469 (2000).

[346] *See* National Report of the United Kingdom, at 26 (noting concerns expressed by foreign registrars about seizures of domain names in non-British name space).

[347] *See id.* at 37 (noting targeting analysis in copyright cases and threshold requirement of substantial publication within the jurisdiction in libel cases).

[348] *See id.* at 38.

[349] *See* Golden Eye (International) Ltd. v Telefónica UK Ltd. [2012] EWHC 723, [117] (Ch), *approved in Viagogo (SC)*, [45] (Lord Kerr JSC).

[350] *See* National Report of the United Kingdom, at 33.

Although Court of Justice case law makes reference to competing fundamental rights and the proportionality test applied by national courts talks of the "values" underlying rights and the "justifications" for restricting rights, the analysis has tended to adopt a pragmatic hue. Thus, analysis in the United Kingdom has focused heavily on the costs of implementing particular measures.[351] And certainly the lowest-cost-avoider principle underlying Article 8(3) of the Information Society Directive, as noted in the recitals to that instrument, encourages that kind of approach. But the relevant principles identified by the Court of Justice appear to call for some for more normative content, and fuller engagement with the competing values.

Thus, the English courts have taken into account enforcement mechanisms in place to combat child pornography in developing orders that are proportionate and no burdensome in costs. But of course, insofar as these orders reflect a balance of competing fundamental rights under the Charter of Fundamental Rights, one might expect a different calculus in cases of intellectual property infringement and child pornography. The weight of the relevant interests might plausibly be quite different.

VI. Assimilation and General Principles

This final Part of the Chapter addresses two final conceptual questions that are raised by the discussion above. First, to what extent does (and should) the secondary liability of online service providers depart from the secondary liability standards generally, and from offline applications in particular? Second, is secondary liability simply the application of general principles of tort law, divorced from the policies of particular areas of law that define the counterpart primary liability?

On the first question, it is typical for a positively constructed standard of secondary liability—for example, the elements of a claim of joint tortfeasorship under English law—to reflect long standing principles of tort law that applied offline.[352] But there may be a difference in the way that courts apply these general standards in the online context. As Jaani Riordan has noted regarding English courts, "[a]lthough there is no formal difference [between the test applicable offline and online intermediaries], in practice courts may be willing to infer the required mental element in ways that extend liability for online service providers or interpret the general standards in a way that leads to broader prima facie liability being imposed.

---

[351] See id. at 35 ("Orders carrying compliance costs in the region of £5000–£10,000 have been held not to infringe [the freedom to conduct business]. That assessment often folds into the proportionality").

[352] See id. at 7–8 ("English secondary liability principles have been developed incrementally by judges since around the mid-sixteenth century, though obviously not by reference to online service providers, or indeed any particular class of defendants... During the twentieth century, English courts sought to bring new technologies of reproduction and dissemination within the established patterns of secondary liability."); id. at 8 ("the same general standards are applied to online service providers as are used to establish secondary liability against other classes of defendants").

This is sometimes fuelled by a perception that because such services enable primary wrongdoing to occur, they should be obliged to provide an enforcement 'solution'.[353]

Riordan argues the sympathy that judges might have for imposing some enforcement costs on online intermediaries engaging in trade that "carries with it a higher risk of infringement than more traditional methods of trade" has not caused them to adapt the general standards, which he suggests might show that "general standards are ultimately more important than any policy considerations thought specially applicable to service providers."[354] Of course, announcing adherence to general standards while actually making online specific adaptation in their applications brings with it other concerns, not least being the transparency of the law.

In contrast, insofar as the secondary liability of intermediaries is effectively defined negatively by immunity provisions that are commonly found in legislation motivated by e-commerce policy, the standard appears on its face to be more online-specific. Of course, if falling outside zones of immunity does not of itself result in liability, then the effective standard is constructed by an amalgam of generally applicable and internet-specific standards. And this might be the best way of accommodating the diverse range of policy goals at issue in these cases.[355]

On the second question, of the relationship between the standards in different bodies of law and reliance on horizontal principles, again, insofar as courts articulate standards that positively delineate secondary liability, they frequently purport simply to be applying general standards of private law that do not differentiate between the bodies of law at issue.[356] Courts in civil law countries are more likely to assess secondary liability by application of a general tort provision, but even

---

[353] See id. at 8 (discussing L'Oréal SA v eBay International AG).

[354] See id. at 8.

[355] Although much of the discussion has revolved around whether offline principles should be extended to the online context, it is intriguing that the principles of Article 11 of the Enforcement Directive—first applied in the online context—have been applied by analogy (and literal text) in the offline context. Thus, in Tommy Hilfiger Licensing LLC v. Delta Center, the tenant of market halls who sublet the various sales points situated in those halls to market-traders, some of whom use [the locations] in order to sell counterfeit branded products, was held to fall within the concept of 'an intermediary whose services are being used by a third party to infringe an intellectual property right' within the meaning of Article 11. See Case C-494/15, Tommy Hilfiger Licensing LLC v. Delta Center EU:C:2016:528 (CJEU 2016). The Court of Justice declined to draw any distinction between online and physical marketplace (although it did not decide the hypothetical question whether suppliers of electricity were intermediaries).

[356] See Sea Shepherd at [40] (Lord Sumption dissenting) ("In both England and the United States, the principles have been worked out mainly in the context of allegations of accessory liability for the tortious infringement of intellectual property rights. There is, however, nothing in these principles which is peculiar to the infringement of intellectual property rights. The cases depend on ordinary principles of the law of tort."). This is not without exception—for example, the "substantial non-infringing use" doctrine announced in Sony is formally found in copyright law. But that doctrine clearly drew from the "staple article of commerce doctrine" in patent law. Thus, at a substantive level, it may be the application of a general principle of private law. And one might questions whether, although developed in articulating positive standards of liability, one should not regard that doctrine as an immunity doctrine that operates to define secondary liability in a negative fashion.

common law courts frequently comment that the relevant standard for secondary liability is an extension of tort principles. This again appears to draw such common law doctrine closer to the civil law approach. And it also presages the idea that standards are in fact horizontally applied throughout private law.

As Mark McKenna has persuasively demonstrated in the context of trademark law, if American courts were true to their frequent claim that secondary liability is grounded in general principles of tort, the entire *method of analysis* might look substantially different.[357] If one took seriously the Second Circuit's suggestion that the standard for secondary liability in trademark law was derived from general principles of tort law,[358] analysis of considerations commonly informing lowest cost avoider analysis might arguably have been relevant.[359] Under such an approach, courts might take into account the fact that while brand owners may be best positioned to determine infringing conduct, intermediaries may be best positioned to implement preventive measures[360]—and, indeed, may have the expertise best to do so in technologically innovative ways, as many have done.[361]

---

[357] *See* Mark P. McKenna, *Probabilistic Knowledge of Third-Party Trademark Infringement*, 2011 STAN. TECH. L. REV. 10, 26 (2011), *available at* http://journals.law.stanford.edu/sites/default/files/stanford-technology-law-review-stlr/online/mckenna-probabilistic-knowledge.pdf. McKenna makes the point that "[i]f … courts were really to take seriously the tort law analogy, they would not treat cases involving probabilistic knowledge as secondary trademark infringement cases at all; they would treat them as negligence cases." *Id.*

[358] *See Tiffany*, 600 F. 3d at 103 ("Contributory trademark infringement is a judicially created doctrine that derives from the common law of torts.").

[359] *See* McKenna, *supra* note 357, at [26]. McKenna makes the point that "[i]f … courts were really to take seriously the tort law analogy, they would not treat cases involving probabilistic knowledge as secondary trademark infringement cases at all; they would treat them as negligence cases." *Id.* So conceptualized, the U.S. approach might look less conceptually divergent from the approaches followed by some courts in continental Europe (as well as the forms of limited relief available under Article 11 of the Enforcement Directive). *See id.* ("[I]f trademark secondary liability really derived from tort law, liability would exist in cases of probabilistic knowledge only when the defendant unreasonably failed to take precautions in the face of the known risk of infringement. Unreasonableness would be measured, as it generally is in tort cases, by evaluating the probability of harm to the plaintiff and the potential magnitude of that harm, and comparing the product to the cost of the foregone precautions. Secondary liability cases also would entail analysis of causation-in-fact and proximate causation…"); *see also supra* text accompanying note 144 (liability under French law) and notes 263–351 (discussion of Article 11).

[360] The involvement of intermediaries (sometimes as a result of mutually beneficial arrangements between the brand owner and intermediaries) can often be very effective. *See* Kathy Chu & Joanne Chiu, *To Woo Lux Brands, Alibaba Purges Resellers*, WALL ST. J., Aug. 11, 2014, at B1 (noting effectiveness of intervention by Alibaba, the Chinese owner of the Tmall retail site, in reducing the number of counterfeit and gray market goods available on the site, and tying Alibaba's intervention to a brand's opening an official store on the Tmall site).

[361] *See* Beebe, *supra* note 114, at 5 (discussing the automated "counterfeit filter" used by Google to assess the bona fides of landing pages). Intermediaries may also be able to implement measures that, while not preventing infringement, make their customers aware of potential trademark issues. *See* Ginny Marvin, *Google Keyword Planner Now Shows Trademarked Terms*, SEARCH ENGINE LAND (Nov. 27, 2013), http://searchengineland.com/google-keyword-planner-now-shows-trademarked-terms-178359.

Of course, in some recent cases, the US courts appear cognizant of the role of the principle of lowest-cost-avoider; for example, aspects of lowest court avoider analysis might in fact have influenced both the trial and appellate courts in *Tiffany*, despite their disavowal (explicit in the case of the lower court).[362] The trial court spent substantial time discussing the sizeable efforts that eBay had made to detect and prevent infringement, and the Second Circuit noted the (purportedly irrelevant) record on this point.[363] Moreover, both courts noted that eBay was hardly in a position to determine which Tiffany products were counterfeit and which were not, because it never saw or inspected the goods.[364]

However, despite frequent claims of derivation of paternity, once general standards are transplanted and applied in the context of subject-specific claims (such as copyright or trade mark), they can to some extent take on the hue and policy concerns of that law. And perhaps this make more sense if secondary liability is simply a device for effective enforcement of discrete primary rights. Thus, in the United States, despite claims of common roots in the common law, the standard articulated by the Supreme Court for trademarks in *Inwood* and copyrights in *Grokster* are different—and the Court has stressed as much. Copyright in particular has developed, through the *Sony* rule, innovation-related doctrinal adjuncts. This is because secondary liability claims bring into focus the effect of liability upon technological development. The U.S. Supreme Court has commented that secondary infringement claims more directly involve the trade-off between "supporting creative pursuits … and promoting innovation in [the] new communication technologies [that make infringement possible]."[365] The U.S. Supreme Court in *Sony* looked to the staple article of commerce doctrine from patent law to fashion the doctrine of substantial non-infringing use.[366]

Within the context of intellectual property law, the US Supreme Court has rejected the idea that the secondary liability standards should be assimilated.[367]

---

[362] *Tiffany*, 576 F. Supp. 2d at 518 ("Certainly, the evidence adduced at trial failed to prove that eBay was a cheaper cost avoider than Tiffany with respect to policing its marks. But even more importantly, even if it were true that eBay is best situated to staunch the tide of trademark infringement to which Tiffany and countless other rights owners are subjected, that is not the law."); *cf. Tiffany*, 600 F.3d at 109 (noting, in response to the argument that mark holders could not bear the cost of enforcement, that, "[w]e could not, even if we thought it wise, revise the existing law in order to better serve one party's interests at the expense of the other's").

[363] *Tiffany*, 600 F.3d at 109.

[364] *Id.* at 98–99.

[365] *Grokster*, 545 U.S. at 928.

[366] *See Sony*, 464 U.S. at 442.

[367] In *Sony*, the U.S. Supreme Court noted the kinship on this topic between copyright and patent law, but rejected the idea that there might be a similar kinship between copyright and trademark law. *Id.* at 439 n.19. Indeed, the Court refused to apply its own standard for contributory trademark infringement, which it clearly saw as harder to satisfy (at least in the fact pattern at issue in *Sony*). *Id.* ("If *Inwood*'s narrow standard for contributory trademark infringement governed here, respondents' claim of contributory infringement would merit little discussion. Sony certainly does not "intentionally induc[e]" its customers to make infringing uses of respondents' copyrights, nor does it supply its products to identified individuals known by it to be engaging in continuing infringement

But these concerns are, to some extent, also present in trademark law. In trademark law, we must balance the rights of the mark owner with enabling legitimate development of innovative technologies that allow new ways of trading in goods (or offering services). If the concern for technological development is cross-cutting feature of secondary liability claims, one might be able to extrapolate from the copyright context to develop principles applicable more generally to assess secondary liability claims. Once the broader significance of online secondary trademark infringement claims are recognized, some of that doctrine may migrate to other fields of intellectual property law such as trademark law.[368]

But even accepting that the standards might in fact be horizontal and dependent upon general principles of law when positively established, a different picture emerges if one examines rules defining secondary liability negatively. There, although there are some immunity provisions that come close to apply horizontally, such as the E-Commerce Directive in the EU or Section 230 of the CDA in the United States, one does find differences across the different immunity rules. Indeed, even when there appears to be a horizontal instrument—such as the E-Commerce Directive—the conditions of immunity might render them more readily applicable to one type of cause of action than in another. This is perhaps inevitable when these rules are more likely to be legislated—the result of lobbying in a specific context—rather than developed by courts who are more prone to be exposed to a range of fact patterns across the breadth of private law and seek to decide case according to some more general standards.

Some European scholars have questioned whether European trademark law should so readily have adopted a system of notice and takedown that mimics that found in copyright law.[369] But to the extent that secondary liability trademark claims raise similar concerns about chilling technological development and about the dangers of over-enforcement of rights where technologies are used for both infringing and non-infringing purposes, a horizontal approach does not seem wholly inappropriate. Mark Lemley has persuasively argued that at least some aspects of liability—such as safe harbours—be standardized to afford the requisite certainty to OPSs and to prevent opportunistic distortion of the claims asserted against them.[370]

In the context of assistance measures imposed on OSPs without liability, the principles have developed most substantially in the context of copyright law. However, as Mr. Justice Arnold indicated in *Newzbin II*, the principles applicable to

---

of respondents' copyrights"). Recent U.S. trademark decisions endorse the suggestion that the standards in copyright and trademark cases are different. *See* Perfect 10, Inc. v. Visa Int'l Serv., Ass'n, 494 F.3d 788, 806 (9th Cir. 2007) ("The tests for secondary trademark infringement are even more difficult to satisfy than those required to find secondary copyright infringement.").

[368] *Cf.* Dogan, *supra* note 58, at 504, 506–07 (suggesting that courts applying the trademark doctrine properly are drawing, and should draw, upon principles developed in copyright cases).

[369] *See* Martin R.F. Senftleben, *An Uneasy Case for Notice and Takedown: Context Specific Trademark Rights* (Mar. 16, 2012), *available at* www.ssrn.com/abstract/=2025075.

[370] *See* Lemley, *supra* note 36.

determine appropriate measures under Article 11/Article 8(3) (and whether such measures comply with European law) transcend intellectual property law. But that does not mean that Article 11 and Article 8(3) will be applied in exactly the same fashion in both types of disputes. There are important differences in both the factual and legal contexts in which the applicable provision would be invoked in the two different regimes.[371]

For example, as a factual matter, file-matching technology is more likely to be able to accurately identify copyright-infringing files, and thus can be implemented at lower cost than the human-intensive assessments that might be required in the case of trademark claims. It may be that the costs of identifying infringements should fall more heavily on mark owners than copyright owners in weighing the variables relevant to a reasonable compromise of duties.[372] Article 11 is based on the normative principle of lowest-cost-avoider, which surely allows for such context-specific assessment. As a legal matter, the range of permissible uses may vary between copyright and trademark, altering the calculus of effect on the conduct of legitimate business.[373]

Nor does it mean that measures appropriate outside the context of intellectual property should, for mere reasons of practical convenience and minimal cost apply to copyright cases under Article 8(3) of the Information Society Directive or trademark cases under Article 11 of the Enforcement Directive. Thus, in reaching the conclusion that some of the orders issued against intermediaries under Section 97A of the Copyright, Designs and Patents Act 1988 were appropriate, Mr. Justice Arnold appears to have been influenced by the fact that the ISPs in question already maintained a list of sites blocked in order to prohibit access to child pornography and it would be minimally burdensome to add to that list sites hosting copyright infringements.[374] As a pragmatic matter, this seems thoroughly sensible. Viewed through the prism of weighing conflicting normative values, as occurs in the context of fundamental rights, it is not so obvious that the weight attaching to the social interest implicated in restricting access to child pornography is the same as that to be afforded to the protection of copyright.

---

[371] There may also be differences between the types of measures appropriate in counterfeiting cases and those appropriate in cases of trademark infringement, again because the factual and legal context may be different, as evident perhaps from the decision of the drafters of the MOU to focus on counterfeiting.

[372] The factual context may also vary over time. Thus, while most trademark cases will now involve some form of payment processing and shipment of a physical article, it is not clear how 3D printing might alter online practices with respect to counterfeit goods.

[373] See Senftleben, *supra* note 369, at 2 (focusing on the more limited nature of trademark rights). In fact, the relationship between the strength of copyright and trademark rights may be less linear than Professor Senftleben suggests. But the basic point that the systems are different is well-taken.

[374] *Newzbin II*, [2011] EWHC 1981.

In short, handling service provider liability by applying doctrinal elements of primary liability *without* adaptation might cause courts to ignore relevant policy concerns that are necessary to protect vital roles played by the intermediary.[375] By the same token, failing to take account of the nature of the underlying primary wrong might fail to vindicate the purpose of the legal regime in question.[376]

Conclusion

The secondary liability of online service providers remains formally governed by somewhat different standards under diverse national laws, in part because the underlying question of "secondary liability" is conceptualised quite differently in a range of countries. Yet, the practicalities of online commerce have forced legislators, courts, and private actors to find some common approaches, conceptual (and terminological) differences notwithstanding. In particular, transnational practice is increasingly converging around systems of notice and takedown (and more or less demanding variants thereof) and on the imposition of some responsibility on intermediaries to stop wrongful conduct of third parties without finding them secondarily liable for the conduct of their users. The former mechanism requires a greater degree of transparency and public oversight in order to ensure its legitimacy, while the latter (if subject to the same public oversight) can not only assist in the prevention of wrongdoing but also help reduce the pressure to subject well-intentioned OSPs to threats of liability and thus to impede the provision of services vital to contemporary society. Both mechanisms are being developed pragmatically, and efficiency suggests that they should ideally be trans-substantive in nature, allowing OSPs a high degree of guidance regarding expected conduct. But effectuating the policies underlying the system of primary liability at issue in such cases requires that attention also be paid to the different public values at play in different legal claims.

---

[375] *See* Stacey L. Dogan & Mark A. Lemley, *Grounding Trademark Law Through Trademark Use*, 92 Iowa L. Rev. 1669 (2007).

[376] Of course, the service provider may have engaged in macro-behaviour that to some extent is independent of particular disaggregated acts by third parties, and which might appropriately be viewed in an autonomous light. *See* Dinwoodie and Janis, *Lessons, supra* note 62, at 1717. There may be certain conduct of the search engine that causes confusion wholly independent of the content of particular ads, such as the way in which search results are presented on the search results page. *Cf.* Dogan, *supra* note 73, at [19]. One may ask, for example, whether there is clear differentiation of purchased advertising and so-called organic listings, and whether advertisements are clearly labelled as such. *Cf.* E-Commerce Directive, art. 6 (a).

Chapter 2
Secondary Liability of Internet Service Providers in Poland

Xawery Konarski and Tomasz Targosz

Introduction

Poland is a country with a high level of Internet piracy.[1] Liability of Internet service providers should therefore be viewed as a crucial component of the legal system, constantly honed by widespread application. This has not been so. Even though some basic tenets of such liability have been repeatedly discussed in legal literature and rather rare court decisions,[2] this area of law remains underdeveloped. There is a rather widespread conviction that under existing law, such potential liability is practically insignificant, in that the conditions of liability are very hard to meet and the remedies available to the injured parties ineffective. Such a state of affairs is not necessarily justified by the legal instruments that Polish law offers. It is rather a consequence of adhering to simple legal solutions and the unwillingness to use creatively various possibilities that Polish law actually offers. Only recently have the

---

This paper is an updated and expanded version of the article "Secondary liability of service providers" published in "Rapports polonais: XIXe Congrès international de droit comparé = XIXth International Congress of Comparative Law: Vienne, 20–26 VII 2014, Łódź 2014, pp. 407–422.

[1] See e.g. "Analiza wpływu zjawiska piractwa treści wideo na gospodarkę w Polsce [Analysis of the impact of piracy with regard to video content on the Polish economy], report prepared by PwC, available at: https://www.pwc.pl/pl/publikacje/piractwo/analiza_wplywu_zjawiska_piractwa_tresci_wideo_na_gospodarke_w_polsce_raport_pwc.pdf

[2] That is, these are rare when compared to the scale of various infringements on the internet and the number of court decisions on other liability issues.

X. Konarski (✉)
Polish Chamber of Information Technology and Telecommunications and Internet Advertising Bureau Poland, Warsaw, Poland
e-mail: xawery.konarski@traple.pl

T. Targosz
Department of Intellectual Property Law, Jagiellonian University, Krakow, Poland
e-mail: tomasz.targosz@traple.pl

© Springer International Publishing AG 2017   73
G.B. Dinwoodie (ed.), *Secondary Liability of Internet Service Providers*,
Ius Comparatum – Global Studies in Comparative Law 25,
DOI 10.1007/978-3-319-55030-5_2

harbingers of change appeared; it is far from certain whether they are just a temporary occurrence, or signal a lasting modification of approach.

The term 'secondary liability' (or its literal translation) is alien to the Polish law. Polish law does not even recognise special regimes of liability for what is termed 'secondary infringements' even in the areas of law where such regulations are particularly common in other jurisdictions, such as in patent law.[3] Before one engages in a more detailed presentation of the legal issues involved, the very concept of 'secondary liability' must be translated into the language that Polish law understands. The most similar notion would be that of an indirect infringement. Indirect infringements of rights or legally protected interests have been known in Polish private law for a long time and have been thoroughly analysed in theoretical works.[4] An indirect infringement, as the name suggests, usually occurs when an action that directly infringes is predetermined (made possible) by another action, more remote in causal nexus, but still indispensable for the infringement to take place.

Despite this general characteristic, 'indirect' infringement may be an ambiguous concept. If we are to assume that this term has any descriptive accuracy, we should understand indirect infringements as a distinct kind of *infringement* of a given right. In other words, an indirect infringement suggests that human behaviour described as such constitutes an infringement of a right (usually a property right and rights following the same design such as intellectual property rights). Indirect infringements are therefore a class of infringements, and therefore closely related to direct (normal) infringements. For an indirect infringement to occur the scope of the right in question has to be sufficiently broad so that it could accommodate not only typical encroachments, but also those that are more remote.

Indirectness of an infringement usually means two things. Firstly, and self-evidently, it cannot be a direct infringement. For example, a direct infringement of an intellectual property right means using the protected subject matter without the right-holder's consent and contrary to law. The law may further require that such use have additional characteristics (a special way of use), which is for example of importance for trademarks.[5] An indirect infringement must be consequently something else than a 'use' in the direct infringement sense, but at the same time must bear some connection to it, as otherwise there would be no justification for expanding the sphere of liability. Indirect infringement is therefore behaviour resulting in direct infringement or at least creating a substantial risk of direct infringement. There must be a causal connection here, between the intermediate element (i.e. an intervention of another factor) and the resulting direct infringement. Indirect infringement in this scenario is human behaviour that makes it possible for others to complete the required elements of a direct infringement. Although this is the most common template, consider another example: A erects a structure on his property (land) obscuring all the light (or draining all the water) so that B may not enjoy

---

[3] Polish patent law does not regulate indirect patent infringements, unlike most other European countries or the US.

[4] Most recently Lackoroński 2013 [Civil liability for indirect infringements].

[5] Kur 2002, 41 ff.

his property to the extent that he should be allowed. There is no third party intervention, but Polish law will consider A's action to constitute an infringement of B's
property.[6] Because A is not using B's property, it will often be called an indirect
infringement.[7]

In both cases the fact that the law considers indirect infringements illegal must
follow from the nature and scope of the protected right. It is especially possible
when such rights are defined broadly, the best example being classic property rights.
Intellectual property rights, and in particular industrial property rights, are however
usually defined in a different way. The balance of interests these rights must reflect
manifests itself not only in the part of the regulation concerning exceptions and
limitations, but often in the very scope of the right and the requirements of
infringement.

Even though such an understanding of indirect infringements could be at least
discussed with regard to infringements in respect of which internet service providers are usually implicated (e.g. copyright law), it is very rarely mentioned in the
context of ISPs' liability, with the exception of some decisions concerning personality rights.[8] What is usually understood as the scope of secondary liability of ISPs
belongs in Poland to the ambit of tort law. Since most infringements on the internet,
such as infringements of personality rights or infringements of intellectual property
rights, are civil law torts, it feels natural to ask whether the general tort law rules
could not accommodate ISP liability. These will obviously be rules pertaining to the
question as to who, other than the direct perpetrator, may be liable for torts. In
Poland 'secondary liability' is consequently primarily an issue of tort law.

The search for liability must begin with finding suitable legal grounds in the
existing provisions of law. The most popular solution is to apply art. 422 of the
Polish civil code (CC), regulating liability of persons inducing a tort, assisting in the
commission of a tort, or consciously taking advantage of the damage caused to
another.[9] The conditions of liability are: (a) an intentional[10] action consisting in
inducing, assisting or exploiting a tort committed by a third party; and (b) according
to the prevailing although probably incorrect view, an adequate causal link[11] (with
the exception of taking advantage of the damage caused to another)[12]. The route
through art. 422 suffers, however, from serious disadvantages, the most conspicu-
ous of which are the requirements that the liable person must act with knowledge of
a third party committing a tort[13] and that this must generally be knowledge of a

---

[6] Picker 1972, 110 f.

[7] Katner 1982. [Protection of Property against Indirect Infringements].

[8] Judgement of the Regional Court in Warsaw, August 8, 2014, II C 489/12.

[9] Barta and Markiewicz 2009; Barta and Markiewicz 2013, 364, Pacek and Wasilewski 2008.

[10] Now the prevailing view: recently for example the Court of Appeal in Warsaw, 13 June 2014, I ACa 1754/13.

[11] Lewaszkiewicz-Petrykowska 1978, 108 and 110 [Causing Damage by Several Persons].

[12] Taking advantage of the damage resulting from a tort cannot *cause* it since it may only take place *after* the tort has been committed. Hence, an adequate causal link cannot be required.

[13] Banaszczyk 2015, art. 422, point 8,

specific person infringing specific rights. Knowledge that some users of an internet provider's services infringe some third party rights is insufficient. As a consequence, the boundaries of liability are quite similar to the so-called safe harbours,[14] so that the latter do not in fact considerably privilege providers in comparison to the general conditions of liability.

There have been some proposals to handle indirect infringements differently (but still from a tort law perspective). One such way would be to rely on the general tort clause of art. 415 CC, modeled after the French art. 1382 code civile.[15] It can be argued that creating a source of danger for third party rights, such as operating a service widely used for infringements, without taking all the necessary and reasonable steps to prevent infringements from happening, is a tort in its own right.[16] If so, then it does not have to be intentional; any fault (including negligence) would be sufficient. Similar in effect is the idea to consider the term "infringes" used in statutes regulating specific IP rights as referring also to indirect infringements (the same way infringement of property in things is understood as comprising both direct and indirect infringements).

With regard to injunctions, it is possible to apply art. 439 CC. This provision states that any person who is threatened directly by damage as a result of another person's behaviour, in particular as a result of lack of proper supervision of the operation of an enterprise or establishment run by that person, may demand that such a person undertakes measures necessary to ward off the peril and where necessary, to provide an adequate collateral.

There are no theoretical impediments for the above named grounds of liability to be applied independently of one another, and they are all 'general' in the sense that they have not been specifically designed to handle issues of secondary liability of ISPs. These laws are therefore not specific to internet providers' liability, but rather general standards of civil law. The very conditions of liability are consequently the same regardless of the cause of action. They should be equally relevant for infringements of personal interests, copyright, trademark rights and other possible infringements committed on the internet or in the bricks and mortar world. However, the nature of the cause of action in question may have an impact on how liability in practice works, because it will for example decide what is considered an infringement. On a basic level, all intellectual property rights and personal interests (e.g. defamation) are protected (in Poland) with the same legal instrument, i.e. a subjective absolute right vested in the right-holder.[17] Any encroachment upon this right is presumed to be illegal.[18]

---

[14] Targosz 2012 [Comment on Supreme Court's Decision of July 8, 2011, IV CSK 665/10], OSP 2012, no. 4, p. 45 ff.
[15] "Tout fait quelconque de l'homme, qui cause à autrui un dommage, oblige celui par la faute duquel il est. arrivé à le réparer".
[16] Longchamps de Berier 1948, 234–235; Sołtysiński 1970., 163–75, Wilejczyk 2013, 55 i ff.
[17] Traple 2013, 136 ff; Skubisz 2012, 59 ff.
[18] Machnikowski 2009, 380.

Since there are no laws creating secondary liability of service providers (including internet service providers), the existing grounds of liability do not define the term "service provider". The law on E-Commerce defines online service providers, but these definitions are only of consequence for the limitations of liability recognised by this law (implementing the EU's E-Commerce Directive). Therefore, based on the existing grounds of liability, it would not be possible to argue that a certain entity is or is not liable because it falls under (or does not fall under) a definition of a service provider. Once liability has been established according to the general rules, e-commerce safe harbors must be taken into account and liability may in this way be excluded.

Thus, the issue of internet providers' liability or, speaking more generally, liability for indirect infringements, has not been settled in Polish law. There is relatively little case law dealing with these problems. The majority of court decisions have dealt with the protection of personal interests (such as reputation or privacy) as it is often the case that the internet provides ample opportunities for defamation and comparable torts. In a vast majority of cases the plaintiffs demand that certain data be removed (or access to it blocked) and/or damages awarded. It is extremely rare for plaintiffs to try to impose any obligations on ISPs, the purpose of which would be to reduce the scale of infringements.[19]

Secondary Liability—Indirect Infringements

One way of dealing with so-called secondary infringements is to start with the scope of the infringed right and to ask what its infringement actually means. Since the problem of liability of service providers is usually closely related to intellectual property rights, it may be noted that intellectual property rights, as far as their theoretical concept is concerned, have been modeled after property rights traditionally known in civil law.[20] Such rights protect a sphere of exclusivity granted to the rightholder and are effective against anyone. In property law the concept of infringement encompasses also indirect infringements[21] and it has been proposed to handle intellectual property rights in the same way.[22] However, the analysis must surely take into account how specific intellectual property rights have been regulated and whether it has been legally defined what infringement means.

---

[19] The exception being the case decided by the Regional Court in Kraków, judgment of May 27, 2015, IX GC 791/12, referred to in more detail below.

[20] This is so even when one argues that there are too many exceptions from protection to justify the use of the term property (Czajkowska-Dąbrowska, vol. XXI, 272.

[21] Katner 1982 [Protection of Real Estate Property Against Indirect Infringements].

[22] Machała 2007 [Specifics of a Claim for Damages under art. 79 Copyright Act]; Machała and Sarbiński 2002. See also critically as to this approach Laskowska, vol. 120 [Legal Ground of Liability for Indirect Copyright Infringements—On Applying Civil Code in Copyright Law].

When one applies this approach in Polish law, a one-size-fits-all solution is unlikely to emerge. A wide understanding of "infringement" could be a feasible theory in copyright law, because the Copyright Act (CA)[23] does not define infringement. Instead it vests the right-holder with exclusive rights with regard to any use of the protected copyright work and declares that infringement of these rights gives rise to the infringer's liability. It may therefore be argued that the very concept of infringement may resemble what has been known in property law, thus including indirect infringements.

There are however many problems resulting from such construction of art. 79 of the CA, i.e. the provision regulating remedies in the case of copyright infringement. This provision does not seem to differentiate legal consequences of infringement depending on their forms; therefore, very strict rules, providing for example for damages in the amount of a double license fee even in the absence of fault, would have to apply also to those secondarily liable. So far neither the doctrine nor the judiciary have chosen this road and therefore have not been able to offer a more nuanced solution resembling for example the German *Störerhaftung*,[24] a ground of liability evolved from property law (§ 1004 BGB).

Industrial property rights such as trademark or patent rights differ from copyright in one important aspect (of course there are multiple differences, but here only the problem of secondary liability is the subject of discussion). Unlike copyright, where there is no exhaustive list of infringing activities, industrial property rights have been regulated in such a way that the law defines what actions are considered to be within the scope of the right in question.[25] To prove infringement one has to show that the defendant has undertaken at least one of such legally named actions (e.g. offering or selling in the case of patent law). With trademarks the law goes even further, because it adds additional conditions of liability, such as the risk of confusion[26] and requires that any infringement constitutes the use of a protected designation as a trademark (in its function of a trademark).[27] One could say that copyright protection in Polish law reflects the property model. Unlike many legal systems, in which the scope of copyright is defined by specific (albeit broad) rights such as rights of reproduction, distribution or communication to the public, Polish copyright law provides for the exclusive right to *use* the work[28] and does not define what 'use' means, including whether a 'use' can be indirect. Since infringement may be explained as using a copyright work illegally, this innate vagueness of the scope of

---

[23] Copyright and Related Rights Act of February 24, 1994 with amendments (Journal of Laws 1994, no. 24 item 81; consolidated text Journal of Laws 2006, no. 90, item 630 with amendments).

[24] See eg. Leistner 2010, 1 ff.

[25] See art. 66 IPL (patents); art. 154 and 296 IPL (trademarks

[26] Art. 296 (2) p. 2 IPL. Even where double identity occurs, it appears that for infringement to be proven at least one function of the trademark must be disturbed.

[27] Skubisz 2012, 1070 ff. [System of Private Law. Vol. 14A. Industrial Property Law.]

[28] See Giesen 2015, 61 ff.

copyright in Poland could be of consequence for developing a theory of indirect infringement.

Persons who could be considered as secondarily liable usually do not use the protected subject-matter in a way the law on industrial property rights defines as "infringing". Therefore, their liability cannot be based on a wide interpretation of the term "infringement" in the relevant provisions of the law in force.

Liability for secondary infringements, if understood as falling under the wide concept of infringement of a particular right, would be usually be tied to an act of direct infringement (most likely committed by a user taking advantage of the service provided to him), because only a direct infringement could be the required encroachment upon the exclusive sphere granted to the right-holder. Liability for secondary infringements under this theory would also follow liability for primary infringements with regard to such requirements like fault. Liability is always defined by legal regulations relating to a specific category of rights, such as various types of IP rights. In Poland it is usually strict, in that any use of the protected subject-matter without the right-holder's consent and outside the sphere of legal exceptions and limitations is considered unlawful and constitutes infringement, regardless of fault. Fault may be required to justify claims for damages, but not in all areas of law. The most notable exception is copyright law, where it is possible to claim damages in the amount of a double license in the case of an innocent infringement.[29]

Polish Tort Law and Secondary Liability

Liability of ISPs must be established under the relevant provisions of tort law before a question of the application of safe harbours may even arise. Otherwise the latter would be redundant. It is unfortunately not uncommon for the Polish judiciary to overlook this logical order.[30] The reason may perhaps be the practical entanglement of the relevant liability provisions and the provisions on safe harbours, so that, when deciding a given case, they all seem to be part of the same mix. The unfortunate effect of such an approach is the failure to work out more precise conditions of tort

---

[29] Polish copyright law used to provide for damages equal to a triple license fee in the case of fault and double license fee in the case of innocent infringements. The triple license fee was declared unconstitutional by the Constitutional Tribunal (Decision of the Constitutional Tribunal of 23 June 2015 r., SK 32/14 (Journal of Laws 2015, it. 932). The double license fee has not been submitted for constitutional review and is still applied by courts (e.g. the judgment of the Court of Appeal in Kraków, 15 October 2015, I ACa 889/15). However, the reasons the Constitutional Tribunal provided to declare the triple fee for infringements committed with fault unconstitutional apply equally strongly (or even more persuasively) to the double license fee. It is therefore likely this regulation will be amended soon.

[30] Supreme Court's Decision of July 8, 2011, IV CSK 665/10. In this decision the Court considered liability 'based on the provisions of the e-commerce act' (implementing safe harbour regulations). These provisions cannot be the source of liability—their purpose is to exclude liability resulting from some other legal ground.

liability, because the courts feel that due to the applicable safe harbours it would be unnecessary to inquire whether and if so on what conditions an ISP could be liable if the safe harbours did not exist.

Art. 422 CC states that "Liability for damage is borne not only by the direct perpetrator but also by any person who incites or aids another to cause damage and a person who knowingly takes advantage of damage caused to another person." There are three separate torts in this provision. Two of them take place before damage is caused (inducing and aiding), the third may only happen after this has been done (knowingly taking advantage of damage caused to another person). In all three cases the prevailing view requires not only fault, but something more, i.e. intentional conduct.[31] This is of course correct, but also difficult to prove. Most importantly, intention must always be based on knowledge (actual) of a specific illegal action to be taken or already committed by a third party. In other words, in order to be liable on the grounds of art. 422 CC, a service provider would have to intentionally cause another person to commit an infringement or knowingly help this other person to commit an infringement (i.e. be conscious of the fact that this person intends to commit a tort and objectively facilitate this deed). Most service providers cannot be charged with this sort of conduct. One can often accuse them of not doing enough to prevent or reduce infringements, but rarely to actively induce or assist their perpetrators.

An interesting consequence of applying art. 422 CC is that it practically makes safe harbour regulations redundant.[32] There may be some small differences, but generally speaking if secondary liability is based on art. 422 CC, service providers are not liable if they do not have knowledge of specific infringements and therefore are not in need of safe harbour provisions that would shield them from liability. Liability based on art. 422 CC is also tied to establishing primary liability, because liability for inducing or assisting a tort depends on this tort being committed and causing damage.

Art. 422 CC only allows a claim for damages if the demanding conditions of liability that it requires could be met in a practical situation.[33] It is presumed that art. 422 CC cannot be used for injunctions of any kind. On closer inspection this interpretation seems too narrow, as art. 422 CC at least makes inducement or assistance illegal and if the law allows for preventive action (art. 439—see below), one should assume that injunction must be available when the infringement continues.

The prevailing view, however, looks for the legal basis to issue injunctions in art. 439 CC. According to this provision: "Anyone who, as a result of another person's behavior, especially due to a lack of proper supervision of the operations of an enterprise or establishment run by that person or of the condition of a building or other facility in his possession, is directly threatened by damage may demand that the person undertake the measures necessary to avert the imminent danger and, if

---

[31] Lewaszkiewicz-Petrykowska 1978, 112; Szpunar 1957, 287; Supreme Court, July 8, 2011, IV CSK 665/10.

[32] Targosz 2012.

[33] Supreme Court, June 3, 2015, V CSK 599/14.

needed, that he give appropriate security." It is a rather modern solution that other legal systems often have to create by interpretation of general legal provisions;[34] however one of the most surprising features of art. 439 CC is that it has never been even modestly popular in court practice.[35] There is practically no case law (especially of appeal courts or the Supreme Court) analysing this provision and its requirements. On the face of it, there is nevertheless a lot of potential, as far as secondary infringements are concerned. The prevailing view of legal scholars requires that the action to be prevented must be unlawful (i.e. if completed and causing damage, there would be objectively speaking a tort),[36] but fault, let alone intention, are not conditions of liability. Since right holders often argue that the most effective way of preventing future infringements would be for the service providers to implement measures reducing such risks, it is worth noting that art. 439 CC explicitly names operation of an undertaking a source of risk to be averted and refers to inadequate supervision of such operations. Preventive liability has been designed in such a way that the person who may sustain damage can demand that dangerous activities be stopped or (more often) modified or adjusted, so that the danger is, if not reversed, then at least pruned to reasonable levels.

The obvious problem with art. 439 CC, often overlooked in legal literature, is the requirement that the actions resulting in danger must be objectively unlawful. Most service providers will say, their operations are anything but and they will invoke safe harbour provisions with particular emphasis on art. 15 of the E-Commerce law,[37] according to which no monitoring obligations may be imposed. Even in the absence of this exonerating provision it would be necessary to find a legal ground for the objective unlawfulness of the lack of supervision.

It is therefore necessary to turn to the third tort law instrument, the general tort clause of art. 415 CC. This provision states: "Anyone who by a fault on his part causes damage to another person is obliged to remedy it." All the complexities of the Polish general tort clause cannot be discussed in this Chapter, but some basic information is worth noting. Art. 415 CC mentions only fault, but it is not contested that conduct giving rise to liability must not only be subjectively reproachable but first of all objectively improper.[38] This requirement is usually referred to as unlawfulness,[39] although some [courts/scholars] prefer "objective fault".[40] Regardless of the name, it is also not contested that this objective unlawfulness does not have to result from an infringement of a specific statutory provision (e.g. prohibiting a certain action) or an absolute right. It suffices that it is contrary to the so

---

[34] Kötz 1974, 158 ff; Peukert 2008, 304.

[35] Śmieja 2009, s. 621–622.

[36] Lewaszkiewicz-Petrykowska, vol. XIII, 52 ff..; Katner 1982, 156.

[37] Court of Appeal in Warsaw, 11 June 2015 r., I ACa 1842/14; Supreme Court, 14 January 2015, II CSK 747/13.

[38] Machnikowski 2009, 377.

[39] Szpunar 1947, 112 ff.; Sośniak 1959.

[40] Dąbrowa 1968, 7 ff.; Lackoroński 2013, 178.

called principles of community life (social co-existence).[41] These are norms, not necessarily explicitly reflected in any binding statutory provisions, comparable to good morals known in other legal systems. The concept of unlawfulness is therefore very broad and may be developed by the judiciary.

In many publications, as well as in some court decisions, it has been proposed to recognise a general rule of conduct relevant for the application of art. 415 CC, i.e. a rule according to which everyone should act with care so as to avoid causing damage to another.[42] Courts have for example used this approach to find liable organisers of social events, during which some participants have sustained damage due to organisational oversight, or in other words lack of due care to have the event organised in such a way that the risk of harm is significantly reduced.[43] What is particularly important for this ground of liability is its flexibility. It does not strive to offer absolute protection, but only requires what can reasonably be expected to reduce the risk of damage.

Following this reasoning, it should be possible to argue that a service provider who has organised his business in such a way that the risk of direct infringements is high and has not taken any reasonable steps to reduce this risk to an acceptable level should be liable for his own tort. What steps should be required in practice must be determined taking into account many circumstances, such as what is technically available, what are the costs of such measures and what industry standards should apply.

This line of reasoning has been recently adopted by the Regional Court in Kraków.[44] The defendant in this case[45] operates a file-sharing platform named Chomikuj,[46] where numerous illegally distributed files with films, music, ebooks, etc. are available to download. The plaintiffs notified Chomikuj that three movies had been made illegally available on the platform and due to the fact that access to these files had not been immediately blocked (one of the reasons being that Chomikuj considered the notification sent by a 'normal' letter insufficient) sued for damages. Apart from damages the plaintiffs also claim that the entity operating the service has not taken the required precautions to limit the scale of infringements and demand that certain filtering mechanisms be implemented. The precedent-setting nature of the court's decision lies in the fact that the court considered the latter demand justified under art. 439 CC. This seems to be the first case in Poland when the court entertained such a possibility and found it reasonable. The court considered various filtering mechanisms suggested by the plaintiffs, weighing their

---

[41] Sośniak 1959, 102 ff., Machnikowski 2009, 381.

[42] Longchamps de Berier 1948, 234–23, Wilejczyk 2013, 55 ff.

[43] See examples provided by Wilejczyk 2013, 56.

[44] Regional Court in Kraków, judgment of May 27, 2015, IX GC 791/12.

[45] The decision is not final at the time of submitting this chapter. It is under appeal before the Court of Appeal in Kraków, case file no. I ACa 1494/15.

[46] "chomikuj" is a verb created from the noun chomik (hamster). As hamsters, when eating, store food in their mouths, the verb means "to store" (for later).

proportionality, and decided inter alia to follow the German Supreme Court[47] in imposing monitoring obligations (limited to the audiovisual works indicated by the plaintiffs), involving checking links to the infringing files generated by Google and Bing searches.

It is worth adding that applying art. 439 CC is not conditional upon an infringement having been committed, since this provision explicitly refers to a situation before damage has been done. It then only requires an imminent danger, not proof of infringement itself.

Remedies

What remedies would be available against secondary infringers depends on the basis of liability. It should be stressed that the consequences of infringement of intellectual property rights (which, as already mentioned, is a category of case that seems to be crucial for service providers) have been regulated separately for each of these rights. In copyright law, art. 79 provides that the right-holder may request the person who has infringed those rights to stop the infringement; remove the effects of the infringement; redress the damage caused, in accordance with general rules (tort liability as regulated in the civil code), or by paying an amount of money equal to a double value of an adequate remuneration (license fee); and surrender of any benefits gained.

Apart from this there are claims such as publication of a press statement or payment to the Fund for the Promotion of Creativity. Article 80 of the Copyright Act also provides for legal remedies concerning securing the evidence and information claims. This provision makes it possible for the right-holder to request the competent court to obligate a person other than the infringer to provide any information relevant to the claims referred to in Article 79 pertaining to origin, distribution networks, quantity and price of the goods or services violating economic rights, if it is ascertained inter alia that he provides services that are used in activities violating economic rights, and the purpose of those activities is to gain, directly or indirectly, profit or other economic benefits.

Remedies available against infringers of industrial property rights are similar, though not as generous in scope. For example, if there is trademark infringement the right-holder may demand that the infringer ceases the infringement, surrender the unlawfully obtained profits and in case of infringement caused by fault also redress the damage in accordance with the general principles of law, or by the payment of a sum of money corresponding to a reasonable license fee (art. 296 IPL[48]).

Measures for securing evidence and information claims also apply (art. 286[1] IPL), including requesting the court to order a party other than the infringer to provide information necessary to enforce protected rights, i.e. information on the origin

---

[47] BGH, 12 July 2012, I ZR 18/11 (Alone in the Dark).
[48] Act of 30 June 2000 Industrial Property Law, consolidated text Journal of Laws 2013, item 1410.

and distribution networks of the goods or services which infringe an industrial property right, where the likelihood of infringement of these rights is substantial. Additional requirements must also be met, such as the fact that the addressee of a demand "was found to provide services used in activities which infringe the patent, supplementary protection right, right of protection or right in registration", where the above activities are intended to directly or indirectly gain profit or other economic benefits.

Information claims against "innocent" third party service providers are therefore the only legal instrument specifically designed to enhance protection of intellectual property rights by involving entities other than the 'direct' or 'primary' infringer. It would be impossible to find a legal ground allowing injunctions against "innocent" service providers, for example ordering them to block access to certain websites, etc.[49] Of course, one may argue that practically it is not as important as it might seem, because once a service provider has been notified of an ongoing infringement, there is information leading to knowledge and eventually (in lack of action) to fault and liability under art. 422 CC. This is however only partly true. Consider an injunction requesting to block access to a 'pirate' website[50]. The ISP in question (access provider) could not be made liable for such infringements even if notified about the illegal content.

The current state of affairs is incompatible with EU law.[51] Poland has failed to implement art. 8(3) of the directive 2001/29/EC[52] and art. 11 3rd sentence of the directive 2004/48/EC.[53] The law does not recognise the clear safe harbour exceptions introduced by art. 12(3), 13(2) and 14(3)[54] of the E-Commerce Directive

---

[49] It seems that such remedies are required under EU law. See ECJ, 27 March 2014, Case C-314/12 (UPC Telekabel Wien).

[50] Recently especially popular in the UK. See e.g. Richemont International SA and others v British Sky Broadcasting Limited and others [2014] EWHC 3354 (Ch), England and Wales High Court (Chancery Division), 17 October 2014.

[51] See Commission Staff Working Document: Analysis of the application of Directive 2004/48/EC of the European Parliament and the Council of 29 April 2004 on the enforcement of intellectual property rights in the Member States, SEC (2010) 1589 final (Dec. 12, 2010), p. 16: "[n]either Article 11 (third sentence) of the Directive, nor Article 8(3) of Directive 2001/29 link injunctions with the liability of an intermediary". See also CJ EU L'Oreal v eBay C-324/09 at 127–134.

[52] Directive 2001/29/EC of the European Parliament and of the Council of 22 May 2001 on the harmonisation of certain aspects of copyright and related rights in the information society, Official Journal L 167, 22/06/2001 pp. 10–19.

[53] Directive 2004/48/EC of the European Parliament and of the Council of 29 April 2004 on the enforcement of intellectual property rights (OJ L 157, 30.4.2004).

[54] These provisions state that safe harbours do not affect the possibility for a court or administrative authority, in accordance with Member States' legal systems, of requiring the service provider to terminate or prevent an infringement. As such they seem to be of a declaratory nature only, and the failure to implement them literally could be excused. However, since the relevant provisions of the copyright and enforcement directives (art. 8 (3), art. 11, 3rd sentence) have not been implemented, either, the overall impression suggests that safe harbours protect ISPs also against injunctions ordering to prevent infringements. As regards an injunction seeking to terminate an ongoing infringement, this cannot be undermined by a safe harbour provision—at least with regard to host-

2000/31/EC[55] either. There is consequently no legal ground to issue an injunction against intermediaries whose services are used by a third party to infringe an intellectual property right, if the intermediary cannot be made liable for illegal actions. In other words, injunctions could be available only when an intermediary is liable himself (e.g. as an entity creating the risk of damage due to improper supervision of its operations under art. 439 CC). It is also far from certain that art. 439 CC allows for injunctions of the same scope as would be available under art. 8(3) of the Information Society Directive[56] and art. 11 (3rd sentence) of the Enforcement Directive.

If right-holders rely on general civil law provisions on inducement or assistance, a secondary infringer would be liable for damages (jointly and severally with the direct infringer). It is, however, controversial whether this encompasses 'special' damages recognised by intellectual property regulations (e.g. a double license fee in copyright law), or only damages according to the general rules.[57]

In principle, it would be difficult to justify the view that remedies available against primary infringers should in any way limit the liability of secondary infringers, if their conditions of liability have been met. The only limitation in such cases would result from general standards of civil law (for example the same damage cannot be compensated twice).

Safe Harbours for ISPs

Safe harbor provisions for internet service providers have been harmonised on the EU level. Polish law should therefore correspond to the European standard. At first glance the degree of convergence would seem to be exceptionally high as even the article numbers exactly match the E-Commerce Directive. The Polish implementation reveals, however, a few more or less pronounced peculiarities, some of which may make it incompatible with EU law.

The Polish E-Commerce Act follows in the directive's footsteps in that it provides for two types of protection for online service providers. Firstly, it offers

---

ing providers—since such an injunction imparts to the ISP the knowledge about the infringement, thus lifting the safe harbour protection.

[55] Directive 2000/31/EC of the European Parliament and of the Council of 8 June 2000 on certain legal aspects of information society services, in particular electronic commerce, in the Internal Market ('Directive on electronic commerce'), Official Journal L 178, 17/07/2000 P. 0001–0016.

[56] Directive 2001/29/EC of the European Parliament and of the Council of 22 May 2001 on the harmonisation of certain aspects of copyright and related rights in the information society, Official Journal L 167, 22/06/2001 P. 0010–0019.

[57] See Supreme Court, 27 June 2014, I CSK 540/13. In this decision the Supreme Court held that the liability of persons falling under art. 422 CC encompasses also the claim for the restitution of profits that cannot be regarded as a typical action for damages, but is a special instrument used in the legislation concerning intellectual property rights, resembling unjustified enrichment.

protection against liability (Article 12–14). Secondly, it protects them against the (general) monitoring obligation (Article 15).

Considering the definitions of a "service provider" (Article 2 point 6) and "provision of services by electronic means" (Article 2 point 4), it should be understood that the safe harbour provisions apply not only to internet services, but to all online services as long as their provision meets the conditions set forth in the E-Commerce Act. On the other hand, the special regime provided for in art.12–14 applies to specific activities rather than to service providers in general. As such, a service provider can conduct various activities, some of which are covered by safe harbour provisions while others are not.

The Polish legislator has chosen a horizontal immunity scheme for services covered by the safe harbor regime. ISPs are exempted from liability regardless of what substantive law is applicable (civil, administrative, criminal). The law does not differentiate between the forms of liability from which service providers are exempted. Therefore, any potential liability is covered by the exemptions. On the other hand, it should be underlined that even though the safe harbour regime constitutes an additional shield for service providers, it does not modify the underlying material law governing liability. It should be understood as an overlay, applied onto the relevant liability rule and modifying the results of its application.

Intermediary online services can be subdivided into four categories: mere conduit, caching, hosting, and provision of information tools (search engine services and hyperlinking). At present, the E-Commerce Act covers the first three categories, while information tool providers are still excluded from its application.

Mere conduit services consist of either network access services or network transmission services provision. Those services are rendered by telecommunications operators.

Caching providers' services consist of temporary and automatic storage of data in order to make the onward transmission of that data more efficient. As in the case of mere conduit, those services are also performed by telecommunication companies, i.e. entities responsible for routing internet traffic.

Host providers store data provided by third parties. When the E-Commerce law was adopted, typical hosting services were those rendered by data centers renting their hard disk space to content providers. Since then, several new storage activities and services have emerged. They are mainly based on the so-called Web 2.0 model. Despite little case law on that subject, it is commonly understood that those activities (and services) are covered by hosting exemption set forth in the e-Commerce Act. Among those services, the most popular are video-sharing sites, online selling platforms, social networks, and discussion fora. After initial hesitation, the case law has established that it does not matter whether the hosted data is accessible to everyone, or just to the person who uses the hosting service.[58]

In litigation involving personality rights, when art. 14 of the e-commerce Act is habitually invoked by website operators (e.g. discussion forums, etc.), sometimes a

---

[58] Court of Appeal in Warsaw, 11 October 2012, VI ACa 2/12.

question arises whether the provider's role is passive or active. However, it has been correctly recognised that implementation of certain filtering mechanisms (e.g. designed to block swearwords) does not deprive the provider of the liability privilege.[59]

Article 12–14 of the E-Commerce Act provides for a number of specific material conditions to be met in order to benefit from liability exemption. They differ depending on what kind of service is rendered by an online service provider.

In the case of *mere conduit* services, the condition for liability exemption is a passive role of the service provider towards the data transmitted. In particular, ISPs cannot make decisions as to whom the data should be transmitted (they cannot initiate transmission nor select the receiver of the transmission) as well as what happens to the transmitted data (they cannot select or modify the information contained in the data). Potential storage of transmitted data may take place as long as it is for the sole purpose of carrying out the transmission in a communication network, and provided that the information is not stored for any period longer than is reasonably necessary for the transmission (Article 12 point 2 of the e-Commerce Act).

In the case of caching services, the precondition for liability exemption is, as in the case of *mere conduit*, a passive nature—towards the data cached—of the service (Article 13 point 1). In particular, service providers may not modify the content of the data and should use information technology techniques that are recognised and commonly used in those types of activities, while specifying the technical parameters of data access. Moreover, a caching service provider is obliged to disable immediately access to any information it has been storing, upon obtaining knowledge of the fact that the information has been removed from the network at the initial source of the transmission, or access to it has been disabled, or that a court or an administrative authority has ordered such disablement.

Most of the cases involving the question of ISPs' liability concern host providers. It should not come as a surprise considering that the most popular internet services are based on a hosting model. Under Article 14 of the e-Commerce Act, host providers can benefit from the liability exemption when they do not have knowledge of the illegal nature of the data stored or related activities and once they are notified that—through official or reliable notice[60]—they have made access to that data impossible.

It should be noted, in the first place, that the e-Commerce Act does not define the terms "illegal data" or "illegal activity". Unlike the legislation of some other EU Member States, it does not require that the content be "manifestly" illegal. As such, it could be understood that the provision of Article 14 applies to any content that is considered illegal under Polish law.

Comparison of the wording of Article 14 of the E-Commerce Directive and Article 14 of the E-Commerce Act leads to the conclusion that the Polish regulation is more generous for host providers. Under Polish law, they may be held liable only

---

[59] Supreme Court, 23 July 2015, I CSK 549/14.

[60] An official notice may for instance be a court preliminary injunction. A reliable notice may be information from any source, as long as it is reasonably credible.

if they have knowledge of the illegal nature of the data stored (related activity). The courts stress that only "positive knowledge" renders the safe harbour immunity inapplicable.[61] The E-Commerce directive provides for the "actual knowledge" requirement only with regard to claims other than claims for damages. As far as damages are concerned, it suffices that the provider is aware of facts or circumstances from which the illegal activity or information is apparent.[62] This introduces a certain standard of care, whereas the knowledge requirement does not ask for any care from the provider. The failure to implement correctly Article 14 of the E-Commerce Directive may constitute a significant limitation of host providers' liability. In practice, they are regarded to have obtained "knowledge of illegal nature of the data stored (related activity)" only by way of official or credible notice submitted to them by interested parties (e.g. right-holders).

The Directive does not refer to "reliable information" as the source of knowledge, but this difference in wording should not be overstated. Practically, the wording of the Polish statute may, however, prove problematic with regard to some categories of infringements. Whereas it is difficult to deny that notification by an IP right-holder is usually reliable (credible) and that the illegality of making the latest movies or music tracks available online for free cannot be seriously put into doubt, things get more complicated when personality rights come into play. The minority view that in such cases reliable information cannot come from the interested (injured) party[63] must be rejected. But it may be difficult nevertheless for the provider to assess whether a given material is legal or illegal.

ISPs often argue that the concept of "reliable information" should also cover the form in which it is provided. Specifically, they would prefer notifications with precise URL addresses. As the law stands, it is difficult to accept this view, since no provision requires any specific format of information.[64] The only requirement that could be taken into account in this context is whether the notification allows identification of the infringing data. If the answer is positive, it must be assumed that the ISP in question has "actual knowledge" and can no longer rely on the safe harbour exemption.

The courts have unfortunately not worked out the exact parameters of the "immediate prevention of access" requirement. The host providers have little guideline as to what will be considered an "immediate" reaction.

The current regulation does not offer to the provider of the allegedly illegal information any opportunity to submit a counter-notice and defend the legality of the information at issue. Any such actions taken by this person could theoretically impact the reliability of the information provided by the allegedly injured party, but for many ISPs it would be too risky to argue that because of the information supplied by the alleged "direct" infringer they consider the notification to be unreliable.

---

[61] Court of Appeal in Lublin, 18 January 2011, I ACa 544/10, Court of Appeal in Kraków, 19 January 2012, I ACa 1273/11.

[62] See Article 14(3) of the E-Commerce Directive as regards damage claims.

[63] Court of Appeal in Gdańsk, 27 November 2013, I ACa 748/13.

[64] Regional Court in Kraków, judgment of May 27, 2015, IX GC 791/12.

Plans to introduce the notice and take down procedure in an amendment to the e-Commerce Act[65] have so far fallen through. At the moment of completion of this chapter there is no legislative initiative pending.

ISPs' obligation to disable access to unlawful content involves a risk that wrongful notices are provided to intermediaries and that intermediaries, acting on such notices, block access to perfectly legal content. Two groups of situations should be distinguished in this context. The first concerns the liability of an ISP towards the party whose content has been wrongly blocked. According to Article 14 (3) of the E-Commerce Act, service providers are not liable towards the data-storing party for any damage inflicted as a result of disabling access to such data if they have notified that party of their intention to deny access to the data.

The second group of situations concerns the issue of liability of persons submitting wrongful notices. The e-Commerce Act does not regulate this. In cases where such notice has been made in bad faith, general provisions of the Polish Civil Code will apply. It is perfectly reasonable to claim that a bad faith notification is a tort according to art. 415 CC.

One of the central issues of secondary ISP liability is the scope of monitoring obligations. Article 15 of the E-Commerce Act (implementing art. 15 of the E-Commerce directive) states that the "service provider providing services specified in Art. 12–14 is not obliged to check data transmitted, stored or made available by him". This provision, read literally, rules out any and all monitoring obligations, whereas the directive only refers to a *general* monitoring obligation. Needless to say, there is a fundamental difference between a prohibition of all monitoring and a prohibition of general monitoring. The current wording of art. 15 of the E-Commerce Act raises questions as to the admissibility of *specific* monitoring obligations, although we are of the opinion that it is possible to interpret the said provision in a way compatible with European law.[66]

Conclusion

Polish law has all the required instruments to provide an adequate legal framework of secondary liability. It has simply rarely put them to good use so far. The general clause of art. 415 CC and the wide concept of infringement allow shaping liability in a very flexible way and this seems to be of utmost importance in this area of law. Because technology and circumstances quickly change, specific statutory provisions soon become obsolete and start to hamper rather than facilitate the attempts to build a reasonable liability standard. It should be noted that, at least in theory, it is possible to apply in Poland almost all the legal instruments that courts in Germany or France have been able to use to a rather satisfying effect.

---

[65] See the bill of 13 July 2012 on amendment to the e-Commerce Act.

[66] Konarski 2004, 146 [Commentary on the law on providing electronic services].

References

Banaszczyk, Z. 2015. In *Kodeks cywilny. Komentarz. Vol. I* [Civil Code. commentary. vol. I], ed. K. Pietrzykowski. Warszawa: Wydawnictwo C. H. Beck.

Barta, J., and R. Markiewicz. 2009. *Przechowywanie utworów na stronach internetowych* [Storing copyright works on websites] ZNUJ. PPWI 2009, vol. 105.

――――. 2013. *Prawo autorskie* [Copyright law]. Warszawa: Wydawnictwo C.H. Beck.

Czajkowska-Dąbrowska, M. *Treść (Elementy Struktury) Prawa Autorskiego a Treść Prawa Własności* [Content (Structural elements) of copyright and content of property rights]. Studia Iuridica, vol. XXI.

Dąbrowa, J. 1968. *Wina jako przesłanka odpowiedzialności cywilnej* [Fault as condition of tort liability]. Wrocław: Zakład Narodowy Imienia Ossolińskich.

Giesen, B. 2015. Własnościowy model prawa autorskiego—analiza koncepcji przyjętej w prawie polskim [Propietary model of copyright—Analysis of the concept adopted in polish law]. *RPiEiS* LXXVII: 61.

Katner, W.J. 1982. *Ochrona własności nieruchomości przed naruszeniami pośrednimi* [Protection of property against indirect infringements]. Warszawa: Wydawn. Prawnicze.

Konarski, X. 2004. *Komentarz do ustawy o świadczeniu usług drogą elektroniczną* [Commentary on the law on providing electronic services]. Warszawa.

Kötz, H. 1974. Vorbeugender Rechtsschutz im Zivilrecht. Eine rechtsvergleichende Skizze [Preventive protection in civil law. A comparative sketch] *AcP* 174: 97–144.

Kur, A. 2002. *Use of trademarks on the Internet – The WIPO recommendations*. IIC/2 33: 41–47.

Lackoroński, B. 2013. *Odpowiedzialność cywilna za pośrednie naruszenie dóbr* [Civil liability for indirect infringements].Warszawa.

Laskowska, E. *Podstawa prawna odpowiedzialności za pośrednie naruszenie autorskich praw majątkowych – uwagi do stosowania przepisów kodeksu cywilnego w prawie autorskim* [Legal ground of liability for indirect copyright infringements – On applying civil code in copyright Law] ZNUJ PPWI, vol. 120.

Leistner, M. 2010. *Störerhaftung und mittelbare Schutzrechtsverletzung* [Interferer's liability and indirect infringements]. (GRUR-Beil.) 112.

Lewaszkiewicz-Petrykowska, B. 1978. *Wyrządzenie szkody przez kilka osób* [Causing damage by several persons]. Warszawa: Wydaw. Prawnicze.

――――. *Roszczenie o zapobieżenie szkodzie* [Preventive action]. SPE XIII.

Longchamps de Berier, R. 1948. *Zobowiązania*. [Law of obligations]. Poznań.

Machała, W. 2007. Specyfika roszczenia odszkodowawczego z art. 79 ustawy o prawie autorskim i prawach pokrewnych [Specifics of a claim for damages under art. 79 copyright act]. *Studia Cywilistyczne* 47: 189–196.

Machała, W., and R.M. Sarbiński. 2002. Wymiana plików muzycznych za pośrednictwem Internetu a prawo autorskie [Exchange of music files on the Internet and copyright law]. *PiP* 9: 70–78.

Machnikowski, P. 2009. In *System prawa prywatnego. T. 6. Prawo zobowiązań – część ogól*na [System of private law. vol. 6. Obligations – General part], ed. Olejniczak A. Warszawa: Wydawnictwo C.H. Beck.

Pacek, G.J., and P. Wasilewski. 2008. Pomocnictwo w ujęciu cywilistycznym a odpowiedzialność dostawców usług hostingowych – dwugłos w sprawie, [Aiding in civil law and secondary liability of host providers – Two opinions] *PPH* 7.

Peukert, A. 2008. *Güterzuordnung als Rechtsprinzip* [Allocation of goods as legal principle]. Tubingen: Mohr Siebeck.

Picker, E. 1972. *Der negatorische Beseitigungsanspruch* [Defensive claim for abatement or removal]. Bonn: L. Röhrscheid.

Skubisz, R. 2012. In *System prawa prywatnego T 14A. Prawo własności przemysłowej* [System of private law. vol. 14A. Industrial property law], ed. R. Skubisz. Warszawa: Wydawnictwo C.H. Beck.

Śmieja, A, 2009. In *System prawa prywatnego. Tom 6. Prawo zobowiązań – część ogólna* [System of private law. vol. 6. Obligations – General part], ed. Olejniczak A. Warszawa: Wydawnictwo C. H. Beck.

Sołtysiński, S. 1970. *Licencje na korzystanie z cudzych rozwiązań technicznych* [Licences for use of technological solutions] Warszawa: Wydawnictwo Prawnicze.

Sośniak, M. 1959. *Bezprawność zachowania jako przesłanka odpowiedzialności cywilnej za czyn niedozwolony* [Unlawfullness as condition of tort liability]. Kraków.

Szpunar, A. 1947. *Nadużycie prawa podmiotowego* [Abuse of right] Kraków: skł. gł. w krięgarniach Gebethnera i Wolffa.

——. 1957. Wyrządzenie szkody przez kilka osób [Causing damage by several persons] *PiP* 2: 284–300.

Targosz, T. 2012. *Glosa do wyroku SN z dnia 8 lipca 2011 r., IV CSK 665/10* [Comment on the supreme court decision of July 8, 2011, IV CSK 665/10] OSP 4.

Traple, E. 2013. In *System prawa prywatnego. T. 13. Prawo autorskie* [System of private law. vol. 13. Copyright law], ed. J. Barta. Warszawa: Wydawn. C.H. Beck.

Wilejczyk, M. 2013. *Dlaczego nie należy chodzić w tłumie ze szpilką wystającą z rękawa? Naruszenie obowiązku ostrożności jako przesłanka odpowiedzialności deliktowej za czyn własny* [Why one should not walk in a crowd with a pin in one's sleeve? Violation of due care as condition of tort liability]. SPP 1.

Chapter 3
Secondary Liability of Internet Service Providers in the United States: General Principles and Fragmentation

Salil K. Mehra and Marketa Trimble

Introduction

Policymakers, private firms and citizens now generally agree on the critical importance of the Internet for economic and technological development in the United States. In recent years, statistics have suggested that the Internet's contribution to U.S. GDP exceeds that of the federal government.[1] In turn, policymakers have taken steps to bolster and secure the Internet as a form of critical infrastructure,[2] as well as to assure its availability and usefulness to the nation's consumers.[3]

In the United States, private Internet Service Providers (ISPs) have played a key role in spreading Internet access and developing its profound economic and social impact. Typically, U.S. ISPs have deployed wired infrastructure to provide business clients and consumers not only with Internet access, but also with related services such as web hosting and hardware or software and consulting related to Internet

---

Portions of this article are based on previously published material, with permission of the American Journal of Comparative Law, found at 62 Am. J. of Comp. L. 685 (2014).

[1] See Annalyn Censky, *Internet Accounts for more than 4.7% of U.S. Economy*, CNN/MONEY, 19 Mar. 2012 at http://money.cnn.com/2012/03/19/news/economy/internet_economy/index.htm (last visited 5 Jul. 2015) (reporting results of Boston Consulting Group study).

[2] See Improving Critical Infrastructure Cybersecurity, Exec. Order No. 13,636, 78 Fed. Reg. 11,739 (19 Feb. 2013), at https://www.whitehouse.gov/the-press-office/2013/02/12/executive-order-improving-critical-infrastructure-cybersecurity (last visited 5 Jul. 2015).

[3] See Report and Order on Remand, Declaratory Ruling, and Order, GN Docket No. 14–28, FCC 15–24 (26 Feb. 2015), at https://apps.fcc.gov/edocs_public/attachmatch/FCC-15-24A1.pdf (last visited 5 Jul. 2015).

S.K. Mehra (✉)
James E. Beasley School of Law, Temple University, Philadelphia, PA, USA
e-mail: smehra@temple.edu

M. Trimble
Samuel S. Lionel Professor of Intellectual Property Law, William S. Boyd School of Law, University of Nevada, Las Vegas, NV, USA
e-mail: marketa.trimble@unlv.edu

© Springer International Publishing AG 2017   93
G.B. Dinwoodie (ed.), *Secondary Liability of Internet Service Providers*,
Ius Comparatum – Global Studies in Comparative Law 25,
DOI 10.1007/978-3-319-55030-5_3

connectivity. The industry is moderately concentrated, with the five largest industry participants accounting for over 70% of U.S. consumers.[4]

Questions of secondary liability of service providers are of crucial importance to U.S. ISPs. Clear rules on liability provide legal certainty that fosters investment in the industry and its critical infrastructure. Moreover, ensuring a business environment with fairly transparent expectations makes it possible to operate in a manner that provides a fair expectation of return on investment, and lowers the complexity cost of day-to-day operations.

This chapter provides an overview of the rules of secondary liability that affect ISPs in the United States. Section "Secondary Liability Standards" explains the theories of secondary liability applied by U.S. courts. Section "Immunity from Secondary Liability and Safe Harbours" discusses the safe harbours and immunities that U.S. law provides to ISPs. Section "Best Practices" discusses best practices that the private sector has adopted in response to the legal environment; this is followed by a brief conclusion.

Secondary Liability Standards

Secondary liability of ISPs (or intermediaries) for the conduct of others using their services typically arises under defamation law or intellectual property law, particularly copyright law and trademark law. However, ISP liability arises only when provisions that immunise ISPs from liability do not apply. The general immunity provisions, discussed in detail in section "Immunity from Secondary Liability and Safe Harbours", cover a wide range of potential liability but do not apply to all sources of liability; most importantly, they do not affect liability under intellectual property laws. Intellectual property statutes do not immunise service providers from secondary liability, but federal copyright and trademark statutes provide safe harbours that limit service providers' secondary liability; the safe harbour provisions are also discussed in section "Immunity from Secondary Liability and Safe Harbours".

The law on secondary liability of service providers is not codified; the standards for secondary liability of service providers have developed in case law. In general, U.S. law recognizes three types of secondary liability: For contributory infringement courts have drawn on general tort law principles:[5] contributory liability arises when the defendant has actual knowledge of infringement and the defendant

---

[4] See Emil Protalinski, *Over 70% of U.S. households now have broadband Internet access, with cable powering over 50% of the market*, TNW NEWS, 9 Dec. 2013 at http://thenextweb.com/insider/2013/12/09/70-us-households-now-broadband-internet-access-cable-powering-50-market/

[5] Fonovisa, Inc. v. Cherry Auction, Inc., 76 F.3d 259 (9th Cir. 1996); Perfect 10, Inc. v. Visa Int'l Serv. Ass'n, 494 F.3d 788, 795 (9th Cir. 2007). For a discussion of the origins, see Mark Bartholomew & John Tehranian, *The Secret Life of Legal Doctrine: The Divergent Evolution of Secondary Liability in Trademark and Copyright Law*, 21 BERKELEY TECH. L.J. 1363, 1366 (2006).

materially contributes to the infringement.[6] Vicarious liability developed as a separate doctrine; it stems from another general tort law principle—the agency principle of *respondeat superior*[7] and requires that the defendant have the right and the ability to supervise the infringing activity and receive financial benefit from the infringing activity.[8] Inducement liability attaches when the defendant takes 'active steps ... to encourage direct infringement',[9] and in copyright it is understood to be a type of contributory liability, while in patent law, for example, it is a doctrine that is legislated separately from contributory liability.[10]

Other related forms of liability rarely concern service providers because plaintiffs do not typically raise them: General tort liability for aiding and abetting is a close relative of the contributory liability standard. However, Bartholomew and McArdle argue that the law of aiding and abetting is underdeveloped and its uncertain metes and bounds make it unsuitable as a guide for claiming and developing contributory liability for copyright infringement.[11] Bartholomew and McArdle mention an approach to a causal relationship between a defendant's actions and the illegal act as a point where courts' approach to aiding and abetting cases differs from their approach in contributory copyright infringement cases. For aiding and abetting, defendant's actions must be a 'substantial factor in causing the resulting tort'[12]; contributory infringement does not explicitly require a causal relationship.[13]

Two other related torts are the tort of conspiracy and the tort of 'substantial[...] assistance to the other in accomplishing a tortious result [where the] conduct, separately considered, constitutes a breach of duty to the third person'.[14] As opposed to

---

[6] See Fonovisa, Inc., 76 F.3d 259 (contributory infringement in copyright law); Inwood Labs., Inc. v. Ives Labs., Inc., 456 U.S. 844, 854 (1982) (contributory infringement in trademark law). *See also* Tiffany (NJ) Inc. v. eBay, Inc., 600 F.3d 93 (2d Cir. 2009) (as regards an ISP).

[7] Fonovisa, Inc., 76 F.3d 259; A&M Records, Inc. v. Napster, Inc., 239 F.3d 1004, 1022 (9th Cir. 2001); Perfect 10 Inc., 494 F.3d at 802.

[8] See Fonovisa, Inc., 76 F.3d 259; Bridgeport Music, Inc. v. WB Music Corp., 508 F.3d 394 (6th Cir. 2007); Perfect 10, Inc. v. Amazon.com, Inc., 508 F.3d 1146, 1173 (9th Cir. 2007).

[9] Metro-Goldwyn-Mayer Studios Inc. v. Grokster, Ltd., 545 U.S. 913, 936 (2005) (quoting Oak Industries, Inc. v. Zenith Electronics Corp., 697 F. Supp. 988, 992 (N.D. Ill. 1988)). In *Grokster*, the U.S. Supreme Court adopted the inducement rule, according to which 'one who distributes a device with the object of promoting its use to infringe copyright, as shown by clear expression or other affirmative steps taken to foster infringement, is liable for the resulting acts of infringement by third parties'. *Id.* at 936, 937. The Court adopted the inducement rule in reaction to *Sony*, which 'barred secondary liability based on presuming or imputing intent to cause infringement solely from the design or distribution of a product capable of substantial lawful use, which the distributor knows is in fact used for infringement'. *Id.* at 933. *See also* Sony Corp. v. Universal City Studios, Inc., 464 U.S. 417 (1984).

[10] 35 U.S.C. §271(b) (2006).

[11] Mark Bartholomew & Patrick McArdle, *Causing Infringement*, 64 VAND. L. REV. 675, 698 (2011).

[12] *Id.* at 695.

[13] *Id.* at 704, 705.

[14] Restatement (Second) of Torts §876(c).

the general tort of aiding and abetting, the tort of conspiracy requires an agreement between the co-conspirators; such an agreement is not required for the secondary liability of a service provider, thus making it unlikely that plaintiffs will assert conspiracy rather than one of the three forms of secondary liability. Material contribution, which is required for contributory liability, is not an element of conspiracy, which does not however make conspiracy more appealing to plaintiffs who typically find a way to argue material contribution.

The tort of substantial assistance in accomplishing a tortious result does not require that the defendant know of the infringing activity; however, it does require a duty owed to a third person and '[i]t is rare that [...] a separate duty will be found in the case of intermediaries in intellectual property cases'.[15] Thus, these related direct forms of liability have played a less prominent role in demarcating the liability of ISPs.

Notwithstanding the general parameters of the individual types of secondary liability, liability standards vary based on the area of law.[16] For example, one commentator has pointed out that courts have treated the requirement of material contribution differently in copyright and trademark law, requiring for contributory trademark infringement that the defendant 'supply or directly control and monitor the tools of infringement'[17]—something that is not required for contributory infringement of copyright.[18] For vicarious liability in trademark cases, courts require 'that the defendant and the infringer have authority to bind one another in transactions with third parties'[19]—something that courts do not require in copyright cases. As for the requirement of financial benefit in vicarious liability cases, '[i]nfringement that acts as a hypothetical draw for consumers to the defendant's business satisfies the requirement for copyright owners, but not for trademark owners'.[20] Likewise, the immunity provisions mentioned above have wide applicability but do not cover all areas of law, and safe harbour provisions are specific to copyright law and trademark law. As a result, standards for the secondary liability of ISPs differ under: (a) defamation law and a wide variety of other laws; and (b) intellectual property laws.[21]

The law of secondary liability of ISPs draws on the developments in defamation law. Defamation law distinguishes among service providers based on whether they are considered to be publishers, distributors, or conduits; these concepts developed

---

[15] Bartholomew & McArdle, *supra* note 11, at 695 n. 110 (citing Lockheed Martin Corp. v. Network Solutions, Inc., 985 F. Supp. 949, 967 (C.D. Cal. 1997) *aff'd*, 194 F.3d 980 (9th Cir. 1999)).

[16] See Bartholomew & Tehranian, *supra* note 5.

[17] Mark Bartholomew, *Copyright, Trademark and Secondary Liability After* Grokster, 32 COLUM. J.L. & ARTS 445, 454 (2009).

[18] *Id.*

[19] *Id.* at 447.

[20] *Id.*

[21] For various opinions on the desirability of uniformity in the standards see Mark A. Lemley, *Rationalizing Internet Safe Harbours*, 6 J. TELECOMM. & HIGH TECH. L. 101, 107–119 (2007); Wu, *supra* note 25, at 347–349.

long before the Internet, and therefore were not intended to be limited to ISPs. A newspaper is an example of a publisher, a bookstore is considered a distributor, and a telephone company is a conduit. According to the Restatement (Second) of Torts, these service providers fall within the definition of 'one who only delivers or transmits defamatory matter'.[22] According to the commentary in the Restatement, 'delivers' means 'the transfer of possession of a physical embodiment of the defamatory matter', and '"transmits" includes ... the conveyance of defamatory words by methods other than physical delivery, as in the case of a telegraph company putting through a call'.[23] However, 'one who merely makes available to another equipment or facilities that he may use himself for general communication purposes' does not 'transmit' under the Restatement.[24] Publishers are deemed responsible for content by others; distributors are 'subject to liability if [they] know or [have] reason to know of [...] defamatory character' of the content published by others,[25] and conduits are not liable for content published by others even if they are aware of the content.[26]

The line between secondary and direct liability might appear blurry at times. Typically, plaintiffs sue service providers instead of users; however, it is not uncommon for plaintiffs to claim both direct and indirect liability of ISPs. For example, in *Religious Technology Center v. Netcom On-Line Communication Services* the court considered (in addition to a claim of secondary infringement) a claim of direct copyright infringement that the defendant was supposed to have committed by creating a copy of the infringing material posted by a user.[27] The court likened the service provider's activity to that of an 'owner of a copying machine' and rejected the theory of direct infringement.[28] In trademark law, recent court decisions suggest that courts are leaning toward the opinion that the federal trademark statute 'contains virtually no limitation on the type of "use" of a mark that can qualify as direct trademark infringement'.[29] As one commentator has pointed out, this trend towards a broad interpretation of the term 'use' could eventually result in a de facto 'abandonment of the direct/contributory infringement distinction'[30] in trademark law. However, courts might not find that there was a likelihood of confusion caused by the actions of the service provider.

---

[22] Restatement (Second) of Torts §581(1) (1977).

[23] *Id.*

[24] *Id.*

[25] Restatement (Second) of Torts §581 (1977).

[26] Felix T. Wu, *Collateral Censorship and the Limits of Intermediary Immunity*, 87 NOTRE DAME L. REV. 293, 309–311 (2011).

[27] Religious Tech. Ctr. v. Netcom On-Line Commc'n Servs., Inc., 907 F. Supp. 1361 (N.D. Cal. 1995).

[28] *Id.* at 1369.

[29] Stacey L. Dogan, *Beyond Trademark Use*, 8 J. TELECOMM. & HIGH TECH. L. 135, 136 (2010); Rescuecom Corp. v. Google, Inc., 562 F.3d 123 (2d Cir. 2009); Network Automation, Inc. v. Advanced Sys. Concepts, Inc., 638 F.3d 1137 (9th Cir. 2011).

[30] Dogan, *supra* note 28, at 154.

Unless a service provider enjoys immunity from secondary liability or a safe harbour,[31] courts may award damages and injunctions against the service provider. Under copyright law, plaintiff may select between: (a) actual damages and profits; and (b) statutory damages.[32] Attorney's fees, prejudgment interest, and costs may also be awarded. These remedies do not differ from the kinds of remedies that a court may award against a party that is primarily liable. A party that is secondarily liable is 'jointly and severally liable with the direct infringer for all general damages'.[33] When determining the amount of damages, a court will consider the actions of other parties, including non-party infringers.[34] Safe harbour provisions protect ISPs from damages but not from injunctive relief, which courts may grant in accordance with safe harbour provisions, as explained in the following section.

In addition to damages, injunctions, and other remedies, service providers might be subject to other court-ordered acts associated with the activities of others that use their services. Service providers may be subpoenaed to provide information concerning the identity of an alleged infringer. One of the provisions of the Digital Millennium Copyright Act (DMCA)—section 512 of the Copyright Act[35]—includes a special provision concerning such subpoenas; it provides for 'a subpoena to a service provider for identification of an alleged infringer'.[36] The provision specifies the contents of the request, the contents of the subpoena, the basis for granting the subpoena, and the actions required of the service provider who receives the subpoena.[37]

Service providers are unlikely to face criminal liability for the conduct of third parties who use their services. In the context of copyright infringement, there is no explicit statutory authorisation for contributory infringement; doctrine in this area is case-law-driven. In the context of defamatory speech, criminal defamation is largely a dead letter in the United States. A number of states still have criminal defamation statutes on their books, but they are rarely applied and widely considered to have fallen into desuetude; it is unlikely that they would be applied to a service provider, particularly in light of the ability of service providers to take advantage of the safe harbour in Section 230 of the Communications Decency Act (CDA).[38]

---

[31] See infra for a discussion of the immunity and safe harbour provisions.

[32] 17 U.S.C. §504(c) (2006 & Supp. 2011).

[33] Crystal Semiconductor Corp. v. TriTech Microelectronics Int'l, Inc., 246 F.3d 1336, 1361 (Fed. Cir. 2001).

[34] Agence France Presse v. Morel, 10 CIV. 02730 AJN, 2013 WL 2253965 (S.D.N.Y. 21 May 2013). See also, e.g., Arista Records LLC v. Usenet.com, Inc., 07 CIV 8822 HB THK, 2010 WL 3629688 (S.D.N.Y. 2 Feb. 2010) (taking into account actions of defendant's subscribers).

[35] 17 U.S.C. §512 (2006).

[36] 17 U.S.C. §512(h)(1) (2006).

[37] 17 U.S.C. §512(h)(2)–(5) (2006).

[38] 47 U.S.C. §230 (2006).

That does not mean that criminal law has not been deployed, arguably in an abusive manner. For example, in *Puerto 80*,[39] U.S. Customs Enforcement (ICE) seized domain names, effectively taking down websites that hosted links or embeds, alleging that such conduct represented criminal contributory copyright infringement, by taking advantage of seizure laws that allow the federal government to take control of property suspected of being used in criminal activity prior to trial. *Puerto 80* is not an isolated case.[40] In several such cases, as they did in *Puerto 80*, the ICE later voluntarily dismissed its complaint.[41] Academic commentators have observed that ICE's claims in making the seizures were poorly supported in evidence and law.[42] The use of such 'search and seizure' mechanisms against websites based on arguments of criminal infringement should be curtailed by appropriate limiting principles such as weighing privacy interests and the consideration of whether less restrictive alternatives exist.[43] Given the tremendous damage that seizures of websites do to online business, reform in this area is critical.

Immunity from Secondary Liability and Safe Harbours

U.S. legislation includes the following provisions that immunise or provide safe harbours to ISPs from liability for the conduct of others: section 230 of the CDA,[44] a section of the DMCA—section 512 of the Copyright Act,[45] and section 32(2)(B) and (C) of the Lanham Act.[46] The least utilised of the three sets of provisions is section 32(2)(B) and (C) of the Lanham Act.[47]

In 1996, the CDA codified an approach to secondary liability that courts had established in the realm of defamation law prior to the enactment of the CDA; the approach held that conduits were not secondarily liable for defamatory statements, and distributors were secondarily liable only under certain circumstances.[48] Court decisions had previously established that conduits were not liable for the content

---

[39] *Puerto 80 Projs.* v. *United States*, Case 1:11-cv-04139-PAC (S.D.N.Y., 4 Aug. 2011) (order denying petition for release of domain names seized by Immigration and Customs Enforcement).

[40] Nate Anderson, *Government Admits Defeat, Gives Back Seized Rojadirecta Domains*, 29 Aug. 2012, *available at* http://arstechnica.com/tech-policy/2012/08/government-goes-0-2-admits-defeat-in-rojadirecta-domain-forfeit-case/ (last visited 26 Sept. 2015) (discussing return by ICE of domains to parent company Puerto 80).

[41] *Id.*

[42] *Id.* (quoting Prof. Eric Goldman).

[43] *Cf. O'Grady* v. *Superior Court*, 44 Cal. Rptr. 3d. 72 (Cal. Ct. App. 2006) (weighing such interests in civil discovery context).

[44] 47 U.S.C. §230 (2006).

[45] 17 U.S.C. §512 (2006).

[46] 15 U.S.C. §1114(2)(B), (C) (2006).

[47] Section 32 of the Lanham Act was applied in Hendrickson v. eBay, Inc., 165 F. Supp. 2d 1082 (C.D. Cal. 2001).

[48] Wu, *supra* note 25, at 310; see also Lerman v. Flynt Distributing Co, 745 F.2d 123 (2d Cir. 1984).

they transmitted,[49] that distributors were not liable for defamatory statements in the absence of knowledge of the defamatory statements,[50] and that distributors had no duty to monitor the content of publications.[51] However, courts did not seem to agree on whether and how the rules should apply to ISPs. In one case, a court held that a service provider was a distributor,[52] and in another a court decided that a service provider was more a publisher than a distributor because of the content control that the service provider exerted.[53] The CDA responded to these court decisions by providing immunity from liability under defamation law and also under a wide variety of other laws. An early case clarified that the CDA immunises service providers from secondary liability, whether they act as publishers or distributors.[54]

The immunity provisions of the CDA are not limited to ISPs; they also cover Internet users, service providers, and users of interactive computer services other than the Internet. The CDA immunises from liability a 'provider or user of an interactive computer service'.[55] An interactive computer service is defined as 'any information service, system, or access software provider that provides or enables computer access by multiple users to a computer server, including specifically a service or system that provides access to the Internet and such systems operated or services offered by libraries or educational institutions'.[56] The immunity provision does not cover a service provider who is also 'an information content provider … responsible, in whole or in part, for the creation or development of' the content.[57] Entities that have benefited from immunity under the CDA include large, prominent firms such as Facebook, Google, and Yahoo!, as well as those with less power over much more modest Internet fora.[58]

When enacted in 1998, the DMCA built on case law concerning secondary liability for copyright infringement.[59] Courts had already established that service providers would not be secondarily liable for infringement by others unless the service

---

[49] Wu, *supra* note 25, at 310; David S. Ardia, *Free Speech Savior or Shield for Scoundrels: An Empirical Study of Intermediary Immunity under Section 230 of the Communications Decency Act*, 43 Loy. L.A. L. Rev. 373, 398–401 (2010).

[50] Restatement (Second) of Torts §581; Ardia, *supra* note 43, at 397, 398.

[51] Lerman v. Flynt Distrib. Co., 745 F.2d 123, 139 (2d Cir. 1984).

[52] Cubby, Inc. v. CompuServe, Inc., 776 F. Supp. 135, (S.D.N.Y. 1991).

[53] Stratton Oakmont, Inc. v. Prodigy Servs. Co., 1995 WL 323710 (N.Y. Sup. Ct. 24 May 1995).

[54] Zeran v. America Online, Inc., 129 F.3d 327 (4th Cir. 1997).

[55] 47 U.S.C. §230(c) (2006). On difficulties with the definition of a 'user', see Wu, *supra* note 25, at 318–324.

[56] 47 U.S.C. §230(f)(2) (2006).

[57] 47 U.S.C. §230(f)(3) (2006). On the difficulties of identifying service providers responsible for the development of content, see Wu, *supra* note 25, at 324–327.

[58] *See, e.g., Barrett v. Rosenthal*, 40 Cal. 4th 33 (Cal. 2006) (holding that immunity extends even to one 'who had no supervisory role in the operation of the Internet site where allegedly defamatory material appeared and who thus was clearly not a provider of an "interactive computer service" under the broad definition provided in the CDA').

[59] Sony Corp. v. Universal City Studios, Inc., 464 U.S. 417 (1984).

providers knew or had reason to know of specific infringing activity.[60] The safe harbour provisions of the DMCA are limited to service providers of 'digital online communications' and providers 'of online services or network access, or the operator of facilities thereof'[61] who fall into one of four categories. The first category of service providers includes providers who engage in transmission or routing, or those who provide connections on a system or network that is controlled or operated by or for service providers ('Transitory digital network communications').[62] The second category of service providers includes those who engage in 'the intermediate and temporary storage of material' ('System caching').[63] The third category includes service providers who provide for space in which third parties store material ('Information residing on systems or networks at direction of users').[64] The fourth category includes service providers who 'refer[...] or link[...] users to an online location' of material 'by using information location tools' ('Information location tools').[65] The DMCA also has a special provision for nonprofit educational institutions acting as service providers.[66]

The safe harbour provision in section 32 of the Lanham Act is not limited to ISPs; it covers a wide range of publishers and distributors. Section 32 covers 'the publisher or distributor of ... a newspaper, magazine, or other similar periodical or electronic communication' with regard to material that is a part of paid advertising.[67] Electronic communication is defined by reference to 18 U.S.C. §2510(12), where the term is defined as 'any transfer of signs, signals, writing, images, sounds, data, or intelligence of any nature transmitted in whole or in part by a wire, radio, electromagnetic, photoelectronic or photooptical system that affects interstate or foreign commerce'.[68] Excepted from the definition of the term are '(A) any wire or oral communication: (B) any communication made through a tone-only paging device; (C) any communication from a tracking device (as defined in section 3117 of ... title [18]); or (D) electronic funds transfer information stored by a financial institution in a communications system used for the electronic storage and transfer

---

[60] Religious Tech. Ctr. v. Netcom On-Line Commc'n Servs., Inc., 907 F. Supp. 1361 (N.D. Cal. 1995).

[61] 17 U.S.C. §512(k)(1)(A) and (B) (2006).

[62] 17 U.S.C. §512(a) (2006); see also Perfect 10, Inc. v. CCBill LLC, 488 F.3d 1102 (9th Cir. 2007); Columbia Pictures Indus., Inc. v. Fung, 710 F.3d 1020 (9th Cir. 2013) (deciding that BitTorrent sites were not covered by 17 U.S.C. §512(a)).

[63] 17 U.S.C. §512(b) (2006).

[64] 17 U.S.C. §512(c) (2006 & Supp. 2011); see also Perfect 10, Inc., 488 F.3d 1102 (9th Cir. 2007); Viacom Int'l, Inc. v. YouTube, Inc., 676 F.3d 19 (2d Cir. 2012); UMG Recordings, Inc. v. Shelter Capital Partners LLC, 718 F.3d 1006 (9th Cir. 2013); Columbia Pictures Indus., Inc., 710 F.3d 1020 (9th Cir. 2013) (deciding that BitTorrent sites were not covered by 17 U.S.C. §512(b)).

[65] 17 U.S.C. §512(d) (2006); see also Perfect 10, Inc., 488 F.3d 1102 (9th Cir. 2007); Columbia Pictures Indus., Inc., 710 F.3d 1020 (deciding that BitTorrent sites were not covered by 17 U.S.C. §512(d)).

[66] 17 U.S.C. §512(e) (2006).

[67] 15 U.S.C. §1114(2)(B) (2006).

[68] 18 U.S.C. §2510(12) (2006).

of funds'.[69] Courts assessed eBay to be more an 'online auction house' than a 'classified advertiser' and therefore the site did not benefit from the safe harbour under section 32 of the Lanham Act.[70]

The CDA immunises service providers and users from liability under any laws that impose liability on a 'publisher or speaker of any information provided by another information content provider'.[71] Although some critics argue that the CDA immunises service providers only for actions in which they assume the role of a publisher or speaker, courts have applied the immunity even in cases where the service provider arguably did not act in that role.[72] The CDA also immunises service providers from civil liability that may be based on the provider's restricting access to certain types of material or on the provider's providing to others the technical means to restrict access to certain types of material.[73] The CDA is a general intermediary liability provision that is without prejudice to intellectual property-specific provisions (the DMCA and section 32 of the Lanham Act), criminal statutes, state laws that are consistent with the CDA,[74] and the Electronic Communications Privacy Act.[75] Although the CDA arose from a line of defamation cases, the provision has been designed to cover other instances of intermediary liability and has been applied to facts involving breach of contract, tortious interference with business relations, and invasion of privacy.

As for the coverage of the intellectual property law statutes, the DMCA safe harbour provision concerns liability for secondary copyright infringement that may arise under the U.S. Copyright Act.[76] Courts disagree on whether the DMCA extends to secondary liability that may arise under state copyright laws,[77] and it is also questionable whether the DMCA covers secondary liability if such liability arises under anti-circumvention provisions of the U.S. Copyright Act.[78] Section 32 of the Lanham

---

[69] 18 U.S.C. §2510(12) (2006).

[70] Tiffany v. eBay Inc., 576 F. Supp. 2d. 463, 507, n. 33 (S.D.N.Y. 2008); Hendrickson v. eBay, 165 F. Supp. 2d 1082, n. 2 (C.D. Cal. 2001).

[71] 47 U.S.C. §230(c)(1) (2006).

[72] Doe v. MySpace, 528 F.3d 413 (5th Cir. 2008). *See also* Wu, *supra* note 25, at 327, 328.

[73] 47 U.S.C. §230(c)(2) (2006).

[74] According to the CDA, '[n]o cause of action may be brought and no liability may be imposed under any State or local law that is inconsistent with' the CDA. 47 U.S.C. §230(e)(3). See also Evans v. Hewlett-Packard Co., C 13–02477 WHA, 2013 WL 4426359 (N.D. Cal. 15 Aug. 2013) (dismissing state law claims against an app store because the state law was inconsistent with the CDA).

[75] 47 U.S.C. §230(e) (2006).

[76] 17 U.S.C. §512 (2006 & Supp. 2011).

[77] Capitol Records, Inc. v. MP3tunes, LLC, 821 F. Supp. 2d 627 (S.D.N.Y. 2011); UMG Recordings, Inc. v. Escape Media Grp., Inc., 964 N.Y.S.2d 106 (2013). See also Report of the Register of Copyrights, *Federal Copyright Protection for Pre-1972 Sound Recordings,* U.S. COPYRIGHT OFFICE, Dec. 2011, at http://www.copyright.gov/docs/sound/pre-72-report.pdf (last visited 22 Aug. 2013).

[78] 17 U.S.C. §1201 (2006). It is disputed whether the provision concerning circumvention of copyright protection systems creates secondary liability. See Lemley, *supra* note 20, at 107.

Act provides a safe harbour for service providers for the acts of others that have led to infringements of trademarks, false designations of origin, false or misleading description of fact, and false or misleading representations of fact under section 43(a) of the Lanham Act.

As noted earlier, the CDA completely immunises providers and users from both direct and secondary civil liability under federal law (with the exceptions of causes of action listed in the statute) and state laws that are inconsistent with the CDA.[79] Providers do not have to satisfy any conditions to benefit from the general immunity that protects them from being treated as a publisher or speaker.[80] However, to benefit from the immunity in cases where the provider restricts access to material or provides the technical means to others to restrict access to material, the provider must act in good faith.[81]

The DMCA safe harbour concerns liability for copyright infringement and protects service providers only against monetary relief,[82] and injunctions are available against service providers only in forms specified in the statute.[83] To benefit from the DMCA safe harbour provision, service providers must comply with the basic conditions of eligibility,[84] and with additional conditions that are specific to each category of service provider. There are two basic conditions for eligibility. First, the service provider must adopt and reasonably implement a policy against repeat infringers, and inform subscribers and account holders of the policy;[85] the policy must 'provide ... for the termination in appropriate circumstances of subscribers and account holders ... who are repeat infringers'.[86] Second, the service provider must 'accommodate ... and not interfere with standard technical measures used by copyright owners to identify or protect copyrighted works'.[87] The statute specifies the conditions that such technical measures must meet.[88]

ISPs must, additionally, to benefit from the safe harbour of the DMCA, meet other conditions that the statute specifies for each category of service provider.[89]

---

[79] 47 U.S.C. §230(c) (2006).

[80] 47 U.S.C. §230(c)(1) (2006).

[81] 47 U.S.C. §230(c)(2) (2006).

[82] According to the legislative history, Congress intended for the DMCA to cover direct and indirect liability, both vicarious and contributory. S. Rep. No. 105–190, at 40 (1998); H.R. Rep. No. 105–551, pt. 2, at 50 (1998).

[83] 17 U.S.C. §512(j) (2006).

[84] 17 U.S.C. §512(i) (2006).

[85] See Perfect 10, Inc., 488 F.3d 1102 (on the meaning of a 'reasonably implemented policy').

[86] 17 U.S.C. §512(i)(1)(A) (2006).

[87] 17 U.S.C. §512(i)(1)(B) and (i)(2) (2006).

[88] 17 U.S.C. §512(i)(2) (2006).

[89] 17 U.S.C. §512(a) (2006) (for 'Transitory digital network communications'); 17 U.S.C. §512(b) (2006) (for 'System caching'); 17 U.S.C. §512(c), (g) (2006 & Supp. 2011) (for 'Information residing on systems or networks at direction of users'); 17 U.S.C. §512(d), (g) (2006) (for 'Information location tools').

Among these conditions is a notification system requirement[90] that service providers must meet if they fall in the category of service providers who provide for space in which third parties store material ('Information residing on systems or networks at direction of users')[91] or service providers who "refer ... or link ... users to an online location' of material 'by using information location tools' ('Information location tools').[92] The DMCA outlines the details of the system, such as the designation of 'an agent to receive notifications of claimed infringement',[93] elements of notification,[94] and the procedure for the replacement of removed or disabled material.[95]

One of the contentious issues arising from the DMCA conditions has been the question of level of 'actual knowledge' or the existence of 'red flag' knowledge that 'the material or an activity ... is infringing';[96] the absence of such knowledge is required for service providers who provide information storage[97] and information location tools.[98] In *Viacom International, Inc. v. Youtube, Inc.* the court explained that the knowledge (or awareness) must be of specific material, that '[t]he difference between actual and red flag knowledge is ... between a subjective and an objective standard',[99] and that 'the willful blindness doctrine may be applied, in appropriate circumstances, to demonstrate knowledge or awareness of specific circumstances of infringement'.[100]

The safe harbour provision in section 32 of the Lanham Act covers instances when 'the infringement or violation ... is contained in or is part of paid advertising ... or in an electronic communication'. The provision of paragraph (B) limits the remedy available against the service provider to an injunction in cases of 'innocent infringers and innocent violators'. Injunctions are limited to cover only future publications or electronic communication,[101] and the availability of injunctions is further limited in some cases when an injunction would lead to a delay in the

---

[90] 17 U.S.C. §512(c)(2),(3) (2006 & Supp. 2011); 17 U.S.C. §512(g)(2)–(4) (2006).

[91] 17 U.S.C. §512(c) (2006 & Supp. 2011).

[92] 17 U.S.C. §512(d) (2006).

[93] 17 U.S.C. §512(c)(2) (2006 & Supp. 2011).

[94] 17 U.S.C. §512(c)(3) (2006).

[95] 17 U.S.C. §512(g)(2)–(4). The notification system is not free of abuses by some copyright holders and others who appear to be using notifications to impede free speech. See Jennifer M. Urban & Laura Quilter, *Efficient Process or "Chilling Effects"? Takedown Notices Under Section 512 of the Digital Millennium Copyright Act*, 22 SANTA CLARA COMPUTER & HIGH TECH. L.J. 621, 631–36 (2006); Jeffrey Cobia, Note, *The Digital Millennium Copyright Act Takedown Notice Procedure: Misuses, Abuses, and Short-comings of the Process*, 10 MINN. J. L. SCI. & TECH. 387, 387–94 (2009).

[96] 17 U.S.C. §512(c)(1)(A)(i), (ii) (2006).

[97] *Id.*

[98] 17 U.S.C. §512(d)(1)(A), (B) (2006).

[99] Viacom Int'l, Inc., 676 F.3d 19 at 31. *See also* UMG Recordings, Inc., 718 F.3d 1006 at 1023.

[100] Viacom Int'l, Inc., 676 F.3d 19 at 35.

[101] 15 U.S.C. §1114(2)(B) (2006).

dissemination of a periodical or electronic communication.[102] To take advantage of the safe harbour, providers must be 'innocent infringers and innocent violators'.[103]

As for remedies, the CDA includes no provision for remedies against ISPs; ISPs are completely immunised from liability under the CDA. The DMCA includes a provision for injunctions that may be issued against service providers. The provision limits the forms of injunctive relief[104] and lists factors that courts must consider when they decide whether or not they will issue an injunction against a service provider.[105] Courts may restrain a service provider from providing access to infringing material or activity[106] and from providing access to a subscriber or account holder who is infringing.[107] Courts may grant other injunctive relief that is necessary to 'prevent or restrain infringement' that is effective and 'least burdensome' to the service provider.[108] The injunctive relief that a court may issue against providers of 'transitory digital network communications'[109] is limited to: a) 'an order restraining the service provider from providing access to a subscriber or account holder'; and b) 'an order restraining the service provider from providing access ... to a specific, identified, online location outside the United States'.[110] The steps that a service provider may take to comply with the last form of injunctive relief will include blocking. The factors that a court will consider are the burden on the provider and its system or network,[111] the harm to the copyright owner if the injunction is not granted,[112] the technical feasibility and effectiveness of the measure and its potential to interfere with access to non-infringing material,[113] and the availability of 'less burdensome and comparably effective means of preventing or restraining access'.[114] The service provider must receive notice of the injunctive relief and be given an opportunity to appear;[115] only orders for the preservation of evidence or 'other orders having no material adverse effect on the operation' of the service provider's network may be issued without notice.[116]

---

[102] 15 U.S.C. §1114(2)(C) (2006).

[103] 15 U.S.C. §1114(2)(B) (2006).

[104] 17 U.S.C. §512(j)(1) (2006).

[105] 17 U.S.C. §512(j)(2) (2006).

[106] 17 U.S.C. 512(j)(1)(A)(i) (2006).

[107] 17 U.S.C. 512(j)(1)(A)(ii) (2006).

[108] 17 U.S.C. 512(j)(1)(A)(iii) (2006).

[109] 17 U.S.C. §512(a) (2006); see also Perfect 10, Inc. v. CCBill LLC, 488 F.3d 1102 (9th Cir. 2007); Columbia Pictures Indus., Inc., 710 F.3d 1020 (deciding that BitTorrent sites were not covered by 17 U.S.C. §512(a)).

[110] 17 U.S.C. 512(j)(1)(B) (2006).

[111] 17 U.S.C. 512(j)(2)(A) (2006).

[112] 17 U.S.C. 512(j)(2)(B) (2006).

[113] 17 U.S.C. 512(j)(2)(C) (2006).

[114] 17 U.S.C. 512(j)(2)(D) (2006).

[115] 17 U.S.C. 512(j)(3) (2006).

[116] *Id.*

Section 32 of the Lanham Act provides for injunctions against a publisher or distributor 'against the presentation of [the infringing material] in future issues of … newspapers, magazines, or other similar periodicals or in future transmissions of … electronic communications'.[117] A court will deny an injunction if it could cause a delay in the dissemination of newspapers, magazines, other similar periodicals, or electronic communications.[118]

Best Practices

The United States does not have legislation that explicitly mandates graduated response of the type that certain other jurisdictions (e.g., France) have adopted. Instead, with some degree of administrative guidance, copyright industry players, including ISPs, and copyright owners, including members of the MPAA and the RIAA, have developed best practices, voluntary codes, and explicit guidelines and industry standards for dealing with conduct by third parties using the services of the ISPs that allegedly amounts to a violation of law. The most prominent example is the code of conduct formulated by the Center for Copyright Information, the MPAA, the RIAA, and several ISPs.[119]

This system of voluntary graduated response focusing on intellectual property rights infringement, sometimes called the 'six strikes' system, was implemented in February 2013. In this set of best practices, ISPs must place in their terms of service that copyright infringement constitutes a violation of the agreement,[120] create and enforce a copyright alert program,[121] and provide monthly reports containing anonymised information about consumers allegedly committing P2P infringements.[122]

The copyright alert program contains four steps. Step one involves the ISP sending a subscriber a notice alleging a P2P infringement and notifying the subscriber that P2P violations are illegal, a violation of the terms of service, as well as other messages that fall along the same line.[123] The next step is called the 'acknowledgement step'. Here, the ISP must send more notices, which the subscriber must acknowledge receiving; however the subscriber is not required to admit guilt at this stage.[124] The next stage is the 'mitigation measures step'.[125] At this point, the ISP

---

[117] 15 U.S.C. §1114(2)(B) (2006).

[118] 15 U.S.C. §1114(2)(C) (2006).

[119] Center for Copyright Information, *Memorandum of Understanding*, 24–25 (6 Jul. 2011) at http://www.copyrightinformation.org/wp-content/uploads/2013/02/Memorandum-of-Understanding.pdf

[120] *Id.* at 7

[121] *Id.* at 7–13

[122] *Id.* at 14–15

[123] *Id.* at 8

[124] *Id.* at 9–10

[125] *Id.* at 10

must send yet another notice to the subscriber, which also requires acknowledgement of receipt. The ISP must also institute a punitive measure within a predetermined number of days.[126] The ISP also has the ability to issue each subscriber one waiver instead of implementing the mitigation measure.[127] The next step, 'post mitigation measure step', involves instituting a different mitigation measure than used in the prior step as well as notifying the user that their conduct may result in a lawsuit.[128] Additionally the memorandum states that if a subscriber does not receive a notification for 12 months that the system should reset to step one.[129]

The Copyright Alert System framework was devised through the negotiation and cooperation of several major industries and firms, notably the industry associations the Independent Film and Television Alliance (IFTA) and the American Association of Independent Musicians (A2IM); Recording Industry Association of American members Universal Music Group, Warner Music Group, Sony Music Entertainment, and EMI Music; Motion Picture Association of America members Walt Disney Studios Motion Pictures, Paramount Pictures, Sony Pictures Entertainment, Twentieth Century Fox Film Corporation, Universal Studios, and Warner Brothers Entertainment; and the ISPs ISPs AT&T, Cablevision, Comcast, Time Warner Cable, and Verizon. The Governor of New York, Andrew Cuomo, facilitated the negotiations. Some NGO groups, including the Electronic Frontier Foundation, have criticised this process for lacking direct representation of user/consumer interests.[130]

The desirability of self-regulation has been recognized in court decisions. In the landmark case of *Zeran v. America Online,* the court grounded its conclusion that section 230 forbids the imposition of publisher liability on a service provider in part to promote AOL's own freedom to self-regulate.[131] The self-governance described above was negotiated in the shadow of section 230, and adopted with the specific intent to shape a set of industry norms to the contours of the immunity provided by the statute. Similarly, the approach to 'dual use' technologies—that is, technologies that are capable of both infringing and non-infringing use—allow bona fide providers of the technologies to benefit from the immunity and safe harbour provisions. As a result, via self-regulation, ISPs may enjoy a form of immunity in the copyright context that resembles that which they have enjoyed in the context of defamation and other non-IP liability.

---

[126] *Id.* at 10–11.

[127] *Id.* at 12

[128] *Id.* at 12–13

[129] *Id.* at 13

[130] Corynne McSherry, "The 'Graduated Response' Deal: What if Users Had Ben at the Table?" 18 Jul. 2011, *available at* https://www.eff.org/deeplinks/2011/07/graduated-response-deal-what-if-users-had-been.

[131] 129 F.3d. 327 (4th Cir. 1997). See also Doe v. Myspace, Inc., 474 F. Supp. 2d 843, 850 (W.D. Tex. 2007): 'This section reflects ... the potential for liability attendant to implementing safety features and policies created a disincentive for ... services to implement any safety features or policies at all'.

Conclusions

U.S. law on secondary liability of ISPs is fragmented and continues to develop. The fragmentation is perhaps surprising given the deep roots of the Internet in the United States and the general admiration for the industry that propelled the Internet to become the engine of economic progress. However, the legislation concerning ISPs' secondary liability was enacted early in the development of the Internet, at a time when legislators could not have anticipated some user activities that would eventually become commonplace. Along with later technological developments, these new types of user activities continue to challenge the framework that the early legislation created. The emergence of a more interactive Internet with Web 2.0, plus the greater potential for defamatory harm made possible by the publication of private photos and video, have led some to question the boundaries of Section 230.[132]

Future developments in the law of secondary liability of ISPs will have to balance a number of matters, including technological developments that will enable ISPs to implement more monitoring and stronger supervision of their services. Other countries are struggling to delineate the appropriate level of ISPs' secondary liability in light of such developments, and indications internationally suggest that more might be expected of ISPs in the future.[133] At the same time, fundamental rights of service providers appear to be emerging based on narrow construction of existing doctrines of contributory infringement. For example, in *FlavaWorks v. Gunter*,[134] a case involving a website that linked or embedded video files hosted elsewhere, renowned appellate Judge Posner ruled that such conduct could not be contributory infringement, since neither the user nor the service provider is in fact creating a copy or doing an infringing 'performance' of the video. Judge Posner analogised the conduct at issue to stealing a book from a bookstore and then reading it; it may be immoral and illegal, but it is not copyright infringement. Narrow constructions such as this are effectively preserving space for efficient and welfare-enhancing activity in the face of secondary liability claims. Through such rulings and via ISP-friendly self-regulation, ISPs may hope to enjoy rules on secondary liability that give them the freedom to operate successfully in the future.

---

[132] *See, e.g.,* Note, *Badging: Section 230 Immunity in a Web 2.0 World,* 123 Harv. L. Rev. 981 (2010); Barnes v. Yahoo!, 570 F.3d 1096 (9th Cir. 2009).

[133] See Alone in the Dark, Bundesgerichtshof, I ZR 18/11, 12 July 2012; File-Hosting-Dienst, Bundesgerichtshof, I ZR 80/12, 15 Aug. 2013.

[134] Flava Works, Inc. v. Gunter, 689 F.3d 754 (7th Cir. 2012).

Chapter 4
ISP Secondary Liability: A Portuguese Perspective on Omissions as the Basis for Secondary Liability

João Fachana

Overview of the Portuguese Legal Framework of ISP's Secondary Liability

The Portuguese legal framework for secondary liability of Intermediary Service Providers (ISP) was established in early January 2004, with the approval of a national statute about certain aspects of society information services (especially e-commerce)—The *Decreto-Lei* no. 7/2004, of 7 January, amended later by *Decreto-Lei* no. 62/2009, of 10 March. This legal framework is the domestic implementation of Directive 2000/31/EC, known as the E-commerce Directive.[1] The legal framework for secondary liability of service providers in Portugal is, therefore, similar to the domestic legislation of other European Union Member States that have implemented the E-Commerce Directive.

The framework deals mainly with the immunity of ISPs from liability, which varies according to the type of service granted by the provider. The legal framework defines four types of intermediary services for which providers can be the subject of immunity, namely, mere conduit, caching, hosting, and content association.[2]

---

[1] Directive 2000/31/EC of the European Parliament and of the Council of 8 June 2000, which concerns certain legal aspects of information society services, in particular electronic commerce, in the European Internal Market [2000] OJ L 178/1.

[2] The scope of each service is defined in Sections 14 to 17 of *Decreto-Lei* no. 7/2004, of 7 January, and will be further discussed in this Chapter. The Portuguese legislator decided to include the content association service type, which goes beyond the type of services addressed in the E-commerce Directive. However, as we will see below, the immunity standard for these service providers does not differ from that applicable to hosting services.

J. Fachana (✉)
Master of Laws, Faculdade de Direito, Universidade do Porto, Porto, Portugal
e-mail: joaofachana@gmail.com

© Springer International Publishing AG 2017   109
G.B. Dinwoodie (ed.), *Secondary Liability of Internet Service Providers*,
Ius Comparatum – Global Studies in Comparative Law 25,
DOI 10.1007/978-3-319-55030-5_4

Section 11 of *Decreto-Lei* no. 7/2004 states the general principle of ISP liability: ISPs are subject to the general liability standard, subject to exceptions where any immunity is applicable. Thus, when these exemptions are not applicable, in order to find an ISP liable for the conduct of others, the ISP conduct must fulfil the five general requirements for civil liability: (i) a voluntary action or omission caused by a legal person or legal entity; (ii) unlawfulness of that action or omission; (iii) guilt of the person or entity responsible; (iv) damages produced; and (v) a causal nexus between the action or omission and the damages. Even when primary liability is established—when the infringer is actually identified—this can coexist with secondary liability of the ISP. In the end, the ISP and the party primarily responsible for the creation of the unlawful content will be found joint and severally liable and ordered to pay compensation for damage.

ISP secondary liability will also arise in cases where the ISP fails to comply with a certain duty to act—and, in this case, liability is characterised under Portuguese law as 'liability for omission'. Liability of this type arises only when there is a legal duty to act and the person or entity fails to do so. We will pay special attention to this kind of liability, as it is the basis on which ISPs are most likely found secondarily liable.

Unlike other legal frameworks, Portuguese liability standards are uniform and applicable to every field of law and do not differ depending on the cause of action that is at stake. As such, ISP secondary liability will be assessed under the same principles relevant whether the infringing content concerns intellectual property rights, defamation, or offensive content, etc. (Minimal differences may occur between fields of law, as to, for example, the evaluation of damages regarding to intellectual property infringement, but the core liability standards remain unchanged.)

Historical Development of the Secondary Liability of Service Providers in Portugal

Due to the shortage of ISPs based in Portugal, secondary liability has not been raised in national courts.[3] As such, case law has not developed around the immunity framework. However, this issue was developed and debated by some Portuguese authors prior to the adoption of the E-commerce directive and its implementation by *Decreto-Lei* no. 7/2004.

---

[3] This is subject to a few exceptions relating to injunctions sought against access ISPs. But even in those cases, as the ISPs clearly complied with the court order, no secondary liability was ever discussed. The most recent of these injunctions was sought against "mere conduit" service providers for the blocking of Pirate Bay's website, a decision awarded by the Portuguese Intellectual Property Court, in March 2015 (unpublished ruling). As a result, the Pirate Bay is, actually, blocked in Portugal (Van der Saar 2015).

The first author to comment on the issue was Carneiro da Frada (2001), in a keynote speech, where he endorsed the notion that ISPs should not have a general obligation to monitor the content accessed or transmitted by third parties (Carneiro da Frada 2001, 7–31). Other authors, such as Vasconcelos Casimiro (2000) or Menezes Leitão (2002), also advocated similar approaches to the liability regime that should be applicable to ISPs for content originated by third parties (Menezes Leitão 2002, 171–192; Vasconcelos Casimiro 2000).

These conclusions were drawn from the general civil liability principles that deal with liability for omission, particularly regarding duties to act. It was a common understanding that an ISP serves as an intermediary in the access and transmission of information and, accordingly, it could not be held liable for third party content, except for cases where it had real control over the content and had actual or apparent knowledge of the unlawfulness of the content and failed to act. In these situations, the offending ISP would be held liable for omission under Section 486 of the Portuguese Civil Code.

These scholarly opinions regarding ISP secondary liability were in accordance with the discussions held when drafting the E-commerce Directive. In fact, the underlying principle was established as a single, standard rule by the E-commerce Directive and enacted in each EU member-state, namely, the absence of a general duty to monitor: ISP's are not obliged to generally monitor or filter information from third parties that are transmitted or hosted by their services.[4]

The Concept of an Intermediary Service Provider

The E-commerce Directive does not define 'Service Provider' or 'Online', 'Internet' or 'Intermediary' Service Provider. The Portuguese framework defines 'Intermediary Service Provider' (ISP) (No. 5 of Section 4 of *Decreto-Lei* no. 7/2004), for the purpose of knowing which types of Information Society services are subject to the immunity regime established therein. ISPs are those that supply technical services for the access, availability and use of information or online services independently from the generation of the information or service itself. This definition includes, but is not limited to, search engines[5] and operators of online marketplaces,[6] as long as they develop an intermediary activity, as stated above. It also includes telecommunication companies when they provide access to the Internet.[7] *ANACOM*

---

[4] See Section 12 of *Decreto-Lei* no. 7/2004, which implemented Article 15 of the E-commerce Directive.

[5] For example, the search engine 'SAPO' (www.sapo.pt), originally developed by a Portuguese University (*Universidade de Aveiro*).

[6] For example, 'OLX' (www.olx.pt), a Portuguese website where users can advertise the selling of their own belongings to potential interested persons.

[7] The three most relevant players who provide 'mere conduit' services in Portugal are *MEO— Serviços de Comunicações e Multimédia, S.A.*; *NOS Comunicações, S.A.* and *Vodafone Portugal— Comunicações Pessoais, S.A.*

(*Autoridade Nacional de Comunicações*)—the Portuguese communications authority—has a list, on its website, of ISPs currently active in Portugal.

The Relationship Between Primary and Secondary Liability

Primary and secondary liability is intertwined. Secondary liability presumes the existence of primary liability, since it only exists if the action or omission caused by a third party is also unlawful and produces damages to something or someone. Nevertheless, it is not necessary, to establish secondary liability, that the person primarily liable is actually identified. Secondary liability can be established with the notional establishment of primary liability.

Primary and secondary liability co-exist in part because it is difficult to conclude which exact part of the resultant damage was caused by the actions of the person primarily liable and which were caused by the ISP that is secondarily liable. In these cases, Portuguese law imposes joint and several liability for damage, which means that the persons primarily and secondarily liable are bound to pay the plaintiff the total sum of damages awarded (no. 1 of Section 497 of the Portuguese Civil Code). However, in these cases, if the party that is held secondarily liable pays the plaintiff all the damages awarded, it can later demand from the primarily liable reimbursement of the damages that were caused by the primary conduct. In Portugal, this mechanism is referred to as *direito de regresso* (Fachana 2012, 112–113). In cases where it is possible to separate each contribution of the liable parties, then compensation for damages will be awarded in a conjunction system (*conjunção*)—each party will only be obliged to pay the sum of damages that were produced by his actual action or omission.

In Portugal, most actions still aim to pursue only primary liability and, thus, secondary liability of service providers has not been properly analysed by the courts.[8] In fact, Portuguese legal actors are still not well aware that primary liability can co-exist with secondary liability; so, in cases where the content provider is positively identified, courts or lawyers fail to see the possibility also to pursue the provider that could be secondarily liable. Also, since ISPs are exempted from much liability, there is not real incentive to try and pursue secondary liability.[9]

---

[8] This is true at least in respect to higher courts, which define Portuguese jurisprudence. This lack of higher court decisions is due, mainly, to the absence of important ISPs with registered offices in Portugal and, secondarily, to the fact that most cases that reach a court of law do involve an identifiable person that is primarily liable.

[9] Also, as many ISPs are situated in another EU member state, Portuguese liability rules do not apply to them, according to paragraph a) of no. 1 of Section 5 of *Decreto-Lei* no. 7/2004.

The Liability Immunity Standard

As discussed above, the legal framework established by *Decreto-Lei* no. 7/2004 of 7 January, provides for certain types of immunities for service providers (Sections 12 to 17). This immunity framework is applicable to all areas of law except for those that are related to the following matters, which are excluded from the scope of *Decreto-Lei* no. 7/2004: tax, competition, personal data management and privacy protection, legal representation, gambling (including lottery and bets with money), and notarial services (Section 2).

This immunity regime only applies to ISPs. But a question arises whether the regime is intended to apply only to online ISPs or also to offline ISPs. Since the scope of *Decreto-Lei* no. 7/2004 (much like the E-commerce Directive) is the regulation of certain aspects of society information services, we must conclude that immunities will only be applicable to ISPs that act online.

The Portuguese legal framework doesn't restrict exemptions or immunities to civil liability, but extends them to every kind of liability. An ISP can be immunised from criminal liability, civil liability or, even administrative liability. However, because of the way the immunity regime is drafted, it only addresses secondary liability exemption.

As explained, the general rule creating immunity is the absence of a general duty to monitor (Section 12), which means that ISPs are not obliged to monitor or filter information from third parties that are transmitted or hosted by their services. This general principle extends to the absence of any requirement to investigate possible illegal deeds practiced throughout the use of the services by third parties. However, there are certain limitations, some applicable to all ISPs and others applicable only to certain ISPs.

First, there are some common duties that ISPs owe to competent authorities. ISPs must:

1. Inform, immediately, the authorities of the practice of illegal activities via their services, as soon and whenever the fact is known by the service provider.
2. Provide identification of the service's users when requested.
3. Dutifully comply with determinations designed to prevent or end an infringement, like the removal of information or disabling access to it.
4. Providing lists of owners of websites that are hosted by the service provider, when asked to do so (Section 13).

There are also some immunity limitation rules only applicable to certain types of service providers. In order to benefit from these immunities, an ISP needs to comply with certain additional conditions, depending on the nature of the services provided:

1. 'Mere conduit' services (Section 14):[10]

    (a) ISPs that merely transmit information or provide access to a network, without being the origin of the information or without having control over the

---

[10]This applies to services of information transport and information access, typically offered by telecommunications companies throughout the world.

content of the messages transmitted, are exempted from any liability deriving from the information transmitted.

(b) This exemption is still active if the ISP also technologically stores the information in the process of transmission, provided that the storage is exclusively for the purpose of transmission and only during the time required to do it.

2. 'Caching' services (Section 15):[11]

(a) Liability exemption is offered for ISPs that engage in automatic and temporary storage of information exclusively for the purpose of making more effective and economical the information's onward transmission to other recipients of the service upon their request, on condition that:

(i) The ISPs do not intervene on the content of the transmitted messages, or its selection, or its recipients.
(ii) The ISPs respect the access to information conditions.

(b) However, immunity will not exist, if the ISP does not proceed according to the normal professional standards which relates to:

(i) Information update; and
(ii) The use of technology, using it to obtain data on the use of the information.

(c) Liability exemption will also not be applicable if the ISP does not act expeditiously to remove or to disable access to the information it has stored upon obtaining actual knowledge of the fact that the information at the initial source of the transmission has been removed, or access to it has been disabled, or that a court or an administrative authority has ordered such removal or disablement.

3. 'Hosting' (Section 16)[12] and content association (Section 17) services:

(i) ISPs that provide server storage of information shall only loose exemption of liability if the ISP has knowledge of the activity or information whose unlawfulness is evident and the ISP does not immediately remove or disable the access to the information.
(ii) Immunity will also not apply whenever the recipient of the service works for, is subordinate to or is controlled by the ISP.

---

[11] This refers to temporary storage. The Portuguese legislator decided to rephrase the expression given in the Directive (precisely, 'caching') to 'Intermediary storage'. This phrasing is unwise because it can lead to misunderstandings; every type of service predicted in the immunity rules is applicable only to intermediary services.

[12] Under the name 'Main storage'.

ISP Omissions as Grounds for Liability

Under the Portuguese liability general framework, liability can be established either by demonstrating a concrete action by the responsible individual or entity which caused the damage (liability for action) or by demonstrating that the damage was caused by the lack of action which should have been done by the responsible individual or entity (liability for omission).

Not all omissions can constitute grounds for liability under the latter basis. Indeed, in accordance with Section 486 of the Portuguese Civil Code, omissions will only give rise to liability if there was, by law or contract, a duty to act.[13] The question is whether by not imposing a general duty to monitor the content and information transmitted or stored under ISPs services, the law excludes the possibility of having omissions as the basis for secondary liability (Martinez 2005).[14]

There is no doubt that, regarding 'mere conduit' services, such liability cannot arise, since the legal framework establishes objective liability 'shields', i.e. the concrete activity of transmission and/or access to information which is exempt from any and all liability. If the ISP provides services that go beyond the definition of 'mere conduit' services, such as by intervening in the content of the messages which are transmitted, it will be liable if the content is unlawful, and such liability shall be based on an action by the ISP (the direct intervention in the unlawful content) (Section 38).[15]

However, when dealing with 'caching', hosting or content association services, the conclusion cannot be the same as with 'mere conduit' services. In these types of services, an ISP can be liable only if, upon due notice that it is storing infringing content, it does not expeditiously remove or disable the access to the infringing content.[16] As such, in these cases, the only ground for secondary liability is liability for omission.

Thus, the liability 'shields' which arise from Sections 15 and 16 of *Decreto-Lei* no. 7/2004[17] are not mere 'filters' to secondary liability of ISPs, as it could be argued. They are true rules of conduct, integrated within the Portuguese legal framework, which impose legal duties to act.

---

[13] This is similar to the common law definition of constructive or apparent knowledge.

[14] Martinez (2005, 275–290) defends this position, stating such liability for omission cannot exist under the Portuguese legal framework for ISPs secondary liability.

[15] The 'mere conduit' ISPs are, however, still bound to comply with the general obligations towards competent authorities stated in Section 13 of *Decreto-Lei* no. 7/2004 (such as compliance with determinations designed to prevent or end an infringement). If the ISP does not comply with such obligations, it is possible it can be held liable for omission towards the complainant if such lack of compliance gave rise to damages produced by the unlawful contents—and without prejudice to the criminal or administrative liability which can arise due to the lack of compliance of the orders issued by the competent authorities.

[16] Additionally, ISPs engaged in caching services are bound to maintain the information updated and the use of technology according to the common rules of the related business sector.

[17] And this is true also of Section 17, for content association services applying the same framework, although in this part of the Chapter I refer only to Section 16.

Finally, in a claim for secondary liability of an ISP which is responsible for 'caching', hosting or content association services, if the plaintiff proves the breach of a legal duty to act, it will only need to establish the damages produced and the causal nexus between the unlawful omission and the damages, since proving a relevant omission is the same as proving the remaining three requirements for civil liability: an omission *per se*, the unlawfulness of it, and the guilt of the ISP.

Ascertaining the Liability for Omission in the Case of 'Caching', Hosting and Content Association Services

As noted in the previous section, when dealing with 'caching', hosting or content association services, if an ISP is notified of content that it stores and which is evidently unlawful, it has a legal duty to act and to remove expeditiously the infringing content in order for it to remain immune from liability. Otherwise, its omission will constitute grounds for liability.

We must clarify some of the concepts that are relevant to establishing that an ISP has a duty to act. First, the concept of notification: what is a valid notification for the purposes of the ISP immunity framework? Unlike the United States,[18] the EU did not provide for a 'notice and takedown' procedure, and the Portuguese domestication of the E-Commerce Directive also lacks such a procedure. As such, the concept of notification in Portugal is wider and less restrictive than the concept of notification under the Digital Millennium Copyright Act.[19]

However, the lack of regulation does not mean that a 'notice and takedown' procedure should not be followed for the purpose of fulfilling immunity standards regarding 'caching', hosting or content association services. In order for an ISP to be adequately notified of unlawful content, the notification itself will have to be serious (i.e. provide sufficient grounds for the ISP to rely on the veracity of the report) and provide sufficient detail of the location of the allegedly unlawful content. Additionally, and taking into account most ISPs act on a global scale, with limited resources to address the multiplicity of reports, it should be reasonable to

---

[18] See 17 USC § 512 (US).

[19] Section 512(c)(3)(a) states that a notification to an ISP for removal of a certain content must comply with the following requirements: it has to have a physical or electronic signature, belonging to the rights' holder or its duly empowered representative; it has to clearly identify the works where the copyright infringement is being observed; it must clearly identify the material(s) which are infringing the copyrights of the works and also provide any other information which can allow the ISP to identify the infringing content; reasonable and correct information that allows the ISP to contact the complainant, such as a valid e-mail or address; a statement in which the complainant warrants that the content or materials subject of the complaint have not been authorized by the rights' holder or its representative; finally, a declaration where the complainant warrants, under penalty for perjury, that the complaint is true and accurate and, if such complaint is not presented by the rights' holder, that it is dully authorized or empowered to present the compliant in such rights' holder behalf.

demand that the notification contains minimal requirements, in order to be addressed by the ISP. Such notification shall include, at least: (a) the complete identification of the complainant and the identification of the person who makes the complaint on his behalf, if applicable; (b) the identification of the unlawful content, the URL where it is located, and a brief description of the right which is being affected by such content and summary proof that such right belongs to the complainant (Fachana 2012, 117).[20] ISPs should act where the notice provides reasonable information about the infringement in order to be able to act. Only a reasonable and serious notification (in connection with the other requirements which discussion follows) will impose on the ISP a legal duty to act.

Another requirement which must be fulfilled in order for the ISP to have a duty to act is related to the unlawfulness of the content. The law requires such unlawfulness to be evident. There is no doubt, at least in Portugal, that content which is related to child pornography, xenophobic or hate messages, terrorism or racism, is self-evidently unlawful, but the same may not be said for other content, which relates to copyright or trademark law, privacy or defamation.[21]

Lack of regulation on these issues can lead to overreaction by ISPs. If unlawfulness is evident, ISPs will lose immunity if they do not remove the content. But if the ISP removes the content, it will only be held liable by the user that owned the content if it was evident it was not unlawful. Since notifications often concern content the legality or illegality of which is not self-evident, ISPs will tend to choose to remove the content upon due notification. Further, the lack of regulation of notice and takedown procedures will contribute to a lack of rights-holders' awareness as regards their right to notify ISPs to remove unlawful content and, as a consequence, there are few complaints from rights-holders against ISPs.

As stated above, an ISP is liable if it removes content whose legality is evident. This rule is intended to prevent ISPs removing or impeding access to any content that is placed under suspicion by a right-holder without a previous analysis of its potential unlawfulness. There are no precise statutory rules to prevent or deter possible right-holders' abuses of this system. However, the common liability regime will still be applicable to any person that issues a notification to an ISP about third party content for the sole purpose of harming the rights of the person who posted the content.

---

[20] For example, in trademark infringement, the complainant shall produce sufficient evidence that it holds the trademark by indicating the trademark registry number and the trademark offices where it is registered.

[21] For example, some years ago, one Portuguese Public Prosecutor closed a case of copyright infringement moved against more than one hundred users of a peer-to-peer service used for the unauthorized download of copyrighted films. The Public Prosecutor took the view that such downloading was legal as it came under the private copying exemption (Van der Saar 2012). Even at the European level, this issue—downloading from unlawful sources—was only resolved (correctly, according to this author's perspective) by the ECJ's ruling in case C-435/12 *ACI Adam v. Stichting de Thuiskopie* (ECLI:EU:C:2014:254), stating that the private copying exemption only applies to copies derived from lawful sources. For the purposes of this chapter, we are addressing copyright infringement of material whose unlawfulness is evident.

It would be better, therefore, in order to ascertain the specific grounds of secondary liability of ISPs based on omission, that there be further regulation of the requirements of the notification of infringing content (by establishing a true notice and take down procedure, in terms similar to the Digital Millennium Copyright Act) and on the specific meaning of evident unlawfulness.

Extension of the Existing Framework: Lack of Compliance with Interlocutory Injunctions and ISP Liability for Omissions

As we saw above, the legal framework for secondary liability of ISPs imposes certain duties to act in specific circumstances. Some of these circumstances relate only to certain intermediary services ('caching', hosting and content association), while there are other circumstances that apply to all intermediary services indicated in *Decreto-Lei* no. 7/2004.[22]

The question we consider here is whether there can be duties to act other than those imposed by *Decreto-Lei* no. 7/2004 which can be imposed on ISPs given the prohibition against imposing a general obligation to monitor the content accessed or transmitted through ISPs. There may be room to extend these duties. The E-Commerce Directive does not impede EU Member-States demanding from ISPs 'duties of care, which can reasonably be expected from them'[23]. Thus, where the unlawful content is introduced into the web due to a failure of the ISP to exercise its duty of care over its own network, the ISP may be secondarily liable for omission and such liability will not conflict with the general rule of non-obligation to monitor. We are talking here about external interferences (*situações de ingerência*), where the source of the danger is created or permitted by the person or entity on whom the burden of care finally resides (Nunes de Carvalho 1999, 222 et seq).[24]

But can there also be liability for omissions arising from the lack of compliance with interlocutory injunctions (*providências cautelares*) granted by the competent courts of law? This might flow from the injunctive relief provided for in the Portuguese civil procedure code, which can be applied against ISPs, especially those injunctions that require an ISP to remove or block access to certain information or content.[25] An interlocutory injunction has to comply with certain requirements to be accepted by a court:

---

[22] Section 13.

[23] See recital (48) of the E-Commerce Directive.

[24] A clear-cut example in which the ISP may be secondarily liable for omission due to an external interference is in the situation where a third party introduces a virus upon the ISP network and it is proven that the ISP's security measures were not updated.

[25] Under Section 13(c) of *Decreto-Lei* no. 7/2004, ISPs are obliged to comply with the determinations of competent authorities (such as courts) in order to prevent or end an infringement, namely the removal or disabling of access to information.

1. The plaintiff must produce circumstantial evidence that it holds a right that is being infringed by the information that is being transmitted or hosted by the ISP—*fumus boni iuris*; and
2. The plaintiff must prove his justified fear that his right is being or can be damaged with the information that is being transmitted or stored by the ISP—*periculum in mora*.

These injunctions do not need to be sought against the person(s) that can be held liable for the damage resulting from the content, which means that an injunction can be pursued against an ISP even if it is not liable for the content.[26] Recall that one of an ISP's obligations, according to Section 13(c) of *Decreto-Lei* no. 7/2004, is to promptly abide by determinations issued in order to prevent or cease an infraction. Such understanding, which is identified in the E-Commerce Directive[27], has been emphasised both in the InfoSoc Directive[28] and, later on, in the Enforcement Directive[29], regarding the application of injunctions against ISPs for copyright or a related right infringement caused by a third party.[30]

The lack of compliance with provisional measures awarded by a court of law may lead to the ISP being secondarily liable for omission, because, in such case, the ISP is disregarding a duty to act imposed by law—the obligation to comply with orders issued by competent authorities under Section 13 of *Decreto-Lei* no. 7/2004. This liability may be imposed on a storage service provider or a mere conduit service provider, since the obligations under Section 13 are applicable to all kinds of ISP services.

---

[26] One common type of temporary measure which can be imposed on an ISP is the obligation to block access to a certain content or to remove it from its database. Although, as the European Court of Justice has discussed in many of its rulings, mostly when dealing with blocking injunctions over copyright infringing content, this obligation cannot disregard fundamental rights, such as the right of access to information, or the right of privacy. The latest ruling on this matter—Case C-314/12 *UPC Telekabel Wien (ECLI:EU:C:2014:192)*—clarified that such obligation could be imposed, although subjected to restrictive limitations which, overall, may impede such blocking injunctions to be effective in practice (Sousa e Silva, 2015, 222).

[27] E-Commerce Directive, art. 15 (2).

[28] Directive 2001/29/EC of the European Parliament and of the Council of 22 May 2001, on the harmonisation of certain aspects of copyright and related rights in the information society [2001] OJ L 167/10.

[29] Directive 2004/48/EC of the European Parliament and of the Council of 29 April 2004, on the enforcement of intellectual property rights [2004] OJ L 195/16.

[30] InfoSoc Directive, art. 8 (3) and Enforcement Directive, art 9 (1) (a), respectively.

Omissions and Contracts: The Duty to Act Imposed by a Contractual Obligation

Duties to act can be found either in statute or can derive from an obligation arising from a contract entered into by the parties. Under a contract entered between an ISP and its customers, there are typically disclaimers for any and all liabilities for any unlawful use of the service by the customer (the 'user') or any third party. However, such exclusions if absolute and unrestricted, are void, under Portuguese legal standards of liability exclusions and limitations[31] (Monteiro 2003, 391 et seq). Thus, even when existing liability exclusions or limitations, ISPs shall not being exempt of some duties to act when entering into an agreement with a user.

As such, certain risks which are within the scope of the service being offered by the ISP and over which the ISP has effective control, or is able to prevent, may generate duties to act. One typical example is the network security of the ISP. If the ISP fails to make its network security because it is negligent in maintaining its security mechanisms, according to the reasonable professional standards of the industry, then it will be liable to its customers for this omission for any malware introduced in the systems due to the failure to update security mechanisms. But duties to act, such as the example given above, must be reasonable, as these are the only ones which the counterpart in the contract (the user) may reasonably expect the ISP to comply with (Fachana 2012, 132).

When basing omission liability on a contract, the remaining analysis will be the same as when assessing omissions under a statute—with a slight, but important, difference. Where the duty derives from contract—and, thus, the failure of fulfilling such duty is regarded as a contract breach—it is presumed that the ISP is liable, and it has the burden of proving it is not liable[32]; under tort, the burden of proving liability rests with the plaintiff[33]. As such, it is easier for a plaintiff to pursue an omissions claim under a contract than under statute.

Remedies Available for Secondary Liability for Omission

Remedies granted by a court of law are the same for primary and secondary liability, with courts granting compensation for damages caused by the infringing activity of others.

In most instances where an ISP is found secondarily liable for omission, it will also be established that it committed negligent conduct (as opposed to a wilful conduct), since omission will be found most certainly by a lack of due care or

---

[31] Portuguese Civil Code, Sections 800 (2) and 809; Section 18 (a), (b), (c) and (d) of Decreto-Lei no. 446/85 of 25 October.

[32] Portuguese Civil Code, Section 799.

[33] Portuguese Civil Code, Section 487.

unacceptable delay upon pursuing its responsibilities upon being notified of the unlawful content. Such misconduct (or, more properly, the absence of conduct) will typically not be done with a specific intent of non-compliance.

The Portuguese civil code establishes compensation limitations in cases where the liable party is negligent[34]. In these cases, the court can reduce, in a fair way, the compensation to a sum that is less than the amount corresponding to the damages caused, provided the level of negligence of the ISP, together with its financial situation and that of the plaintiff and other circumstances surrounding the case justify that reduction. Since ISPs usually have financial wellbeing and most of time secondary liability will arise due to a failure to comply with a duty to act, the courts will rarely be inclined to reduce the damages.

Although the secondarily liable party will need to pay compensation for the damages caused to the plaintiff, it is also possible, though, for it to recover the amount paid, by demanding the primarily liable person (if he or she can actually be identified, which is also a problem in these cases) reimburse it for the compensation paid as damages[35], in cases where the damages were exclusively produced by the action of the person primarily liable.

A Brief Reference to Criminal Liability

This Chapter has addressed secondary liability mainly from a civil liability perspective, since criminal liability for the conduct of third parties will happen only in a very few and exceptional cases. Portuguese criminal standards (and the Portuguese penal code) do not generally allow for the imposition of criminal liability for the action of others. That is to say, there is no vicarious liability in criminal law.

Still, someone can be held criminally liable for the conduct of others if he or she voluntarily (that is, with *malice*) helped the execution of the criminal offense. Thus, regarding particular cases of criminal offenses perpetrated through web services, an ISP will only be liable if it knowingly aided someone to execute a criminal offense. In this case, the ISP will be punished as an accomplice (Section 27 of the Portuguese penal code).

Because liability is only applicable where the ISP acts with malice, accusation that an ISP had knowledge of the commission of a criminal act through use of the ISP service and failed to remove/impede the criminal content will be judged as a negligent omission, and thus giving rise only to civil liability. Of course, if the ISP wilfully fails to remove or impede the criminal content, then criminal liability may arise.

As such, liability for omission nowadays has largely been addressed as a matter of civil liability; criminal secondary liability of ISPs is of little or no practical interest.

---

[34] Portuguese Civil Code, Section 494.

[35] The statute in Portugal is called *direito de regresso*.

Possibilities of Improving the Current Framework

As seen above, liability for omissions derives from the failure to pursue a duty to act. Such duties, as can ground the secondary liability of ISPs, arise mainly from the immunity framework stated in *Decreto-Lei* no. 7/2004. But that framework can and should be improved.

New developments in the usage of World Wide Web services, in what is called the *web 2.0.* phenomenon, should justify a reform in the immunity regime. The actual immunity regime resides on three key pillars:

1. The technical impossibility of ISPs exercising effective advance control over the contents uploaded and transmitted through their services.
2. The typical neutrality of ISP's with respect to third-party contents.
3. The consequences for Internet access (and its expansion) and the efficiency of e-commerce that would result if a liability regime without immunity shields was to be adopted.

However, nowadays some of these key pillars have lost most of their relevance. On the one hand, there are available various programs/technologies that facilitate content filtering and monitoring—although they can imply an unacceptable limitation of free-speech and privacy rights—see cases *Promusicae* C-275/06 (ECLI:EU:C:2008:54), *Scarlet Extended* C-70/10 (ECLI:EU:C:2011:771) and *UPC Telekabel Wien* C-134/12 (ECLI:EU:C:2014:192). On the other hand, ISPs, when using a profit model that is based on advertising, give incentive for more upload and diffusion of user generated contents—since it helps increasing their profit margin—which surely does not fit with neutrality arguments. Finally, Internet and web services are fully implemented in modern societies, both in the USA and EU countries. Whilst immunity and liability standards are still adequate solutions for most of the ISPs' activity in the *web 2.0.* paradigm, the system can be improved by some small changes that will, in my opinion, adequately adapt the immunity regime to this new paradigm.

One useful change would be the standardisation of what should be contained in a valid notice to the ISP about an unlawful content. The current regime does not provide any information about the requirements of a notice in order for it to constitute the legal awareness requiring an ISP to act on it. Setting out the requirements for a valid notice would also help ISPs to know whether they should or should not investigate the facts contained in the notice.

There should also be developed an effort, probably made by the competent regulatory authorities, to give guidelines which help define the meaning of 'evident unlawfulness', in order to help dissipate subjective doubts about the (un)lawfulness of content which generally is more discussable, like copyright and personal rights issues (as image or honour rights). One good addition could be the development of "type-scenarios" which could help cease doubts (both of ISPs and of right holders) about certain activities (and the scope of such) which one can reasonably understand to be (un)lawful.

These improvements would not only clarify the scope of liability immunity, but also would clarify the grounds on which an ISP can be held liable for omission, since it would make objective, in a much clearer way, the specific duties to act which can be imposed on an ISP when dealing with notices of possible infringing content.

Finally, ISPs should have the burden of proving that their actions fall within immunity provisions. Currently, the plaintiff is required to prove that the ISP is not protected by the immunity shields; that is to say, the plaintiff has to prove that the hosting ISP was aware of the illicit content and did nothing to remove it. This is unfair since the plaintiff will not have the best access or even technological tools to be able to access information that proves this proposition. ISPs are in a much stronger position to demonstrate to a court that they had no knowledge of the unlawful content or, knowing it, did everything in their power to remove it, so it is fair to place the burden of proof in this matter on them.[36] As such, as is the case under contract, when addressing liability for omission under statutory law, ISPs should have the burden of proving that they did fulfil their duties to act—leaving the plaintiff only with the burden of establishing the concrete duties applicable in the case and with the proof of the remaining requirements of civil liability.

In the Information Society we all are living in, which is getting more digital and less physical each day ahead, the role of ISPs is of the uttermost importance for persons and companies pursuing their day-to-day life and businesses. As a result, ISPs omissions and failures to act will have greater impact in our society, as technological developments grow and, together, our need for their services. As such, and although secondary liability has not been discussed in Portugal with the same detail and analysis as in other EU countries, we can reasonably expect that the subject will get the relevance it deserves and, hopefully, take into account some suggestions for improvement stated in this Chapter.

References

Carneiro Da Frada, Manuel. 2001. "Vinho novo em Odres Velhos"? A responsabilidade civil das "operadoras de Internet" e a doutrina comum da imputação de danos. *Direito da Sociedade da Informação* II: 7–32.

Casimiro, Sofia de Vasconcelos. 2000. *A Responsabilidade Civil pelo Conteúdo da Informação Transmitida pela Internet*. Coimbra: Almedina.

Fachana, João. 2012. *A responsabilidade civil pelos conteúdos ilícitos colocados e difundidos na Internet: Em especial da responsabilidade pelos conteúdos gerados por utilizadores*. Coimbra: Almedina.

Leitão, Luis Manuel Teles de Menezes. 2002. A Responsabilidade Civil na Internet. In *Direito da Sociedade da Informação* III: 171–192.

---

[36] This change would work only if it was accompanied by the adoption of legal standards regulating the requirements for a valid notice from the injured party; this is one more extra reason to regulate the requirements that a notice must have in order to oblige an ISP to act.

Martinez, Pedro Romano. 2005. Responsabilidade dos prestadores de serviços em rede. In *Lei do Comércio Electrónico Anotada*, ed. Portugal Ministry of Justice, 271–289. Coimbra: Coimbra Editora.

Monteiro, António Pinto. 2003 (reedition). *Cláusulas limitativas e de exclusão da Responsabilidade Civil*. Coimbra: Almedina.

Nunes De Carvalho, Pedro. 1999. *Omissão e dever de agir em Direito Civil*. 222 et seq. Coimbra: Almedina.

Sousa E Silva, Nuno. 2015. A Perspectiva do Equilíbrio entre a Propriedade Intelectual e (Outros) Direitos Fundamentais. *Revista de Direito Intelectual* 1: 209–223.

Van Der Saar, Ernesto. 2012. File-sharing for personal use declared legal in Portugal. *TorrentFreak*. https://torrentfreak.com/file-sharing-for-personal-use-declared-legal-in-portugal-120927/. Accessed 25 July 2015.

———. 2015. The Pirate Bay will be blocked in Portugal. *TorrentFreak*. https://torrentfreak.com/pirate-bay-will-blocked-portugal-150302/. Accessed 25 July 2015.

Chapter 5
The Legal Framework Governing Online Service Providers in Cyprus

Tatiana Eleni Synodinou and Philippe Jougleux

Introduction

Fifteen years have passed since the enactment of the E-Commerce Directive.[1] The Directive established for three main types of OSPs a regime of exoneration from liability for illicit activities or unlawful content that passes through their services, or is stored on their servers. Nonetheless, the question of OSP liability is far from being harmonised. This is natural, since the Directive regulates only the question of the immunity provided to OSPs, leaving the thorny issue of liability to be regulated by the national law of Member States (Walter 2010, 1088). Indeed, as Christina Angelopoulos notes "the veneer of approximation that the safe harbours supply masks the persisting fragmentation of substantive liability law along European borders" (Angelopoulos 2013, 254).

The grounds of liability of OSPs, as a result, have to be discerned from the general tort law of each EU member state. National courts in many jurisdictions have been struggling for many years to clarify the landscape. This has often led to a case by case application of general tort law principles or of specific regimes, such as the German disturber liability regime (Störerhaftung) (Hoeren and Yankova 2012, 501). It is noteworthy that even similar statutes have generated different interpretations,

---

[1] Directive 2000/31/EC of the European Parliament and of the Council of 8 June 2000 on certain legal aspects of information society services, in particular electronic commerce, in the Internal Market (Directive on electronic commerce), OJ L 178, 17.7.2000, p. 1

T.E. Synodinou (✉)
University of Cyprus, Panepistimiou Avenue 1, Aglantzia, 1678, Nicosia, Cyprus
e-mail: synodint@ucy.ac.cy

P. Jougleux
School of Law, European University of Cyprus,
6 Diogenis Str., Engomi, 1516, Nicosia, Cyprus
e-mail: P.Jougleux@euc.ac.cy

© Springer International Publishing AG 2017   125
G.B. Dinwoodie (ed.), *Secondary Liability of Internet Service Providers*,
Ius Comparatum – Global Studies in Comparative Law 25,
DOI 10.1007/978-3-319-55030-5_5

not only because of the circumstances and/or the status of the parties involved in the dispute, but because of the cultural and/or economic background of the jurisdiction issuing the decision (Farano 2012).

Nevertheless, the question still remains completely unexplored in certain jurisdictions, such as in Cyprus. As such, every analysis of the issue of the liability of OSPs in Cyprus is based on the hypothetical application of general tort rules and of the E-Commerce Directive's safe harbour by Cypriot courts. Tracing the legal path that a court could take in future litigation is undoubtedly a hazardous intellectual operation. At the same time, since the tort law of Cyprus is based on UK tort law, the exploration by common law courts of similar questions could in some instances serve as useful guidance.

The Liability of OSPs in Cyprus

The question of liability, either direct or secondary, of OSPs has not yet been raised before the Cypriot courts. Consequently, any future litigation will be decided on the grounds of general tort and criminal law. More precisely, the horizontal standards based on the general law of tort (sections 12, 13 and 14 of Cap. 148[2]) and on general criminal law (section 21 of Criminal Code about aiding and abetting/criminal accomplice rules) will apply.

Cypriot legislation has not been more explicit than necessary in order to implement the EU Directives in respect of the legal issues related to OSPs. In this context, no statutory definition of the concept of service provider exists. The Cypriot legislator avoided dealing with the issue of exoneration of liability of service providers, such as search engines or linking providers, other than the main three categories of OSPs expressly covered by the E-Commerce Directive's safe harbor.[3] Consequently, the problematic question of the application of the E-Commerce Directive's immunity regime to other categories of OSPs has been left to be decided by the courts.

Copyright Law 59/1976, which implemented Article 8(3) of the Information Society Directive,[4] uses the term "intermediaries" without providing any definition or further explication. Of course, Article 8(3) itself does not deal with the issue of intermediaries' liability, since injunctions against intermediaries can be ordered

---

[2] Civil Wrongs Law, Chapter 148 (Cap.148).

[3] Some Member States, such as Spain, Austria, and Portugal, included the additional question of the liability of providers of hyperlinks and search engines in their national laws. See Report from the Commission to the European Parliament, the Council and the European Economic and Social Committee—First Report on the application of Directive 2000/31/EC of the European Parliament and of the Council of 8 June 2000 on certain legal aspects of information society services, in particular electronic commerce, in the Internal Market (Directive on electronic commerce) COM/2003/0702 final.

[4] Article 8 par. 3 states as following: "Member States shall ensure that rightholders are in a position to apply for an injunction against intermediaries whose services are used by a third party to infringe a copyright or related right".

regardless of their liability (Husovec 2013a).[5] So, even if Law 59/1976 had provided a definition of the term "intermediary", this definition could have applied only in respect of remedies provided by Article 8(3) and not in general with respect to the question of intermediaries' liability. Article 11 of the Enforcement Directive would also normally have been a legal basis on this matter, but the Cypriot legislator has not specifically transposed the Directive.

Nonetheless, the general provisions of the law of civil procedure could theoretically be used as a legal basis for the granting of such injunctions, even though this has not yet been confirmed in case law.

Primary Liability of OSPs

Civil Liability

It should be noted first of all that Cypriot tort law is based on common law, which is the English private law in force until the date of the independence of the Republic of Cyprus (1960). Even after that date, English case law continues to exert persuasive force when applying the traditional rules of common law. Ruling on the civil liability of the OSPs, it is probable that the Cypriot judge will use the classical common law mechanism of joint tortfeasor liability, as codified in the Cypriot legal system.[6] Under that approach, each tortfeasor is independently liable for the total damages sustained by the victim. It has been ruled that the Section 11 of the Cap 148 applies both in cases of one common tortuous act and when two or more tortuous acts contribute to the creation of a single wrong.[7]

The joint tortfeasor must be so involved as to be guilty of a tort himself. This would be the case if he induced, incited, or persuaded the primary infringer to engage in the act, or if there is a common design or concerted action or agreement on a common action to secure the doing of the act.[8]

Specific legislation, such as Law 59/1976 (copyright law),[9] provides a specific legal basis for the liability of intermediaries that contribute to copyright infringement. In this respect, Section 13 of that law prohibits direct copyright infringement, but also prohibits inciting and authorising copyright infringement. There is no Cypriot case law on the concept of inciting or authorising copyright infringement

[5] Husovec, Injunctions against Innocent Third Parties, para. 116.

[6] Section 11 of Civil Wrongs Law (Cap.148) states that "When two or more persons are respectively liable under the provisions of this Law for any act and such act constitutes a civil wrong such persons shall be jointly liable as joint civil wrong doers for such act and may be sued therefor jointly or severally (…)".

[7] This was decided in Mouharem and Another ν Paνlides (1983) 1C.L.R. 526.

[8] Newzbin [2010] EWHC 608 (Ch), [2010] F.S.R. 21 at [108]; Dramatico v Sky [2012] EWHC 268, [2012] E.C.D.R. 14 at [82]; EMI v Sky [2013] EWHC 379 (Ch); [2013] E.C.D.R 8 at [71]–[74].

[9] Available at http://www.cylaw.org/nomoi/enop/non-ind/1976_1_59/index.html. (in Greek).

under Law 59/1976 and, therefore, the relevant standard has not been elaborated by the courts. But it can be deduced from the text that this prohibition of infringement by incitment creates for OSPs a statutory duty of care.

UK case law[10] and, in general, case law of the commonwealth countries could be used as guiding precedent. Even if authorisation under Section 13 presupposes primary infringement by the person to whom authorisation was given and, consequently, is an area where primary and secondary liability overlap (Angelopoulos 2013), authorisation must be considered as a separate and distinct tort and as a distinct cause of action.[11]

In the relevant UK case law neither the level of control required nor the level of prevention methods that should be taken in order to determine the "authorisation" is explicit and currently the standard is largely decided on a case by case basis (Yan 2012).[12] Normally, a mere enablement or facilitation of the infringement shall not constitute "authorisation".[13] A less restrictive approach (interpreting a similar, but slightly different, statutory test) has been followed by Australian courts.[14]

Since both the UK and the Australian approaches could serve as guiding precedent, Cypriot courts could opt for a more restrictive or more flexible position depending on the circumstances of each case. Alternatively, they may avoid these approaches and apply the general common law principles on joint tortfeasance,[15] or attempt to combine all the possible legal grounds, namely direct infringement, authorization and joint tortfeasance.[16]

Holding the OSP directly liable would require clear and direct assistance to the infringer. The most common source of direct liability in Cypriot civil law, as in common law in general, is the tort of negligence. However, merely letting the user infringe, without any intervention, should be characterized as an omission, as a "nonfeasance". According to case law, the mere passive conduct cannot provide the

---

[10] Infringing copyright by authorization is provided in the UK Copyright, Designs and Patents Act 1988, s. 16(2).

[11] Hanimex (1987) 17 FCR 274 at [32]; Australasian Performing Right Assn Ltd. v Jain (1990) 26 FCR 53; 96 ALR 619; 18 IPR 663 (Jain).

[12] Yan 2012, 123.

[13] See CBS Inc. v Ames Records and Tapes [1981] 2 All ER 812.

[14] University of New South Wales v Moorhouse, [1975] HCA 26; (1975) 133CLR 1 (High Ct. Australia, 1 Aug. 1975)

[15] Even though recent UK case law seems to be oriented towards a more flexible approach, this is dependent on the facts of the specific cases and the law is still unclear on this point. Indeed, for some commentators, UK law has not really advanced in this area since the Amstrad case and it is still difficult to claim under authorisation of infringement in the UK, especially when compared to Australia. See Hocking 2012, 87. Opting for the general common rules of joint tortfeasance is a more sensible and straight forward option in the view of Angelopoulos. See Angelopoulos 2013.

[16] In the UK case, The Football Association Premier League Ltd. v British Sky Broadcasting Ltd. and others [2013] EWHC 2058 (Ch), that concerned a website that provided links to sports video streams submitted by users, the website was held directly liable for communication to the public but the judge repeated the "backup" grounds of authorisation and joint tortfeasance. See Savola 2014, 279.

basis for liability.[17] Exceptions apply when a special relationship exists with the victim or with the infringer, or when the person took part in the creation of the danger or knew the existence of this specific danger.[18]

This principle, however, does not apply when a positive obligation to act exists. Such an obligation could be found in specific legislation. For instance, it could be argued that the OSP as data controller has a statutory duty of care towards the subject of personal data, since the protection of the security of the personal data has been explicitly added to the list of obligations of the data controller in the Cypriot law.[19] Similarly, in our view, a failure to make a data breach notification, when doing so is mandatory, should also trigger direct liability of the OSP.

Criminal Liability

Service providers might be held criminally liable under specific legal provisions establishing criminal liability, such as article 14 of Law 59/1976 (on the protection of copyright),[20] or Law 138 (I) 2001 (on personal data protection),[21] or section 21 of the Cypriot Criminal Code (on aiding and abetting).

Rules on criminal accomplice liability could be applied to a third party who knowingly incites ("solicits, encourages, pressurises, threatens or endeavours to persuade"), or aids, abets, counsels or procures another to commit a criminal copyright offence (Strowel 2009). A necessary prerequisite for establishing criminal liability is the "actual knowledge" and consequently the *mens rea* of the OSP, which would be difficult to establish in practice, since OSPs are often legal entities and classic questions related to the establishment of mens rea appear. General common law principles in respect of the criminal liability of companies apply; in particular, the criterion of "directing [the] mind and will of the corporation, the very ego and centre of the personality of the corporation" is used to determine liability.[22] In this context, corporations in Cyprus can be criminally liable if the criminal act can be attributed to persons who can be characterized as the corporation's "directing mind and will".

---

[17] *Stovin v Wise* (1996).

[18] *Smith v Littlewoods Organisation Ltd.* (1987).

[19] See Article 98A of the Law no. 156 (I) 2004 on electronic communications.

[20] Article 14 (3) of law 59/1976 provides for criminal sanctions also for those who authorize the unauthorized communication of protected works to the public. Nevertheless, there is not any specific case law concerning on line service providers.

[21] Article 26 of that law provides for criminal sanctions in case of illegal data processing by the data controller. According to the opinion of the Working Group of article 29 social networks service providers can be considered as data controllers. See: Article 29 Working Party's Opinion 5/2009 on online social networking, adopted on 12 June 2009 (WP 163).

[22] Supreme Court, *Suphire Securities and Financial Services Ltd. v the Republic of Cyprus* 2008 2 AAΔ 486.

Secondary Liability of OSPs

As a principle, Cap. 148 on torts[23] provides only four instances of possible vicarious liability (in Sections 12 and 13): the liability of the master for the acts of the servant, the liability of the represented for the acts of the agent, the liability of the employer for the acts of the employee and, in certain circumstances, the liability of the contractor for the acts of the independent worker.

None of these forms of liability concern service providers. However, the case law—inspired by UK case law on common law issues—could add by exception in certain cases other forms/cases of vicarious liability, justified by some special or unique circumstances. In this domain, Cypriot courts will almost certainly refer to relevant English case law. For instance, a hospital is liable for the doctor's wrongful act whether it pays him directly[24] and the prison is liable for a prisoner's assault to an employee.[25]

According to the latest developments in English case law, establishing vicarious liability involves a two stage evaluation. First, it has to be examined whether the relationship between the tortfeasor and the third party is one which is capable of giving rise to vicarious liability. Second, an examination of the connection between the third party and the tortfeasor's wrongful act or omission.[26] The first stage is established in case a closed relationship involving control of the tortfeasor's action is proved. The second stage is traditionally related to the third party's authorization of the wrongful act, but there can be found some recent case law evolutions on this point. Two new criteria have also been accepted as a kind of sufficient connection: the "relative closeness" connecting the tort and the nature of an individual's employment establishes liability[27]; liability is also established when the relationship's nature "significantly enhanced the risk" of a tortious action.[28]

In conclusion, vicarious liability does not differ so much from direct liability: a certain degree of implication of the OSP to the realisation of the infringement, which is an intervention related to the infringer, has to be found. It is also interesting to note that the mechanisms of vicarious liability have recently been completed by new developments in the field of direct liability related with "the non-delegable duty of care"[29]

---

[23] http://www.cylaw.org/nomoi/enop/non-ind/0_148/index.html (in Greek).

[24] Cassidy v Ministry of Health [1951] 2 KB 343.

[25] Cox v. Ministry of Justice; CA 19 Feb 2014 [2014] EWCA Civ 132, [2014] ICR 713, [2014] PIQR P17, [2014] ICR 713, [2015] 1 QB 107, [2016] UKSC 10.

[26] The Catholic Child Welfare Society and others (Appellants) v Various Claimants and The Institute of the Brothers of the Christian Schools (Respondents) [2010] EWCA Civ 1106 [2012] UKSC 56.

[27] Lister v Hesley Hall Ltd. [2001] UKHL 22.

[28] The Catholic Child Welfare Society and others (Appellants) v Various Claimants and The Institute of the Brothers of the Christian Schools (Respondents), op.cit., par.86.

[29] Dickinson J. & Nicholson A., "Supreme Court Closes Another Vicarious Liability Loophole: Woodland v Swimming Teachers Association", (2015) 21(2) EJoCLI. The authors explained that :

Nonetheless, it seems very unlikely that a judge applies the rules of vicarious liability to found liability of OSPs for the acts of their users.

As a rule and, provided that the mechanism of secondary liability is used, the OSP will be liable in the same way as the main tortfeasor for the whole of the damage, according to Section 11 of Cap.148.[30] The only restriction concerns the sum of damages, which cannot exceed the actual loss of the victim. In case of successive actions, therefore, the liability is limited to the damages not covered in the first instance. In case of a compromise, it has been ruled that the judge is not bound by the result of the compromise, and can take as relevant the sum that a court would have given but for the compromise.[31]

The Safe Harbours of OSPs

When implementing the E-Commerce's Directive OSPs' safe harbours, Cyprus has followed closely the relevant provisions of the Directive. Articles 12, 13, 14 and 15 of the Directive were transposed in sections 15, 16, 17 and 18 of Law 156 (I) 2004. The safe harbours provided in this law cover mere conduit, caching and hosting activities. Law 156 (I) 2004 does not distinguish specific categories of OSPs depending on the specific area of law.

The law does not expressly distinguish between primary and secondary liability. But under the E-Commerce Directive, immunity will normally be applicable in cases of alleged secondary liability. In the case of direct liability, immunity could be invoked in any action based on the tort of negligence. The immunity's application to actions based on a statutory duty of care (this is the case when the legislator imposes a positive obligation to the OSP on a specific matter) is more problematic. Indeed, in the case of the violation of a statutory duty of care the OSP does not act anymore as an intermediary, but as a principal tortfeasor. For instance, under European legislation on personal data protection, Google is directly liable in cases of violation of the right to be forgotten.[32]

"there is now a duty upon hospitals, schools and other institutions entrusted with the custody, charge or care of vulnerable people, not merely to take reasonable care when delegating the performance of its core functions to independent contractors, but to ensure that reasonable care is taken by those contractors".

[30] It has been ruled in the Supreme Court case *Mouharem and Another v Pavlides* (1983) 1 C.L.R. 526, that this article normally *stricto sensu* about joint tortfeaser liability applies both when more than one person is liable of a tortious act and when several independent actions contribute to the realisation of one damage.

[31] Supreme court, *Επίσημος Παραλήπτης ως Trustee της περιουσίας του Κύπρου Κυπριανού and others v Δημητρίου and others* (*Episimos paraliptis os Trustee of periousiass of Kyprou Kyprianou and others v Demetriou and others*), (1997) 1 A.A.Δ. 336.

[32] CJEU, Google Spain SL, Google Inc. v Agencia Española de Protección de Datos, Mario Costeja González (2014), Case C-131/12. In this context, it has been argued that search engines have a special civil liability on the grounds of the Directive 95/46. See Bruguière 2015, 15–23.

Law 156 (I) 2004 repeats for these purposes the general definition of service provider in article 2 of the E-Commerce Directive, according to which "service provider" is any natural or legal person providing an information society service. The conditions for immunity vary depending on the activity of the provider and are identical to those established by the E-Commerce Directive.

No case law has interpreted or applied any of these provisions to date in Cyprus. But Cypriot legislation has followed closely the provisions of the E-Commerce Directive and establishes a safe harbour regime for mere conduit, hosting and caching activities. It has not extended this regime to other activities, such as for the activities of search engines and of providing links or online market places.

The relevant provision implementing Article 12 of the E-Commerce Directive is Section 15 of Law 156 (I) 2004. According to this provision, where an information society service is provided that consists of the transmission in a communication network of information provided by a recipient of the service, or the provision of access to a communication network, the service provider is not liable for the information transmitted, on condition that the provider: (a) does not initiate the transmission; (b) does not select the receiver of the transmission; and (c) does not select or modify the information contained in the transmission. The acts of transmission and of provision of access referred to in paragraph 1 include the automatic, intermediate and transient storage of the information transmitted in so far as this takes place for the sole purpose of carrying out the transmission in the communication network, and provided that the information is not stored for any period longer than is reasonably necessary for the transmission. This provision does not preclude a court or administrative body from imposing on the service provider appropriate measures to terminate or prevent the infringement.

Section 16 of Law 156 (I) 2004 implements Article 13 of the E-Commerce Directive. According to this section, where an information society service is provided that consists of the transmission, in a communication network of information provided by a recipient of the service, the service provider is not liable for the automatic, intermediate and temporary storage of that information that is performed for the sole purpose of making more efficient the information's onward transmission to other recipients of the service upon their request, on condition that: (a) the provider does not modify the information; (b) the provider complies with conditions on access to the information; (c) the provider complies with rules regarding the updating of the information, specified in a manner widely recognised and used by industry; (d) the provider does not interfere with the lawful use of technology, widely recognised and used by industry, to obtain data on the use of the information; and (e) the provider acts expeditiously to remove or to disable access to the information it has stored upon obtaining actual knowledge of the fact that the information at the initial source of the transmission has been removed from the network, or access to it has been disabled, or that a court or an administrative authority has ordered such removal or disablement. This provision does not preclude a court or administrative authority from requiring the service provider to terminate or prevent an infringement.

Section 17 of Law 156 (I) 2004 implements Article 14 of the E-Commerce Directive. More specifically, according to this provision, where an information society service is provided that consists of the storage of information provided by a recipient of the service, the service provider is not liable for the information stored at the request of a recipient of the service, on condition that: (a) the provider does not have actual knowledge of illegal activity or information and, as regards claims for damages, is not aware of facts or circumstances from which the illegal activity or information is apparent; or (b) the provider, upon obtaining such knowledge or awareness, acts expeditiously to remove or to disable access to the information. This provision does not apply when the recipient of the service is acting under the authority or the control of the provider.

Section 18 of Law 156 (I) 2004 implemented Article 15 of the E-Commerce Directive. According to this provision, intermediaries are not subject to a general obligation to monitor information which they transmit or store, nor a general obligation actively to seek facts or circumstances indicating illegal activity. Nonetheless, information society service providers have to inform promptly the competent Public authority (the Ministry of Industry and Tourism) of alleged illegal activities undertaken, or information provided by recipients of their service, or obligations to communicate to the competent authorities, at their request, information enabling the identification of recipients of their service with whom they have storage agreements.

Law 156 (I) 2004 has thus implemented almost verbatim the provisions of the Directive. And as a result the Cypriot legal framework on the immunity of OSPs naturally presents the pitfalls which have been highlighted in respect of the Directive. No specific notice and take down procedures or counter notification mechanisms to detect fraudulent notifications and put back content (Farano 2012; Baistrocchi 2002), can be found in the statute or in codes of conduct or any Cypriot courts' guidelines.

The reluctance of Cypriot courts to deal with the question of liability of OSPs and the safe harbour regime established by Law 156 (I) 2004, has been highlighted by the recent preliminary reference posed to the CJEU by the District Court of Nicosia in the case *Papasavvas and Others*.[33] The case is apparently outside of the scope of application of the intermediary's immunity regime, since it concerns defamatory content published by journalists (and not by readers) in the defendant's printed newspaper and on the defendant's website. The District Court of Nicosia asked the following preliminary questions:

Bearing in mind that the laws of the Member States on defamation affect the capacity to provide information services by electronic means both at national level and within the European Union, might those laws be regarded as restrictions on the provision of information services for the purposes of applying Directive 2000/31/EC?

If the answer to Question 1 is in the affirmative, do the provisions of Articles 12, 13 and 14 of Directive 2000/31/EC, on the question of liability, apply to private civil matters, such

---

[33] *Papasavvas and Others*, Case C-291/13, lodged at 27/05/2013.

as civil liability for defamation, or are they limited to civil liability in matters concerning business to consumer transactions?

Bearing in mind the purpose of Articles 12, 13 and 14 of Directive 2000/31/EC relating to the liability of information society service providers and the fact that, in many Member States, an action must exist in order for a prohibitory injunction to be granted which will remain in force pending full completion of the proceedings, do those articles create individual rights which may be pleaded as defences in law in a civil action for defamation, or must they operate as an obstacle in law to the bringing of such actions?

Do the definitions of 'information society service' and 'service provider' in Article 2 of Directive 2000/31/EC and Article 1(2) of Directive 98/34/EC, as amended by Directive 98/48/EC, cover online information services the remuneration for which is provided not directly by the recipient, but indirectly by means of commercial advertisements posted on the website?

Bearing in mind the definition of 'information service provider', laid down in Article 2 of Directive 2000/31/EC and Article 1(2) of Directive 98/34/EC, as amended by Directive 98/48/EC, could the following, or any of them, be regarded as a 'mere conduit' or 'caching' or 'hosting' for the purposes of Articles 12, 13 and 14 of Directive 2000/31/EC:

- a newspaper that operates a free website on which the online version of the printed newspaper, with all its articles and advertisements, is posted in pdf format or another similar electronic format;
- an online newspaper which is freely accessible but the provider obtains money from commercial advertisements posted on the website, where the information contained in the online newspaper comes from the newspaper's staff and/or freelance journalists;
- a website which provides (a) or (b) above for a subscription?[34]

As Martin Husovec notes, most of these questions lead to obvious answers (Husovec (1) 2013b). Nonetheless, the questions from the Cypriot court could be seen as a first step towards a rising awareness of the question of liability of intermediaries on the Internet. The factual background, as described in the reference, raises the issue of classic media content provider liability and not the question of the liability of a hosting provider that stores content supplied by third parties. At the same time, the Opinion of the Advocate General and the judgment of the CJEU could have raised, discussed and clarified related legal issues, even as *obiter dicta*, and this would be more than welcome. In this vein, in its ruling on 11 September, 2014, the Court affirmed that that "article 2(a) of Directive 2000/31/EC must be interpreted as meaning that the concept of 'information society services', within the meaning of that provision, covers the provision of online information services for which the service provider is remunerated, not by the recipient, but by income generated by advertisements posted on a website" and that "the limitations of civil liability specified in Articles 12 to 14 of Directive 2000/31 do not apply to the case of a newspaper publishing company which operates a website on which the online version of a newspaper is posted, that company being, moreover, remunerated by income generated by commercial advertisements posted on that website, since it has knowledge of the information posted and exercises control over that information, whether or not access to that website is free of charge". As such, the court clarifies the

---

[34] See Sotiris Papasavvas v O Phileleftheros Dimosia Etairia Ltd., Takis Kounnafi, Giorgos Sertis, (Case C-291/13), Judgment of 11 September 2014.

conditions of the non-application of the safe harbor to an OSP: to be liable, the OSP must have knowledge of the information and must exercise control over the information. At the same time, it rejects the third condition about the existence or not of a remunerated activity proposed by the Cypriot judge. Here, both knowledge and control were presumed since the information had been uploaded by the journalists working for the newspaper and it is assumed they worked under the supervision and control of their employer.

Injunctions Against OSPs

General Framework

Injunctions against OSPs in civil proceedings for taking down/removal of illegal Internet content could theoretically be obtained on the grounds of general provisions permitting the Court to order an injunction on the basis of Article 32 of the Courts of Justice law 14/60, whose justification lies in the law of equity. To date, the conditions for the application of Article 32 have been defined by Cyprus case law mainly in the context of cases unrelated to the Internet.[35] In general, courts are cautious in granting such orders. Article 32 of the Courts of Justice Law confers power on the Court to grant an injunction in all cases in which it appears to the Court just or convenient to so do. However, the justice and convenience of the case is not the sole consideration to which the Court shall pay attention and certain supplementary conditions have to be met: (a) a serious question arises to be tried at the hearing; (b) there appears to be a "probability" that the plaintiff is entitled to relief; and (c) unless it shall be difficult or impossible to do complete justice at a later stage without granting an interlocutory injunction. The injunction mainly concerns stopping the use, disclosure or exploitation of illegal or harmful content by the defendant in the frame of a civil action, while the adjudication of the main civil action is still pending. The purpose of the interim order is to prevent certain action or to order the defendant to act in a specific way until the end of the adjudication of the main civil action. If the plaintiff succeeds in the main civil action, the order becomes absolute.

Recently, a demand for an injunction on the grounds of Article 32 of Law no.14/60 was filed against Facebook asking the District Court of Paphos to order the company to remove a number of offensive comments aimed at a local man posted on a local business profile and to take steps to ensure that future related comments were taken down immediately. The Court accepted in principle the granting of the

---

[35] *Odysseos v Pieris Estated Ltd. and Others* (1982) 1 CLR;· *Constantinides v Makriyiorghou* (1978) 1 CLR 585; *Karydas Taxi Co Ltd. v Andreas Komodikis* (1975) 1 C.L.R. 321; *A. Kytala v Anna Chrysanthou and others* (1996) 1 A.A.Δ. 253;· *T.A. Micrologic Computer Consultants Ltd. v Microsoft Corporation* (2002) 1(Γ) A.A.Δ. 1802.

order, but before issuing the order gave guidelines that Facebook be notified of the claim; the issue is still pending.[36]

Since there is not yet any relevant case law, it is unclear whether the general provisions of Article 32 of the Courts of the Justice Law no. 14/1960 about interim orders could be applied to oblige ISPs to block and/or filter sites containing illegal content. One the one hand, it could be argued that blocking orders are within the authority and discretion of the civil courts to grant injunctions, relying on, among others, UK case law, which can be used as a persuasive non-binding precedent. Indeed, Mr. Justice Arnold in *Cartier International and Others v. BSkyB and others*[37] had to consider whether a power to order ISPs to block trademark infringing sites could be found in the general jurisdiction to grant injunctions in section 37(1) of the Senior Courts Act 1981, which is the counterpart of Article 32 of the Courts of the Justice Law no. 14/1960. In the UK, as in Cyprus, there is a special provision in the UK Copyright, Designs and Patents Act 1988,[38] which enables the Court to order such measures in copyright cases, but there is no equivalent statutory provision in relation to trade mark infringement. Mr. Justice Arnold accepted that the power of the Court to grant injunctions is unlimited and can be exercised in new ways. On the other hand, it could be argued that site-blocking by ISPs requires special legislation, and could be ordered only in the specific cases clearly provided by law. This was the stand of the District Court of Tel Aviv on 1 July 2015 in the *ZIRA* case,[39] where it was held that there is no authority for the courts to order ISPs to take any actions in disputes between plaintiffs and third party defendants. So, while it is probable that the courts in Cyprus can use Article 32 of Courts of the Justice Law no. 14/1960, which concerns interim orders, as a legal ground for obliging ISPs to block and/or filter sites, this possibility has not yet been confirmed by case law.

Specific Provisions

Furthermore, injunctions against OSPs may be ordered on the grounds of specific provisions, such as copyright law, gambling legislation and legislation dealing with child abuse and child pornography.

Section 13 (5) of Law 59/1976 (about the protection of copyright) implements Article 8(3) of the Information Society Directive, and provides that right holders are in a position to apply for an injunction against intermediaries whose services are used by a third party to infringe a copyright or related right. This provision does not presuppose or have as a prerequisite the intermediary's liability. The provision has

---

[36] District Court of Paphos, *Neophytos Georgiou v Facebook Inc., and others*, no. of the action 569/2015, 23/4/2015, available on line at: http://www.cylaw.org/cgi-bin/open.pl?file=apofaseised/pol/2015/4120150130.htm&qstring=facebook.

[37] [2014] EWHC 3354 (Ch).

[38] Section 97A of the Copyright, Designs and Patents Act 1988.

[39] Civil file (Tel Aviv) 37,039-05-15 ZIRA v John Doe et al. (July 1, 2015).

not yet been applied by Cypriot courts, but it could be used in order to stop or prevent copyright infringements by various means. Normally all the appropriate remedies will be considered by courts, including information disclosure, or DNS and IP address blocking. In the light of the CJEU's decisions in *Scarlet*[40] and *Netlog*,[41] deep packet inspection or URL blocking should be excluded, since it is extremely intrusive of the right to privacy, data protection and the secrecy of communications. It is noteworthy that in Cyprus, even if there is (as in the UK) no specific privacy tort, case law has recognised that violation of privacy is a tort under Article 15 of the Constitution.[42] The Cypriot Constitution also guarantees the secrecy of communications, while personal data protection is established by law that has implemented the relevant EU Directives. So, every preventive blocking method which requires a general monitoring of the content of packets will normally be deemed unconstitutional.

Sections 15, 16 and 17 of Law 156 (I) 2004[43] provide that these provisions do not affect the power of a court or administrative authority to order the service provider to terminate or prevent the infringement. There is no case law specifying what measures could be ordered by a court in order to prevent or terminate the infringement.

Online gambling, which is outside of the scope of application of the E-Commerce Directive,[44] is regulated by Law 106/2012. According to section 65 of Law 106/2012 about online betting, Internet providers have an obligation to block websites that do not have the permission to provide services of online betting. Indeed, betting services in Cyprus via electronic means are offered legally only if the betting operator is authorised by the National Betting Authority. The National Betting Authority provides a list with sites that have to be blocked by ISPs. Access to sites that offer electronic betting services that are licensed by Member States of the European Union has not been barred by reason of the transitional provision of Article 91 (3) of the Law 106 (I) 2012. According to this provision, legal persons which, during the entry into force of this Law, are licensed to conduct online betting on the basis of a license issued by a Member State of the European Union may continue to provide such services only for the type of betting services which are authorised by this law. They are considered to operate under this law until they are granted a license from the National Betting Authority, provided they have submitted an application for a license pursuant to the provisions of this Law. However, it has been ruled

---

[40] CJEU, *Scarlet Extended SA v Societe Belge des Auteurs, Compositeurs et Editeurs SCRL (SABAM)*, Case C-70/10, Judgment of 24 November 2011, par. 50 : "Moreover, the effects of that injunction would not be limited to the ISP concerned, as the contested filtering system may also infringe the fundamental rights of that ISP's customers, namely their right to protection of their personal data and their freedom to receive or impart information, which are rights safeguarded by Articles 8 and 11 of the Charter respectively".

[41] CJUE, *Belgische Vereniging van Auteurs, Componisten en Uitgevers CVBA (SABAM) v Netlog NV*, Case C-360/10, Judgment of 16 February 2012.

[42] *Τάκης Γιαλλουρος v Ευγένιου Νικολαου (Takis Yiallouros v Evgeniou Nikolaou)* (2001) 1A Α.Α.Δ 558.

[43] Available at http://www.cylaw.org/nomoi/indexes/2012_1_106.html. (in Greek).

[44] See article 1 par. 5 (d) of Directive 2000/31.

recently that such a blockage when used against a legal person that is established in another EU member State violates primary European law (sections 43 and 49 EC). More precisely, the blockage of a website was suspended for the reason that the order to block it was found to be illegal, since the Republic of Cyprus has not yet issued the appropriate regulations and has not officially opened the procedure so that companies established in other member states can ask for a permission from the National Betting Authority to operate in Cyprus.[45]

Besides, article 11 of Law 91(I)/2014 implemented article 25 of the Directive 2011/92/EU of 13 December 2011 on combating the sexual abuse and sexual exploitation of children and child pornography. More specifically, article 11(1) provides that the Court in any stage of the procedure can order the prompt removal of web pages containing or disseminating child pornography. Article 11(2) empowers the Court to order the blocking of access to web pages containing or disseminating child pornography towards the Internet users who reside in Cyprus. Internet service providers offering services or access to the Internet within the territory of the Republic of Cyprus are obliged, when they gain knowledge or when they are informed by the competent department of the existence of child pornographic content on any site, to immediately take appropriate measures for the interruption of access by Internet users.

OSPs in the Eye of the Storm: Balancing of Fundamental Rights

Independently of the primary or secondary liability of OSPs and of their obligation to respond to court injunction, a final issue shall be raised in respect of the legal framework regulating OSPs. OSPs, by definition, play a fundamental role in the law enforcement in the digital era, since they typically possesses the only means by which to determine the identity of the perpetrator. It would be too simple to just declare that the denial to give access to their data constitutes an obstruction to justice. Indeed, OSPs have a duty to protect the confidentiality and the security of the personal data they process.

There are no voluntary codes or best practices in Cyprus that have been concluded or followed by all service providers on this issue. However, the terms of services of each provider normally contain certain clauses that permit the service providers to terminate the provision of services or disclose data to third parties in case of a violation of law. For example Article 8.3 of the general terms of use of "primetel" services provides that "the Company and the Subscriber are bound against each other to treat in strict confidence and secrecy any information or data provided by the Company to the Subscriber and vice versa, for the purposes of or pursuant to the Agreement, unless the specific information is already in the public

---

[45] Supreme Court, *Betfair International Plc's v. Republic of Cyprus*, case 492/2013, 30/4/2013.

domain or their disclosure is necessary in accordance with a court judgment or order or with the applicable legislation for the time being."[46]

However, it is important to note that to date Cypriot case law has been very cautious on issues of protection of privacy and of the secrecy of communications. Under articles 15 and 17 of the Constitution (protection of privacy and of the secrecy/confidentiality of communications), it is contrary to the Constitution to use the IP address of an accused as a means of proof of an illegal activity (violation of privacy, defamation etc.).[47] If the obligation to obtain a warrant in order to oblige the ISP to reveal the identity of his user is well established by the Cypriot case law, a lot of uncertainty prevails as regards the collection of the IP address itself. In a case where a Facebook account had been modified in a defamatory way, the plaintiff had by himself collected the perpetrator's IP address through the Facebook advanced options and had given it to the police. The Cypriot judge in first instance decided that the procedure was illegal: since the IP address is a personal datum, a judicial supervision of the collection process should have occurred. This decision aimed to force the police to obtain a warrant against the ISP for the identification of the owner of the IP address, which is logical, but also another warrant before this one, for the collection of the IP address itself. It provoked, however, so much confusion that recently the Cypriot judges in the same case on appeal ruled otherwise and decided that the IP address by itself—that is without connection to the ISP log—is not a personal datum.[48]

Conclusions

The existing Cypriot legal framework for the regulation of intermediary liability should be modernised. The legal framework is simply the result of the implementation of EU Directives and, therefore, it has lacunas since it does not deal with all the pressing issues. Indeed, the very concept of "intermediary" or "service provider" must be defined. The liability of search engines, online market places, and providers of links must also be regulated. "Notice and take down" procedures must be established. Since the legal basis for liability is general tort law, the standard of liability in the specific case of service providers should be clarified (possible criteria to be taken into account: extent, kind, degree of control of the infringing activities, profit of the provider, etc.).

However, a legislative framework, as perfect as it can be, remains useless without effectiveness. The particularity of the Cypriot situation resides in the—until now—

---

[46] See: http://www.primetel.com.cy/en/terms.

[47] See for example: Supreme Court, Criminal appeal 135/2008 (http://www.cylaw.org/cgi-bin/open.pl?file=apofaseis/aad/meros_2/2011/2-201107-135-08.htm&qstring=IP) and Supreme Court, civil appeal 134/2011 (http://www.cylaw.org/cgi-bin/open.pl?file=apofaseis/aad/meros_1/2012/1-201209-134-11.htm&qstring=IP).

[48] Supreme Court, Isaia, 17 July 2014, appeal number 402/2012.

under-development of collective rights management societies handling copyrights. Without any clear policy of enforcement of copyright on the Internet, the Cypriot courts cannot play the traditionally important role (as a common-law country) that they should play in the construction of a legal framework concerning the OSPs. Even if on the European level (and contrary to the US), the safe harbour is horizontal, affecting not only liability under copyright law but all forms of civil liability, it is very important to notice that in practice most applications of the statute concern IPR enforcement. In Cyprus, therefore, it can be asserted that the lack of development (with very few collecting societies) of the IPR management in general contributes mostly to the absence of discussion about the liability of OSPs.

References

Angelopoulos, Christina. 2013. Beyond the safe harbours: Harmonising substantive intermediary liability for copyright infringement in Europe. *Intellectual Property Quarterly* 3: 254.

Baistrocchi, Pablo. 2002. Liability of intermediary service providers in the. EU directive on electronic commerce. *Santa Clara High Technology Law Journal* 19: 111. http://digitalcommons. law.scu.edu/chtlj/vol19/iss1/3.

Farano, Béatrice Martinet. 2012. *Internet intermediaries' liability for copyright and trademark infringement: Reconciling the EU and U.S. Approaches.* TTLF Working Paper No. 14. http://www.law.stanford.edu/organizations/programs-and-centers/transatlantic-technology-law-forum/ttlfs-working-paper-series.

Hoeren, Thomas, and Silviya Yankova. 2012. The liability of internet intermediaries—the German perspective. *IIC* 43(5): 501–531.

Husovec, Martin. 2013a. Injunctions against innocent third parties: The case of website blocking. *JIPITEC* 2013–2.

———. 2013b. (1). New preliminary reference on safe harbors from Cyprus. http://www.husovec. eu/2013/07/new-preliminary-reference-on-safe.html.

Savola, Pekka. 2014. Blocking injunctions and website operators' liability for copyright infringement for user-generated links. *E.I.P.R* 36(5): 279.

Strowel, Alain. 2009. Introduction: Peer-to-peer file sharing and secondary liability in copyright law. In *Peer-to-peer file-sharing and secondary liability in copyright law*, ed. Alain Strowel, 18. Cheltenham: Edward Elgar.

Walter, Michel. 2010. Information society directive-article 8. In *European copyright law, a commentary*, ed. Michel Walter, Silke von Lewinski, 1088, no 11.8.13. Oxford: Oxford University Press.

Yan, Min. 2012. The law surrounding the facilitation of online copyright infringement. *E.I.P.R* 34(2): 123.

Chapter 6
Analysis of ISP Regulation Under Italian Law

Elisa Bertolini, Vincenzo Franceschelli, and Oreste Pollicino

Introduction

In Italian Law, liability for unlawful acts is regulated by Title IX of Book 4 of the Italian Civil Code. The ground rule is contained in Article 2043, which is entitled 'Compensation for unlawful acts' and states: 'Any intentional or negligent act that causes an unjustified injury to another obliges the person who has committed the act to pay damages'.

In the Code, the concept of 'secondary liability' is not clearly expressed in a specific rule or article. If we distinguish, as in the common law tradition, between 'vicarious liability' and 'contributory liability', we find rules that could be catalogued in the former or latter category. As for vicarious liability, for example, the rule of *respondeat superior* may be found in Article 2049 as regards

Elisa Bertolini, Assistant Professor of Comparative Public Law, School of Law at Bocconi University, Milan, wrote paragraphs 4–8 and 12; Vincenzo Franceschelli, Full Professor of Private Law, University of Milan Bicocca, wrote paragraphs 1–3; Oreste Pollicino, Full Professor of Constitutional Law, School of Law at Bocconi University, Milan, wrote paragraphs 9–11.

E. Bertolini • O. Pollicino
School of Law, Bocconi University, Via Roentgen 1, Milan 20136, Italy

V. Franceschelli (✉)
Private Law, University of Milan Bicocca, Via Festa del Perdono n. 14, Milano 20122, Italy
e-mail: lexfran@tin.it

© Springer International Publishing AG 2017
G.B. Dinwoodie (ed.), *Secondary Liability of Internet Service Providers*,
Ius Comparatum – Global Studies in Comparative Law 25,
DOI 10.1007/978-3-319-55030-5_6

employer-employee relations[1] and in Article 2058 as regards the liability of parents, guardians, teachers, and masters of apprentices[2]. It is more difficult to find a specific article of the Code that expressly refers to contributory liability, if by this we mean a form of liability on the part of someone who does not directly damage someone or something but nevertheless makes 'contributions' to an intentional or negligent act perpetrated by somebody else. Such a contribution would be probably be evaluated through the general principle of joint and several liability, which is set forth in Article 2055: 'If the act causing damage can be attributed to more than one person, all are liable *in solido* for the damages'.

The issue of civil liability on the Internet is highly complex and has been one of the most controversial issues of the last two decades. The birth and massive development of the Internet have presented many difficult challenges for the liability regime. The issue potentially involves every actor on the Internet, namely, both providers and users, and infringements can take place in all fields of law (privacy, intellectual property and copyright, contracts etc.).

The focus of this Chapter is the civil liability regime applicable to Internet Service Providers ('ISPs'). ISPs offer users access to the Internet and related services. Due to their role as intermediaries, it has become increasingly difficult to set a clear framework for their potential liability. Across the world, in the last two decades IPSs have been sued many times for infringing activities (i.e. copyright, trademark, privacy infringement). As intermediaries, ISPs are seldom directly engaged in the infringing activity; however, they are sued as indirect infringers in that they operate equipment and services that facilitate an infringing activity. Therefore, when dealing with ISPs, what is relevant is not liability *tout court*, but secondary liability.

Out of this arises one main question arises: to what extent should ISPs be liable for the actions of those who use their services? Besides the problem of having to define the concept of secondary liability, a great number of ISPs are not confined to their national borders and therefore they have to face different legislative regimes and different courts of justice. The issue of Internet governance should be dealt with at a supranational level, thereby avoiding situations where ISPs are responsible in one jurisdiction, but not another, for the same activity (from the Italian perspective see Costanzo 2013). Thus, the choice by the European Union (EU) to deal with ISPs' secondary liability by issuing a directive, which needs to be implemented in each Member State, was not ideal. In fact, the application of the resulting national legislation has led to conflicting results in the Member States' courts (Smith 2011).

---

[1] Liability or masters or employers: 'Masters and employers are liable for the damage caused by an unlawful act of their servants and employees in the exercise of the functions to which they are assigned'

[2] 'The father and mother, or the guardian, are liable for the damage occasioned by the unlawful act of their minor emancipated children, or of persons subject to their guardianship who reside with them. The same applies to a parent by affiliation. Teachers and others who teach an art, trade, or profession are liable for the damage occasioned by the unlawful act of their pupils or apprentices while they are under their supervision'

Courts have usually resorted either to vicarious liability or to the general principles of civil liability (in Italy, set by Article 2043 of the Civil Code) (Barazza 2012).

However, this Chapter will focus on ISPs' secondary liability in Italy, in particular after implementation of the EU E-Commerce Directive through Legislative Decree 70/2003, identifying the evolutionary pattern of the regime and outlining the most controversial case law on the matter. The Chapter highlights the most distinctive features of the Italian liability regime for ISPs.

Generally speaking, the concept of secondary liability consists of holding the ISP liable for not stopping the infringing use of its services by users and it is usually possible to identify two types of secondary liability:

1. Vicarious infringement. Vicarious liability arises when ISPs: (a) possess the right and ability to supervise the infringing conduct of the direct infringer; and (b) have a direct financial interest in the exploitation of the copyrighted materials.
2. Contributory infringement. Contributory liability arises when ISPs: (a) have knowledge of the infringing activity; (b) induce, cause, or materially contribute to the infringing conduct of another.

In this Chapter, it will be argued that the Legislative Decree seems to provide for contributory liability, whereas the courts of justice, through the distinction between 'passive' and 'active' provider, seem to contemplate vicarious liability, as an alternative way of assessing providers' liability.

ISPs' Civil Liability Before 2003

The key legislative provision shaping ISPs' secondary liability is Legislative Decree 70/2003, introducing exemptions from liability (Alvanini 2010; Bocchini 2003; Bugiolacchi 2005; Cassano and Cimino 2004a; Tosi 2003; Di Ciommo 2003). However, analysing the issue diachronically, we have to identify two different phases: the first prior to implementation of the E-Commerce Directive 2000/31 (Van Eecke 2011) and the second following that implementation through the above-mentioned Legislative Decree 70/2003 (Ortaglio and Zingales 2011).

Prior to the enactment of Legislative Decree 70/2003, lacking a specific legislation, it was up to the courts to identify the elements required to establish ISPs' liability, usually by applying the general standard for the establishment of secondary liability in civil and criminal law. Mainly, courts held ISPs liable by applying the publisher liability set out in Law 47/1948 (the Press Law). Such an application of the Press Law—broadly or by analogy—to the Internet was problematic for a number of reasons in terms of ISPs' liability, mainly from a criminal standpoint, considering the difficulty in defining the nature and role of ISPs (De Cata 2010). Moreover, Article 57 of the Italian Criminal Code provides for an offense based on a failure to act; more precisely, it imposes on a newspaper editor the obligation to control what is published, in order to prevent breaches of the law. Therefore the liability of the publisher is founded on the idea that, having a prominent role, he has an obligation to monitor all material published. Thus the key problem is whether or not it is possible to impose that same obligation to monitor all material published to the ISP and in so doing comparing the ISP to a newspaper editor. Defamation is, on the contrary, punished according to Article 595 of the Italian Criminal Code, stating defamation can be a civil or criminal offense (the latter, punishable by imprisonment ranging between six months and three years). But publication of information on the Internet lacked the character of the type of information regulated by the Press Law because of the definition of 'printed material' found in that law. Moreover, the Italian Criminal Code distinguishes penal offences on account of any act or omission (or failure to act).

A body of case law grew up, including these notable examples.

1. Court of Cuneo, 23-6-1997: In 1997, the Court of Cuneo analysed this new form of media in depth, focusing in particular on its technical working. According to the Court, considering the large amount of information published every second on the Internet, it was impossible to consider the ISP to be responsible for such information. The technical features of the Internet and the nature of hosting activity prevented the possibility that an ISP could preventively monitor information—and, even if that would have been possible, it was too costly and would compromise the speed of the net. Therefore, the ISP could be liable only in cases in which it had knowledge of the unlawfulness of the information and failed to act.

2. Court of Naples, 8-8-1997: In some cases, starting from the Court of Naples in 1997, IPSs were held liable by analogy under the Press Law and thus publisher liability, the courts comparing the activities carried out by ISPs with the activities of newspapers and journals (Riccio 2002).

3. Court of Teramo, 11-12-1997: In another case in the Court of Teramo, ISP liability was excluded on the basis of the contract between the ISP itself and its customer.

4. Court of Vicenza, 23-6-1998: The Court of Vicenza, in the same year, issued an order to seize the servers of an ISP that hosted defamatory conduct pursuant to Article 321-bis, par. 3 of the Code of Penal Procedure.

5. Court of Rome, 4-7-1998: Almost the same reasoning was used a year later by the Court of Rome, in a case involving defamation perpetrated through newsgroups with no administrator. Here the Court affirmed that the ISP could not be deemed liable for any defamatory message conveyed through the newsgroup and moreover that the ISP had no obligation to monitor all messages published in the newsgroup.

6. Court of Oristano, 25-5-2000: The Court of Oristano, in 2000, clearly opposed analogous application of the Press Law to the Internet, holding that information published on the Internet did not meet the definition of 'printed material' in Article 1 of the Press Law and that allegedly defamatory conduct on the Internet should be regulated under Article 595 of the Criminal Code.

7. Court of Monza, 14-5-2001: The reasoning of the Courts of Cuneo and Rome was applied by the Court of Monza in 2001. Here the Court, ruling in a case on counterfeiting and unfair competition, excluded the possibility of the hosting provider's liability, relying on general principles of law and on the basis of foreign experiences (namely, the German one). Considering that *ad impossibilia nemo tenetur*, the Court held that it was not feasible, from a technical standpoint, to impose on an ISP an obligation to monitor all the information uploaded by users on the Internet.

8. Court of Florence, 21-05-2001: In a similar case on counterfeiting, a week after the Monza decision, the Court of Florence affirmed criminal liability for hosting a website whose name violated the trademark of a third party because the ISP registered the corresponding domain name. The Court held that ISPs could also be held liable on the basis of Article 2043 of the Civil Code,[3] setting up the responsibility for an unlawful act (that shares some common features with the Anglo-Saxon notion of contributory liability). In other cases, ISPs' liability was based on Article 2050 of the Civil Code,[4] regulating liability arising from the exercise of dangerous activities and Article 2051 of the Civil Code,[5] providing for damage caused by things in the defendant's custody. In the case of both Article 2050 and 2051, the ISP has the burden to prove that it had no active part in the injury (Article 2050) or that the injury is the result of a fortuitous event.

The Legislative Decree 70/2003

Legislative Decree 70/2003 implemented EU Directive 2000/31 on E-Commerce in Italy. The wording of the Italian Decree is almost a literal translation of the Directive and, thus, similarly lacks the procedural structures like those provided in the US Digital Millennium Copyright Act (DCMA). In particular, the Italian legislation does not specify 'notice and takedown' procedures for ISPs to follow where they host unlawful content.

Moreover, the Decree does not adapt the liability regime of the EU-Directive to the principles of the Italian legal system (De Cata 2010). Thus, there is a shift from the general standard of liability in tort, which is based on guilt and/or criminal intent (Article 2043 of the Civil Code) and on semi-objective or aggravated liability (Articles 2050 and 2051 of the Civil Code), to a special standard based on an

---

[3] Article 2043. Compensation for unlawful acts: Any intentional or negligent act that causes an unjustified injury to another obliges the person who has committed the act to pay damages.

[4] Article 2050. Liability arising from exercise of dangerous activities: Whoever causes injury to another in the performance of an activity dangerous by its nature or by reason of the instrumentalities employed, is liable for damages, unless he proves that he has taken all suitable measures to avoid the injury.

[5] Article 2051. Damage caused by things in custody: Everyone is liable for injuries caused by things in his custody, unless he proves that the injuries were the result of a fortuitous event.

exemption from liability for those providing information society services (as the ISPs are). It is the ISP's failure to act according to a legal provision that founds such liability. As we will see, this exemption cannot be applied to all ISPs but only to 'passive' ones, which have to be distinguished—as the courts and the literature have done—from 'active' ones.

Since the Legislative Decree provides for a special regime of exemption of liability, which differs from the general regime, the cases of exemption are strictly predetermined by the law, with no possibility of broader interpretation. Furthermore, there is also a need to identify clearly what procedures an ISP must follow to qualify for the safe harbour from liability. Finally, the absence of a 'notice and takedown' system does not help clarify the framework of ISPs' secondary liability.

The New Liability Regime: Articles 14–17 of the Legislative Decree 70/2003

Legislative Decree 70/2003 introduced a secondary liability regime quite different from the previous one—which had been based, as mentioned above, on the civil law tradition—depending, on the contrary, on the activities performed by the ISP. Thus, exemptions addressing three different types of ISP activities were created—mere conduit (Article 14), caching (Article 15) and hosting (Article 16)—to which correspond specific duties, obligations and liabilities. Another key provision is Article 17, which provides that ISPs have no general obligation of surveillance.

This partition of the ISPs' activities, which reflects the one embraced by the EU Directive, is no longer able to capture adequately the activities of modern ISPs. The activities in which an ISP can be engaged are not traceable to conduit, caching or hosting. There is a higher grade of detail and differentiation in the activities, meaning that the activities of ISPs are both far more varied, and less discretely differentiated, than this classification suggests. In particular, the hosting activity has undergone many changes due to technological development and a hosting provider can now offer a range of services to its customers. The definition given by the Decree is generic and outdated when compared to the services now provided by a hosting provider. This lack of correspondence between present ISPs' activities and the ones provided by the Decree has led the courts to adapt the Decree's provisions to the new Internet panorama, creating a new classification of hosting activities (search engine, linking, suggesting and so on) and thus a sort of 'new' liability regime based on a bipartition of the hosting provider concept, distinguishing the 'passive' hosting provider (reflecting the hosting provider covered by the EU Directive and of the Decree) from the 'active' hosting provider (to which a particular liability regime has to be applied). Therefore 'active' and 'passive' provider are roles developed by case law through a factual reconstruction of the functions performed by the ISP. Basically, neither 'active' nor 'passive' providers have an obligation to control or filter content posted by users. Nevertheless, when an unlawful conduct occurs, active providers – being non-neutral with respect to content – are expected to be more responsible than passive providers, i.e., to do something more. But the law in force offers no legal grounds for such a construction of the relevant provisions. As we will see, the courts decide each situation on a case by case basis, leading to case law which is not always consistent even when dealing with the same type of activity. Besides, this distinction has been rejected by the Court of Milan (7-1-2015[6]), in the appeal decision of the case *RTI v Yahoo*, pointing out that such a distinction cannot be found in the E-Commerce Directive, and EU case law clarified that automated activity do not change the nature of the hosting provider (decision C-324/09, *L'Oreal v eBay*).

However, we will consider first the provisions of the Legislative Decree and then turn to the key role played by the courts. As noted above, the Decree, repeating the content of the EU Directive, distinguishes three main activities: mere conduit (Article 14), caching (Article 15) and hosting (Article 16). According to Article 14:

1. The ISP performing a mere conduit or access provider activity, shall not be liable for the transmitted information, on the condition that:

    (a) it does not initiate the transmission;
    (b) does not select the receiver of the transmission; and
    (c) does not select or modify the information contained in the transmission.

2. The act of transmission and the provision of access referred to in (1) includes the automatic, intermediate and transient storage of the information transmitted, provided that this aims at the sole purpose of carrying out the transmission in the communication network and that the information is not stored for a period longer than reasonably necessary for the transmission.

3. Courts or administrative authorities with supervisory functions (one of the three administrative authorities—with responsibility for communication, data protection and antitrust), may require the ISP, performing one of the activities mentioned in (2), also by means of an interim order, to terminate or prevent an infringement.

According to Article 15:

1. The ISP performing a caching activity shall not be liable for the transmitted information on the condition that it:

    (a) does not initiate the transmission;
    (b) does not modify it;
    (c) complies with the specific conditions on access to the information;
    (d) complies with the rules regarding updating of information specified in a manner widely recognised and used in the industry.;
    (e) does not interfere with the lawful use of technology widely recognized in the industry to obtain data on the use of the information and

---

[6] See *infra*, para 1.10.

(f) acts expeditiously to remove or disable access to the information it has
    stored upon obtaining actual knowledge of the fact that the information at
    the initial source of the transmission has been removed from the network, or
    access to it has been disabled, or that a court or an administrative authority
    has ordered such a removal or disablement.

According to Article 16:

1. The ISP performing hosting activities shall not be liable for the stored information at the request of a recipient of the service, on the condition that:

   (a) it does not have actual knowledge of illegal activity or information and, as
       regards claims for damages, is not aware of facts or circumstances from
       which the illegal nature of the activity or information is apparent;
   (b) upon obtaining such knowledge or awareness, he acts expeditiously to
       remove or to disable access to the information.

2. Courts or competent administrative authorities may require the ISP, including by
   issuance of an interim order, to terminate or prevent an infringement action or to
   disconnect the access.

Finally, Article 17:

1. provides for the immunity of ISPs, stating that, within the limits set in arts.
   14–16, there is not a general obligation:

   (a) to monitor the information which they transmit or store; and
   (b) to actively seek facts or circumstances indicating illegal activity.

2. However, the ISP has the duty:

   (a) to inform without delay the competent public authority of alleged illegal
       activities undertaken or information provided by recipients of its services,
       and
   (b) to communicate to the competent authorities, at their request, information
       enabling the identification of recipients of their service with whom they have
       storage agreements.

Whenever the ISP fails to comply with the above-mentioned obligations, it will be
held liable for the consequent damages. This is not mandated by the directive, but if
flows from the case law. Therefore, the ISP has to prove that it had no actual knowledge of any illegal activities undertaken by the recipients of his service. Besides,
whenever the right-holder informs the ISP of an unlawful content and asks it to
remove it, the ISP cannot invoke the exemption because the communication of the
rights-holder gives the ISP actual knowledge of the infringing activity (see for
example Court of Milan, 19-5-2011, para 1.10).

Some Critical Issues: Actual Knowledge and Notice and Takedown

The absence of a general obligation to monitor is not without its critics. In particular, two issues that are connected to the lack of a general obligation to monitor seem problematic:

1. the determination on whether an activity is illegal; and
2. the meaning of actual knowledge.

As regards (1), it seems to be quite inappropriate to have the ISP judge whether an activity is illegal. It is as if an ex post obligation to monitor for unlawful content was introduced by Article 17 (Tosi 2012). The ISP has no obligation to control the content uploaded by the recipient of the service, nor to look for unlawful content. However, it must intervene whenever it has actual knowledge of the unlawfulness of content.

Point (2) is related to (1). It is not possible to infer clearly from the wording of the Decree the cases in which an ISP can be considered to have actual knowledge of an illegal activity. Given the absence of an obligation to actively monitor or search for illegal content/activity, is it sufficient to trigger liability that the ISP have direct knowledge or is it enough to have indirect knowledge? Or must an ISP be sent notice by a judge or an administrative authority to remove the unlawful content? There is a lexical asymmetry in the text of the Decree as regards the knowledge of the ISP. 'Effective knowledge' is mentioned by Article 15 (caching activity) and Article 16 (hosting activity and criminal liability). Article 16 also refers to simple 'awareness' when dealing with civil liability. Moreover, 'simple' knowledge is mentioned by Article 17, when affirming that the ISP has to inform the competent authority of illegal activity. The lack of clarity has, however, been mitigated by case law.

According to some cases[7], the ISP has actual knowledge of an illegal activity when it receives a notice from one party, usually a rights-holder[8]. Thus, according to these courts, it is not necessary to confer actual knowledge on an ISP that a specific notice be received from a judge or an administrative authority. But, given that a mere notice from one party is enough, there is still a critical issue as regards the detail of the notice. On this issue, in the '*Yahoo! Italia link*' or '*About Elly*' case, the Court of Rome[9] stated that the notice should be specific and detailed in order to impose actual knowledge upon the ISP of the unlawful conduct and thus lose the safe harbour privilege. Therefore it is possible to identify two steps. The first one:

---

[7] See the '*Yahoo! Italia link*' case: Court of Roma, Sez. proprietà intellettuale, 20-3-2011, *PFA Film v Yahoo! Italia* and 11-7-2011, *Yahoo! Italia v FPA Film.*

[8] Moreover, the courts are not unanimous on these issues, meaning that according to the Court of Milan in '*Google Autocomplete*' (25-3-2013) the ISP has a legal obligation to remove the unlawful content only when receiving a notice from the competent public authority, namely a judge or an administrative body.

[9] In the order of 11 July 2011.

whenerver the ISP is informed and can be considered to have an actual knowledge, he automatically loses the safe harbour privilege. And the second one: the ISP will then be considered liable and the general rules on liability will apply. However, even if this can be considered as a solution to the problem of defining actual knowledge, it creates a new issue, namely, the problem of the liability of the ISP that takes action against illegal activity/content after receiving a notice from a party, but it transpires that the notice in question was completely groundless (Sammarco 2011; Ricolfi 2013).

According to some scholars, Article 16(a) identifies two different types of liability: criminal and civil. The ISP should be criminally liable every time it does not act to remove the unlawful information despite having knowledge. Contrarily, it should be deemed liable from a civil standpoint whenever it fails to act against information which is manifestly illicit (Riccio 2003). Therefore, the difference would be that in the criminal field the ISP should investigate the unlawfulness of the information, thus being responsible in the cases of failure to remove unlawful information after having investigated it. According to other scholars, Article 16 provides for a single type of responsibility, namely, a civil one (Nivarra 2003; Cassano and Cimino 2004a, b).

The notion of actual knowledge could have been modelled on DMCA and on the 'notice and takedown' system set up in the United States (Riccio 2002). The DMCA provides a safe harbour whenever the ISP takes action on the basis of a notice of a third party; the ISP cannot be liable for having removed content which was not unlawful. However, in case of a valid counter notice, ISPs must wait 10-14 days before they re-activate or allow access to the claimed infringing content. Also, the ECJ, in *Scarlet v SABAM* (Case C-70/10, *Scarlet Extended SA v Société Belge des auteurs, compositeurs et éditeurs* SCRL (SABAM)) affirmed that Article 15 of the E-Commerce Directive prevents Member States from imposing on ISPs an obligation to monitor all the information on their servers. (Siano 2011). Likewise, in *SABAM v Netlog* (ECJ, 16-2-2012, case C-360/10, *SABAM v Netlog*) (Bellia et al. 2012) the ECJ reaffirmed the prohibition against imposing on ISPs a duty to provide a preventive filtering mechanism on the communications of all their customers for an unlimited period at the ISP's expense[10] (Franceschelli 2012). Such a mechanism would be unrealistic, mainly for reasons of costs but also because it undermines the capacity of ISPs to conduct their own business. Moreover, this may undermine competition, favouring the providers that can bear the costs of the mechanism and thus might create an oligopoly. The issue of competition is a key one; the new regulation on the protection of copyright issued by the Communication Authority, in Italian Autorità per le garanzie nelle comunicazioni (AGCOM), stresses the importance of promoting fundamental freedoms, including economic freedom and competition (Article 2).[11] Lastly, there are also doubts about the effectiveness of the mechanism.

---

[10] Also the Court of Milan in the 'Sky' case, *RTI v Yahoo! RTI v ItalianOnLine* (see below) expressed a similar view.

[11] On the regulation, see infra, par. 12.

Article 17, as well as the whole Legislative Decree, tries to strike a balance between opposing interests, namely freedom of the Internet, of expression and privacy on the one hand, and control and surveillance of contents that may infringe someone's rights on the other (Pizzetti 2011; Bellezza and Pollicino 2012a, Bellezza 2012; Bertolini 2010, 2013; De Minico 2011). Also the courts have tried to strike such a balance case by case; relevant in the Italian case law are two decisions, 'Peppermint'[12] and 'FAPAV'[13], both focusing on the role played by ISPs in the enforcement of copyright law, namely, whether an ISP should transmit to rights-holders the personal data of the customers associated to the relevant IP address, thus making it possible for the rights-holder to identify the person violating the copyright law (Musso 2010; Scorza 2007; Caso 2007; De Cata 2008; Franceschelli 2007).

However, Member States, in the implementation of the E-Commerce Directive in their respective domestic legislation, are free to impose on ISPs a duty to report to the competent public authority possible illegal activity conducted through their services or the transmission or storage within their services of unlawful information and to communicate, at the request of the national competent authorities, information that identifies the recipient of the service that allegedly carried out an illegal activity (Article 15.2). Moreover, the E-Commerce Directive leaves the Member States free to:

1. grant to courts or to administrative bodies the power to order the hosting provider to remove unlawful content;
2. introduce specific national regulations governing the notice and takedown procedures in respect to the unlawful content stored by users (Article 13.2).

Here, as regards the duties that may be imposed on ISPs, there is the main difference between the E-Commerce Directive and the Legislative Decree 70/2003 (Apa and Pollicino 2013). While the Directive requires hosting providers to remove the unlawful content, even in absence of a specific order issued by the competent public authority, the Decree (namely arts. 16 and 17) expressly states that hosting providers shall promptly inform the competent public authority of the presence of the unlawful information or content and remove it only after the competent public authority issues a specific order. The courts have many times, although not always (as in the '*Yahoo! Italia link*' case), stressed the need for a public authority to order the removal of the content (see also '*Google Autocomplete*'). Therefore, under the Italian Decree, the hosting provider shall be deemed liable for damage suffered by third parties only when the hosting provider does not:

1. promptly report the information or content to the competent public authority;
2. promptly remove the relevant information or content after receiving a specific order to do so by the competent public authority.

---

[12] Court of Rome, sez. feriale, 18-8-2006; sez. IX, 22-9-2006; sez. IX, 9-2-2007; sez. IX civile, 5-4-2007; sez. IX civile, 20-4-2007; sez. IX civile, 26-4-2007
[13] Court of Rome, decision of 15-4-2010, '*FAPAV v Telecom*'.

The Decree grants broader protection to ISPs than does the EU Directive. Moreover, the Decree does not create a 'notice and takedown' system like the one provided in the DMCA. Such a system operates in Italy by virtue only of self-regulation. Indeed, the new AGCOM regulation at Article 5 reaffirms that 'notice and takedown' procedures can be provided only through ISPs' self-regulation. The introduction of a system modelled on the American one could solve some of the problems mentioned above, in particular the involvement of the ISP in determining the unlawfulness of the content or the ownership of the right infringed. Indeed, the system set up in the DMCA does not provide such an involvement, being an automated one. The ISP has neither the knowledge nor the tools to be able to monitor the content and to decide whether it is illegal. Moreover, in an automated system the ISP could not be deemed liable in case of removal of content based upon a groundless notice. However, as stated above, Article 16 of the Decree prevents the provision of a 'notice and takedown' system modelled on the DMCA. It would be possible to overcome this provision in two ways:

1. interpreting the Italian provision in light of the supremacy of European Union law, which means that a Member State cannot broaden the immunisation to cases that do not conform to its text or purpose; or
2. referring for a preliminary ruling to the ECJ on the conformity of Article 16 with European Union law, in particular Article 14 of the E-Commerce Directive.

However, in a 'notice and takedown' system, the ISP can be considered as a sort of 'censor' on the Internet. This poses some problems regarding the freedom of expression guaranteed by Article 21 of the Constitution. It is also noteworthy that the Decree does not clarify which Italian public authority is deemed competent to issue removal orders addressed to ISPs (Bellezza and Pollicino 2012b).

Codes of Conduct (Article 18 of the Legislative Decree 70/2003)

Article 18 of the Decree encourages the adoption of codes of conduct (as does Article 3 of the AGCOM regulation), copies of which have to be transmitted to the European Commission and to the Ministry of Productive Activities. Moreover, the codes have to guarantee the protection of minors and human dignity.

The Scope of the Legislative Decree 70/2003

The scope of the Legislative Decree, like the EU Directive, is information society services, which include E-Commerce. However, excluded from the scope of the special liability regime are, according to Article 1 of the Legislative Decree:

1. the following services:

   (a) telecommunications;
   (b) radio broadcasting;
   (c) medical and legal advisory.

2. the following fields:

   (a) tax law;
   (b) privacy;
   (c) competition law;
   (d) services offered by subjects not established within the European Union;
   (e) the activities of notaries or equivalent professions to the extent that they involve a direct and specific connection with the exercise of public authority;
   (f) representation of a client and defence of his interests before the courts;
   (g) gambling activities which involve wagering a stake with monetary value in games of chance, including lotteries and betting transactions.

3. Paragraph 3 affirms that the Decree does not affect measures taken at Community or national level, in order to protect public health and consumers, to authorise detective activities and private security, preserve public security, to prevent money-laundering, drug and weapon trafficking.

Nowadays, the civil liability of ISPs is usually traced back to extra-contractual liability (rather than to the contractual one[14]) and it is therefore possible to distinguish the following regimes of liability:

1. the special regime established by Legislative Decree 70/2003;
2. the general rules of tort law contained in Articles 2043, 2050 and 2051 of the Italian Civil Code;
3. the regime for unlawful processing of personal data, in violation of the provisions set forth in Legislative Decree 196/2003;
4. the regime for violation of copyright found in Law 633/1941 (Copyright Law).

The Definition of ISP

Another important issue is the definition of an 'ISP'. As noted above, Legislative Decree 70/2003 gives a definition of ISP connected to the activities involved that has been now overcome by the development of the Internet and of the services offered by ISPs.

According to Article 2 of both the Directive and the Legislative Decree 70/2003, ISP is defined as any natural or legal person providing an information society

---

[14] Some scholars, however, trace back the actual liability regime to contractual liability. See Di Majo 2012, La responsabilità del provider tra prevenzione e rimozione. *Corr. giur.* 4: 553–560.

service, with an information society service being any economic activity carried out online or offering a paid service. The definition is quite broad and is no longer useful in identifying which providers can enjoy immunity. On this issue, it is important to recall Recital 42, which narrows the scope of the Directive, limiting the immunity to circumstances where the service provided is a mere technical process of operating and giving access to a communication network.[15] Recital 42 is the basis of the distinction discussed above between 'passive' and 'active' hosting providers.

Since the enactment of Legislative Decree 70/2003, in particular in the field of copyright law, there has as a result emerged a need to identify a new subject, the 'active' ISP, who will fall beyond the safe harbour of Legislative Decree 70/2003 and therefore be potentially liable for unlawful activities such as copyright infringements. This need has arisen mainly with reference to the activities of search engines and so-called 'UGC Platforms' (user generated content) (Osbourne 2008; Bassini 2012).

Indeed, a provider can perform other activities that may be qualified as active, such as:

1. indexing of contents;
2. selection and organisation of contents;
3. filtering of contents; and
4. advertising.

All these activities, not being merely automatic but active, cannot fall within the safe harbour regime. In other words, when an ISP consciously engages in indexing and organisation of contents from which he derives an economic gain, he loses the exemption from liability. In this case the ISP is not neutral with respect to the content, meaning that it cannot fall under the safe harbour protection and therefore such an activity can be considered to be similar, in a broader sense, to the publisher liability (Tosi 2012)[16].

This distinction between 'passive' and 'active' provider is based on recognition that it would be impossible to charge the ISP with an obligation of surveillance as regards the contents of all data transmitted on the net. Besides, some ISPs provide no more than the mere access to the net to their customers (Frosini 2011, 2013; Costanzo 2012; Passaglia 2011, De Minico 2010; Duni 2007, 2008; De Marco 2008; Pisa 2010). This idea, embodied in the Recital 42 of the E-Commerce Directive, echoes the DMCA, according to which not all providers can be held liable, but mainly the ones who contribute to the editing of the content, the so called content providers (ICPs). However, the notion of 'active' ISP, based on Recital 42

---

[15]Recital 42: 'The exemptions from liability established in this Directive cover only cases where the activity of the information society service provider is limited to the technical process of operating and giving access to a communication network over which information made available by third parties is transmitted or temporarily stored, for the sole purpose of making the transmission more efficient; this activity is of a mere technical, automatic and passive nature, which implies that the information society service provider has neither knowledge of nor control over the information which is transmitted or stored'.

[16]See para 1.10 on the most relevant case law.

and developed by the courts, is different from the ICP. If the ICP can be considered as a sort of publisher, the same does not go for the 'active' ISP, which performs slightly different activities, such as operating a search engine, linking or uploading UGC. Courts have been very careful on this matter, proceeding case by case in order not to render meaningless the exemptions set in Legislative Decree 70/2003. Therefore, courts started to treat extrajudicial notice on the basis of Article 16 of the Decree sufficient to confer actual knowledge of the infringement on an ISP and thus to exclude the application of the safe harbour. In truth, this is a stretched interpretation of Article 16, which literally states that the ISP has to remove the unlawful content only after the notice of the competent public authority (courts have recognised that this is a stretched interpretation). This problem could be solved in light of the atypical nature of the 'active' ISP, to which the Decree cannot be applied. According to courts, Article 16 can be applied also to a non judicial or administrative authority based notice in all the cases when the provider can be considered an 'active' one. This leads to a sort of less formalized notice compared to the one required for 'passive' providers. This issue of the nature of the notice, following which the ISP has to remove unlawful content, is strictly linked to the one mentioned above, relating to the problem of the liability of an ISP that removes content on the basis of a groundless notice sent from a third party rather than a judge or administrative authority.

All the subjects acting on the Internet may incur some sort of liability, depending on the activity they carry on. Trying to summarise these different activities and the connected liability, we can distinguish:

1. special regime of liability for ISPs, which provides for safe harbour in special cases for mere conduit, caching and hosting ('passive' provider) (principle of liability deriving from a failure to act);
2. commune liability for ICPs and 'active' ISPs, on the basis of Article 2043 of the Civil Code, deriving from an intentional or negligent act;
3. primary liability for ICPs and 'active' ISP comparable to a publisher, deriving from infringements of the Legislative Decree 196/2003 on privacy.

Primary Liability

The special regime of liability provided by the Legislative Decree concerns only ISPs' secondary liability, meaning that ISPs are primarily liable for their own activities that constitute an infringement of a particular statute. Obviously it is fundamental in this case that the unlawful conduct can be directly ascribed to the ISP. The Legislative Decree provides for safe harbours only in terms of secondary liability, meaning that when we are not dealing with secondary liability the Decree does not apply.

Case Law

In the Italian context, the courts play a fundamental role in the development of the ISPs' liability regime.

It was previously mentioned that the courts tried to adapt the liability regime provided by the Legislative Decree to the new kinds of services offered by the ISPs. In analysing case law of the last decade, the services offered by ISPs are seldom traced back to the three activities identified in the Decree, meaning that the liability issue is considered case by case according to the type of service offered by the ISP in that particular case, which is usually not a simple hosting activity, but rather something more complex and difficult to define, like a search engine or linking. Moreover, Italian case law on the liability issue has dealt mainly with requests for precautionary measures issued by intellectual property rights holders against hosting providers in connection with the unlawful dissemination of protected creative works through the services provided by the latter.

This section will focus on the analysis of the most relevant case law in order to show how the courts have interpreted the Legislative Decree. The issue of the distinction between 'passive' and 'active' ISPs is perhaps the most relevant issue with which the courts have dealt. Facing a legislative provision that has not been amended, even after the main transformations undergone by ISPs' activities, the courts had to alter their interpretation of that provision, in order to make it suitable to regulate the new scenario:

1. Court of Milan, 20-1-2011 and 19-5-2011: The distinction between active and passive ISPs was drawn for the first time by the Court of Milan in two proceedings brought by RTI (Italy's main private broadcaster and part of the Mediaset Group), for copyright infringement against IOL-ItaliaOnLine (Court of Milan, 20-1-2011, *RTI v ItaliaOnline Srl.*) (Bellan 2012) and Yahoo! Italia (Court of Milan 19-5-2011, *RTI v Yahoo! Italia Srl.*) (Barbieri and De Santis 2011a, b), whose online video-sharing platforms displayed RTI's TV programmes. In these cases, the Court held that the safe harbour provided by the Decree did not apply to the platforms, as they played an active role in organising the videos uploaded to their platform, with a view to financial gain, despite having been provided by the rights-holder with sufficiently detailed notices on the unlawful conduct of the platform. However, despite such an active role, the court held that it is still not possible to consider the 'active' ISP a content provider. The Court therefore introduced the notion of active hosting provider which is not an ICP (which usually uploads its own content or a third part's one on the Internet) beside the one of mere hosting provider, so called passive hosting, performing just a neutral role.

    Yahoo! Italia appealed against the 2011 decision and on 7th January 2015 the Court of Milan issued a landmark decision on ISPs' liability, overruling the decision of first instance. One of the main findings of the decision is the irrelevance of the distinction between 'active' and 'passive' provider. The Court held that Yahoo simply offered a video sharing services and therefore it is a hosting provider within the meaning of the E-Commerce Directive and thus subject to the liability exemption. No obligation of prior verification or ex-ante control, or monitoring, of the hosted content can fall on the ISP. In particular, the Court of Appeal stressed that the current technologies are not per se enough to make the services offered by the hosting provider 'active'.

Moreover, the Court confronted the issue of the ISP's removal obligation, clarifying that the right holder must expressly and specifically identify the allegedly illicit content before the hosting provider is liable to remove that content. It is necessary that the right holder expressly indicates the URLs allegedly in violation of his or her rights. Therefore, a generic warning with only the title of the programs is not enough. Besides, the Court clarifies that it is not possible to ask either for an injunction, forcing the ISP to set up a filtering system or at least a monitoring obligation with respect to future violations, both obligations considered to be against the freedom of enterprise. Similarly, the hosting provider is not subject to any "stay down" obligation in order to prevent that the same illicit content already removed being uploaded again, being an obligation in contrast with the provisions of the E-commerce Directive.

Surely this decision seems to overcome the distinction between 'active' and 'passive' hosting provider and to clarify the duties of the rights holders asking for the removal of the unlawful content which has to be clearly identified.

2. Court of Rome, 20-10-2011: The same distinction between 'active-passive' provider was acknowledged by the Court of Rome in the 'RTI/Choopa' case (Court of Rome, 20-10-2011, R TI/Choopa) (Apa and De Santis 2011). Here the Court held that anyone who merely provided hosting services to a video-sharing website could not be deemed liable for copyright infringement in a case where video had been uploaded by users on the video-sharing website or directly uploaded by the website owner. On the contrary, the Court held directly liable the provider that operated the video-sharing website in question as an 'active' hosting provider, which was deemed aware of the unlawfulness of the relevant content. Therefore, a distinction has to be made between ISP's liability in case of linking to contents protected by copyright and indexed by the search engine and when the provider himself operates the video-sharing website (as do *Google Video* or *Youtube*, so called UGC platforms).

Moreover, *the RTI/Choopa* court confronted another important issue, namely the duty of the ISP to act when it receives a notice. According to the Court, such a duty for 'passive' hosting providers may derive only from a specific order sent by the competent authority (a judge or an administrative authority) and not from a notice sent by the rights-holder (like the one sent in the 2011 *RTI v Yahoo! Italia Srl.* mentioned above).

Given the growing importance of this distinction, Italian courts have tried to identify the key elements to qualify an ISP as an active hosting provider (Apa and Pollicino 2013):

(a) The insertion of clauses within the Terms and Conditions of the service which might reveal the exercise of control of the information stored by the customers. In the 'RTI/Choopa' case, the Court of Rome held that the fact that the provider of a social media platform inserted a clause that granted the platform a non-exclusive license on the content uploaded by its users was fundamental in order to qualifying the provider as an active hosting provider.

(b) The organisation and selection of the information provided by customers that gives the provider greater profits than those generated by the hosting service. For instance, in different cases the insertion by the provider of promotional messages associated with the content posted by users on its service was considered an element which qualifies the provider as an active hosting provider. See 'RTI/Yahoo!', mentioned above.

(c) In other cases, Italian courts have held that the general liability exemptions for hosting providers do not apply to providers considered to be active (and therefore not to be neutral with respect to the content) because they:

(i) provided a search tool that enabled users to search for content by keyword;

(ii) indexed and selected videos and offered a related videos search function, which automatically displayed content related to the user's search results;

(iii) offered a notice and takedown mechanism for the notification of alleged infringing content; and

(iv) directly uploaded some content.

(See the decision by the Court of Milan, 24-2-2010 and 21-12-2012).

3. The leading case on ISPs' liability in Italy is *Google v Vididown*, known as 'RTI/IOL-ItaliaOnLine' case (Court of Milan, 24-2-2010 and 21-12-2012) (Camera and Pollicino 2010; Franceschelli 2010; Apa and Pollicino 2013; Apa and De Santis 2013). The case was about a video showing four students at a Turin school teasing an autistic boy, which was uploaded to Google Video. Three Google executives were then charged with defamation and unlawful data processing. According to the public prosecutor, Google acted as content provider in connection with the processing of the boy's sensitive personal data (here, the health condition), because Google allegedly handled the boy's data and profited from the disputed video due to the sponsored links that appeared on it (Ad Words). The defendants were acquitted on the defamation charge, given that no duty to monitor the uploaded content can be imposed on ISPs, according to Article 15 of the E-Commerce Directive and its Italian implementing provisions. However, the Court of First Instance found them liable for unlawful data processing according to Article 167 of the Italian Data Protection Code, since the privacy notice for data subjects made available on the Google Video platform provided no information about the need for the express consent of third parties appearing on any uploaded video before they were uploaded.

On 21st December 2012, the Court of Appeals of Milan reversed that decision, ruling in favour of the three Google executives, finding them not guilty of unlawful data processing. The main points of the decision can be summarised as follows:

(a) ISPs are not charged with a preventive duty to monitor the information and content uploaded by users on their platforms, as affirmed both in the EU Directive and in Legislative Decree 70/2003. The reason is that such a monitoring activity would be impossible from a technical standpoint and could undermine freedom of expression.

(b) Failure to provide data subjects with a proper privacy notice does not result in unlawful data processing, punished as a crime under section 167 of the Italian Data Protection Code.

(c) The data controller of the sensitive data was not Google, but the person uploading the video. It was therefore up to him to obtain the boy's parents' consent before the upload.

(d) No sponsored link appeared on the disputed video. Thus, Google had no intent to profit from it. Moreover, there was no wilful intent since the executives were not aware of the content of the disputed video before the upload. Therefore, both the profit and the wilful intent requirements, which must exist for liability to arise for unlawful data processing under the Italian Data Protection laws, were lacking.

(e) As regards the core issue in the case, namely, whether Google acted as a 'passive' or 'active' provider, the Appeals Court made reference to this distinction as drawn by the First Instance Court. The Court supported a view that the provisions regarding ISP liability must be construed and enforced in accordance with the evolution of the Internet, and also in light of new circumstances as compared to those pertaining when the E-Commerce Directive was implemented. In particular, the Court underlined that in today's world, the services that an ISP offers are not limited to the technical process that simply sets up and provides access to the network: as in the case of the content provider, they extend to making it possible for users to submit their own content and other people's content on the network. They cannot, therefore, escape the duty to comply with the standard regulations governing liability for data processing.

The Court, despite acknowledging that, due to the increasing number of services offered by ISPs, it is not possible to understand them as being simple Internet access providers anymore and that a 'hybrid' category of ISPs has emerged, namely, the 'active' hosting providers, failed to present a clear model of liability (Apa and Pollicino 2013). Despite it being held that 'active' ISPs lack the character of neutrality and that they are more responsible than 'passive' providers, it is not clear to what extent this is the case. Furthermore, an active ISP is not considered a content provider and therefore it is not per se excluded from the safe harbour privilege. Thus it is unclear under what conditions an active hosting provider qualifies as such and, accordingly, is liable for the unlawful content posted by users. Here the core issue is that although it can be reasonable to expect 'active' ISPs to have more control over content than 'passive' ones, such a *quid pluris* in terms of control cannot amount to a legal obligation to monitor content posted by users and this tension is extremely difficult to translate into legal terms (Apa and Pollicino 2013). The relevant provisions do not require any provider to act in order to remove unauthorised content unless they are advised of the same or requested to by the competent authority, as a consequence of (and in accordance with) the lack of a preventive obligation to control.

The decision of the Court of Appeals was confirmed by the Court of Cassation (Cass. pen., sez. III) on 17th December 2013. Court of Milano, 25-1-2011 and 31-3-2011.

4. When considering suggestion services offered by Google, we have to recall the two decisions of the Court of Milan in 2011 in the 'Google Suggest' case (Court of Milano, I sez. civ., 25-1-2011, *Bardolla v Google* and 31-3-2011, *Google v Bardolla*). The case involved a man in connection with whose name the Google search engine suggested the words 'fraud' and 'swindler', which thus was considered to be defamatory. In this case, the issue of ISP liability was at stake—a search engine is considered an ISP—but Legislative Decree 70/2003 was considered in the end not to be relevant because the ISP could not be considered liable for not having removed unlawful content, but instead for a fault in programming the software which was an unlawful act carried out by the same ISP. However, the Court considered the connection between the name of the man and the word 'fraud' made by the Google Suggest software defamatory.

Google appealed, due to lack of motivation, both on its liability and on the defamatory character of the connection. Moreover, Google claimed not to be an ICP. Again, it has to be underlined that to ICPs is applied publisher liability and therefore no exemption is possible in this case. The Court rejected the appeal, underlining that in the case Google, as a search engine, had to be considered as an ISP, on the basis that a search engine is basically a database with software, thus operating as an intermediary in the information society (as a host provider). However, the main point of the case was not whether Google was a hosting provider, but about its liability in terms of the suggestions made by the Google Suggest service. This suggestion made between the name of the man and the word fraud, the Court affirmed, is a result of the software created by Google expressly to speed up and make easier the search process and, therefore, Google is primarily liable for any injury that results from the application of the software. As noted above, here the Decree is not relevant because the issue was not whether Google was liable for defamation, but for the improper programming of the software.

5. Court of Pinerolo 30-4-2012: The same conclusion was reached by the Court of Pinerolo in an analogous case where the Google Suggest service was deemed not to be defamatory (Court of Pinerolo, 30-4-2012, *X v Google*). According to the Court, such an activity falls under the hosting provision, for which the Legislative Decree provides a safe harbour provision, and moreover the suggestion was made through an automatic software, thus lacking a defamatory intent by Google.

6. Court of Milan, 25-3-2013: Again the Court of Milan held that Google, despite offering a suggest service, could not be considered as a content provider on the basis that the service merely shows the results according to the most frequent research carried out by Internet users; therefore, Google in this case is a mere caching provider, falling under the safe harbour provisions. In so doing, the Tribunal of Milan overruled a previous decision on the same issue, finding Google liable (Court of Milan, 24-3-2011). To be or not to be a content provider is relevant because ICPs, being considered as publishers, cannot fall under the safe harbour privilege.

7. Another case involved the Google Autocomplete service (cited *supra*), but here the focus of the Court of Milan was notices and specific orders by a public authority. The case was linked to a civil action brought by two non-profit associations against Google alleging that Google's Autocomplete and Related Searches results included defamatory words. The Court ruled that ISPs have a duty to remove illicit content only if so required by a specific order from the competent judicial authority (Bellezza and De Santis 2013). The Court of Rome in the 'RTI/Choopa' case, mentioned above, affirmed this interpretation of the Decree.

8. Court of Florence, 25-5-2012: The Court of Florence—in a case involving Google (*Meneghetti/Google*) that, as a caching provider, had indexed a website whose content violated a trademark—held that Italian law requires a specific court order to remove the infringing content, a notice of a party being not sufficient (Apa and De Santis 2012).

9. Another important service offered by ISPs is linking. In the 'Yahoo! Italia link' or 'About Elly' case referred to above, Yahoo! Italia was sued for having linked to an audio-visual content (on the movie *About Elly*), whose rights were held by PFA Film and for not having removed it even after a notice sent by the rights-holder, thus infringing Article 156 of the Copyright Law (Law 633/1941). In this case, two points are most relevant: (a) whether or not Yahoo! Italia could be considered liable for such an activity (that is, whether the safe harbour of the Legislative Decree applies or not); and (b) the required detail of the notice sent by the right-holders to the ISP, requesting the removal of the unlawful content. In the decision of first instance, the Court of Rome found that the exemption from liability in Legislative Decree 70/2003 could not be applied to Yahoo! Italia on the basis that—despite the provider having no legal obligation of preventative monitoring and having not taken an active part in the selection of the content— once it had actual knowledge of the unlawful content identified by the URL, it could have acted to prevent the indexing and linking of the content. The general notice the rights-holder sent to Yahoo! Italia reporting the unlawfulness of the content liked by the provider was considered by the Court enough to vest Yahoo! Italia with actual knowledge and thus oblige it to remove the content. The Court of Rome issued an interim measure because it found that the search engine bore contributory liability for infringement of copyright in the film by illicit websites, which allowed the streaming or downloading and peer-to-peer sharing of the film without the consent of the film's distributor.

Yahoo! Italia appealed and the Court overruled the injunction. The Court affirmed that in this case:

(a) the failure by Yahoo! Italia to provide data subjects with a proper privacy notice did not result in unlawful data processing, punished as a crime under section 167 of the Italian Data Protection Code.

(b) Yahoo! Italia had not taken an active part in selecting or indexing the content and was acting as a conduit and therefore had immunity under arts. 14-17 of the Legislative Decree 70/2003 (so the provider could not be considered to be an 'active' provider).

(c) Yahoo! Italia had no general obligation of surveillance, on the basis of Article 17 of the Decree.

The Court also expressed the same reasoning in the *Yahoo! Italia v Alfa Films*, of the same year (Court of Rome, 16-6-2011). The Court focused on the complex issue of what constitutes actual knowledge of the unlawful act/content. It is only after having actual knowledge that the ISP loses its exemption from liability. Therefore it is fundamental to establish when an ISP can really be considered to have an actual knowledge. In this case the right-holder, PFA Film, sent a notice containing only a general request to remove all links to the movie *About Elly*. Here the Court did not consider this notice sufficient to confer actual knowledge. Rather, according to the Court, a notice has to be very detailed in order to allow the provider to clearly identify:

(a) the link to the website with the unlawful content;
(b) the right violated, which may differ from case to case;
(c) proof that the one sending the notice is actually the right-holder; and
(d) the infringer.

The Court affirmed again that is not up to the ISP to control and to investigate in order to find the unlawful content in absence of clear and detailed information by the right-holder.

10. Court of Milan, 19-5-2011: A different conclusion was reached by the Court of Milan in the 'RTI/Yahoo! Italia' case (Barbieri and De Santis 2011a, b) The Court held that the right-holder's notices were sufficiently detailed, as the rights-holder (a leading Italian broadcaster) had 'specified each program from which the reported files had been extracted which are transmissions with popular appeal, and even a superficial and quick control should have proved at least the ownership of plaintiff's rights'.

11. Court of Rome, 17-8-2011: Another important linking case is the 'Rojadirecta' case, decided by the Court of Rome. The case involved the website www. rojadirecta.es, which linked to a foreign website sharing audio visual content (mainly sports) free of charge. According to the Court, the linking service offered by the ISP was itself lawful, but it was the aim of the provider—to

secure financial benefit through advertising—that made such an activity unlawful. Moreover, the provider took an active part in organising and indexing contents, and providing the customers with instruction on how to use the platform, and therefore the linking activity could not fall under the safe harbour of Legislative Decree 70/2003. The Court of Rome upheld this conclusion in another case involving a provider which hosted a website streaming copyrighted material (Court of Rome, 20-10-2011). Here the decision focused both on the hosting provider and on the one operating on the website-ICP.

12. Court of Milan, 7-1-2010: In 2010, the Court of Milan decided another linking case in *Sky Italia v D.B and Telecom Italy*. Here a website shared links to soccer matches whose rights were owned by Sky Italia. The website provided its users with instructions on how to use the linking service and therefore it was held liable by reference to Article 41 of the Criminal Code.[17] The ISP, Telecom Italia, was considered to have played a neutral role, providing only access to the website and thus fell within the safe harbour protecting mere conduit activity.

ISPs' Criminal Liability

It is also worth considering ISPs' criminal liability (Ianni 2011; Gambuli 2005). Here we can distinguish two main categories of criminal offense:

1. the one deriving from an action; and
2. the one deriving from a failure to act.

As regards failures to act, it is controversial whether a hosting provider may be deemed criminally liable for omission as regards a criminal offense performed by a third party, namely, for not having removed a content which is unlawful from a criminal standpoint. Moreover, neither the EU Directive not the Legislative Decree 70/2003, both focused on E-Commerce, provide for criminal liability to be associated to the three different activities of the ISPs.

Italian criminal law distinguishes two categories of criminal offense for omission or failure to act:

---

[17]Article 41. *Concurrent causes*: 'The presence of pre-existing, simultaneous or supervening causes, even though independent of the act of omission of the offender, shall not exclude a causal relationship between his act or omission and the event. Supervening causes shall exclude a causal relationship when they were in themselves sufficient to bring about the event. If, in that case, the act of omission previously committed itself constitutes an offense, the punishment prescribed therefore shall be applied. The preceding provisions shall apply even when the pre-existing, simultaneous or supervening cause consists of the unlawful act of another person'.

1. Proper: based on a criminal provision, which provides for an action to be done in order to avoid an offense to happen.
2. Improper: there is not an express criminal provision, but the offense can be inferred from case-law. According to Article 40, para. 2 of the Italian Criminal Code,[18] failing to prevent an act which one has an obligation to prevent is the same as having caused it.

Further, some courts have tried to found ISPs' criminal liability by analogising the ISP to a publisher. In so doing, it would have been possible to apply Article 57 of the Criminal Code, which deals with press-related crimes and, therefore, the ISP would have a legal duty to monitor the lawfulness of all materials uploaded on its server, even by third parties. However, the Court of Cassation (Cass. pen., sez. V, 6-7-2010, n. 35511 and Cass. pen., sez. V, 28-10-2011, n. 44126) excluded the applicability of Article 57 to access providers, service providers, hosting providers, and forum and blog administrators.

It is worth noting that when considering cyber offenses it is convenient to distinguish 'traditional' offenses that can be perpetrated also on the Internet and cyber offenses *stricto sensu*, namely, offenses that can be perpetrated only on the Internet. It is quite clear that ISPs can be perpetrators of the latter category of offenses, whereas it is controversial whether they can liable for the first category. There are a series of so called hard cases where the courts have held ISPs criminally liable or not liable on the basis of broad or narrow interpretations of the criminal offence (Ingrassia 2012).

It is still controversial whether it is possible to hold the ISP criminally liable when it does not conform to the order of the competent public authority to remove unlawful content or fails to report to said authority infringements of which it has become aware. In such cases, where the potential criminal liability of ISPs is still not clear, it is convenient to make reference to the main case law of Italian courts since the enactment of the Legislative Decree 70/2003 (as far as it concern the previous case law, see above):

1. Court of Milan, 24-2-2004: The Court of Milan in 2004 held that a website had no obligation, from a criminal point of view, to monitor the published materials to which it linked for unlawful content. The case was about a website that published links to other websites and one of them shared child pornographic content, thus violating Article 600-ter, par. 3 of the Criminal Code. Cass. pen., sez. III, 29-9-2009, n. 49437.
2. The courts have also dealt with whether linking and sharing copyrighted materials can give rise to criminal liability in two main cases, the first one involving Sky Italia in 2006 and the second one The Pirate Bay in 2009 (Cass. pen., sez. III, 29-9-2009, n. 49437) (Cuomo 2011; Merla 2010):

---

[18] Failing to prevent an event which one has a legal obligation to prevent shall be equivalent to causing it.

(a) Court of Milan, 8-2-2006: The 'Sky-calcio libero' case involved a website which linked Chinese websites making available a stream of Italian football matches whose copyright was held by Sky Italia. The Court of Milan rejected a request by Sky Italia to seize the website, affirming that only the direct uploading of copyrighted material was a criminal offense and not the linking to an upload made by a third party.

To the contrary, the Court of Cassation (Cass. pen., sez. III, 4-7-2006, n. 33945) (Sammarco 2006), affirmed that it could rule out the cooperation of the Italian website in the criminal offense, in the sense that the Italian website facilitated the copyright infringement. Therefore, the Court quashed the order of the Court of Milan that rejected the seizure of the website. Here the Supreme Court seems not to have considered that the criminal offender is the one uploading the unlawful content, meaning that it is when the copyrighted material is uploaded for the first time that the offense is perpetrated and therefore it is not possible for a third party to contribute to the criminal offense (Flor 2007).

(b) In the 'The Pirate Bay' case, the Court of Cassation had to deal with a provider that allowed its customers to upload files on its website. According to the Court, the ISP not only prepared the communication protocol to share the files (for which it could not be held criminally liable) but it indexed, updated the files and arranged a search engine; thus the Court held the ISP criminally liable on the basis of Article 171-ter, par. 2, lett. a)-bis of the Copyright Law of 1941, because its conduct allowed its customers to share unlawfully copyrighted materials. Therefore, the provider was deemed criminally liable for having cooperated in the infringing activity.

The AGCOM Regulation

No reform of Legislative Decree 70/2003 is presently envisaged. However, at the end of July 2013 the AGCOM issued a 'Regulation on copyright on the electronic communication networks and implementing measures pursuant to Legislative Decree 70/2003' (www.agcom.it; Bellezza 2013a, b) which entered into force on 31 March 2014 and affects ISPs' secondary liability.

The main chapter of the regulation is the third, 'Provisions on the Protection of Copyright on the basis of Legislative Decree 70/2003' (arts. 5-9), providing for a new administrative procedure before AGCOM that right-holders may follow in order to have the broadcasting on audio-visual media services of infringing works stopped. According to Article 6, a right-holder may file a complaint before AGCOM (after completing a prescribed form), requesting an order against the Audiovisual Media Services Providers (MSP) to stop the broadcasting of the relevant programme whenever there is an infringement of copyright or of an allied right. The rights-holders should provide all the details and documentation of the alleged infringement. The complaint before the Authority cannot be filed if the rights-holder has already taken legal steps before a court of law. Within 7 days of receiving the complaint, AGCOM's Direction for media services (one of the six directions compositing the Authority[19]) might decide to:

1. Dismiss the case if:

   (a) the right-holder has not used the AGCOM compliant form or failed to provide key information on the alleged infringement of its right;
   (b) the complaint is related to a subject matter which is outside of the scope of the Regulation;
   (c) the right-holder has already filed a legal complaint before a court of law;
   (d) the MSP conforms voluntarily to the request of the rights-holder;
   (e) the complaint seems manifestly groundless.

2. Start a formal investigation by sending a notice to the MSP informing him on the details of the case (Article 7). Under Article 7, the MSP might then file a counter-complaint before AGCOM, explaining its defence within 5 days of the receipt of the AGCOM notice. When receiving the notice, the MSP can decide to voluntarily conform to the request of the rights-holder. Within 20 days from the sending of the notice the Board ends the investigation and suggests to the Board for services and products (one of the two boards composing AGCOM) the adoption of the appropriate measures towards the MSP. During the investigations, the direction for media services might require third parties to disclose information and documents which are relevant to the case, which must be provided within 5 days of the request. According to Article 8, the Board within 35 days shall issue its decision on the case:

   (a) It can dismiss the case for groundlessness of the complaint;
   (b) It may adopt one of the following measures:

      (i) if the MSP is a hosting provider and has servers within the national borders, the Board may issue an order requiring the removal of the infringing material; in case of violation of copyright, the MSP is required to disable access to the infringing digital works;
      (ii) if the MSP is a caching provider and has servers outside the national borders, the Board will issue an order requiring to disable the access to the website. In its first year of implementation, the regulation proved to be quite effective. The Authority received 209 notices, 55% of which has been closed following the spontaneous removal of the violating content. However, before entering into force, the regulation was challenged before the Regional Administrative Tribunal of Lazio, in Italian Tribunale amministrativo regionale (TAR) Lazio, by consumers and small business associations (Altroconsumo, Movimento di difesa del Cittadino, Assoprovider and Assintel) for violating freedom of expression, economic freedom and principle of proportionality. Following this challenge, the TAR decided on 26 September 2014 to refer a question

---

[19] For the organisation of the Authority, see the website www.agcom.it.

regarding its constitutionality to the Constitutional Court (www.federalismi.it Focus TMT—24 novembre 2014). More precisely, the TAR referral lamented:

1. the violation of the principles of statutory reserve and judicial protection which is provided in relation to the exercise of freedom of expression and economic initiative;
2. the violation of criteria of reasonableness and proportionality in the exercise of legislative discretion;
3. the violation of the principle of natural justice, because of the lack of legal guarantees and moreover the violation of judicial safeguards for the exercise of freedom of expression online, which should have been at least equivalent to those laid down for the press.

Moreover, the TAR questioned the constitutionality of the entire notice and take down system, and asked the Constitutional Court to review whether the EU Directive which served as the legal basis for the regulation, allowing the administrative enforcement of online copyright infringements, was constitutional.

The Constitutional Court will give the answer to the question referred to it, but this may take up to two years. Since the TAR did not suspend the regulation and the case is still pending, AGCOM, as stated in October by its President Cardani, shall proceed cautiously, addressing only cases of real urgency.

References

Alvanini, Sara. 2010. La responsabilità dei service providers. *Il Diritto Industriale* 4: 329–337.
Apa, Ernesto, and Federica De Santis. 2011. *Active or passive? Court of Rome rules on role and liability of ISPs.* www.internationallawoffice.com. Accessed 15 Dec 2011.
———. 2012. *Ancora sulla responsabilità del motore di ricerca: l'ordinanza del Tribunale di Firenze nel caso Meneghetti contro Google.* www.portolano.it/2012/07/ancora-sulla-responsabilita-del-motore-di-ricerca-lordinanza-del-tribunale-di-firenze-nel-caso-meneghetti-contro-google. Accessed 25 July 2012.
———. 2013. *ISP liability for user-uploaded content – an Italian perspective.* www.medialaws.eu. Accessed 30 May 2013.
Apa, Ernesto, and Oreste Pollicino. 2013. *Modeling internet service provides' liability: Google vs vividown.* Milan: Egea.
Barazza, Stefano. 2012. Secondary liability for IP infringement: converging patterns and approaches in comparative case law. *Journal of Intellectual Property Law & Practice* 7 (12): 879–889. doi:10.1093/jiplp/jps164.
Barbieri, Antonella, and Federica De Santis. 2011a. *Court rules on television programmes on video-sharing platforms.* www.internationallawoffice.com. Accessed 28 July 2011.
———. 2011b. *Protection of television programmes on video-sharing platforms.* www.internationallawoffice.com. Accessed 6 Oct 2011.
Bassini, Marco. 2012. Commercio elettronico e tutela dei segni distintivi. Responsabilità degli intermediari e trend giurisprudenziali. In *Tutela del Copyright e della privacy sul web: quid iuris?* ed. Andrea Maria Mazzaro and Oreste Pollicino, 45–86. Rome: Aracne.
Bellan, Alberto. 2012. Commento. *Il Diritto Industriale* 3: 253–261.

Bellezza, Marco. 2012. Privacy e diritto d'autore nell'era digitale: alla ricerca di un bilanciamento. In *La tutela dei dati personali in Italia 15 anni dopo. Tempo di bilanci e di bilanciamenti*, ed. Giuseppe Franco Ferrari, 93–114. Milan: Egea.

———. 2013a. AGCOM e diritto d'autore: un rapporto ancora difficile. Note a caldo sul nuovo regolamento #ddaonline. in www.medialaws.eu. Accessed 25 July 2013.

———. 2013b. *On the newly proposed Italian Regulation on Copyright Protection in Audiovisual Media Services*. www.medialaws.eu. Accessed 30 July 2013.

Bellezza, Marco, and Federica De Santis. 2013. *Google not Liable for Autocomplete and Related Searches Results, Italian Court Rules*. www.portolano.it/2013/04/google-not-liable-for-autocomplete-and-related-searches-results-italian-court-rules. Accessed 5 Apr 2013.

Bellezza, Marco, and Oreste Pollicino. 2012a. *Privacy versus diritto d'autore: tra bilanci e nuove sfide nell'era digitale*. www.medialaws.eu. Accessed 23 Jan 2012.

———. 2012b *Mercato dei servizi internet ed ambiti di possibile competenza dell'Autorità Garante della Concorrenza e del Mercato*. Riflessioni a margine del caso Private outlet. www.medialaws.eu. Accessed 3 August 2012.

Bellia, Marco, Gaia Anna Maria Bellomo, and Margherita Mazzoncini. 2012. Commento. *Il Diritto Industriale* 4: 346–362.

Bertolini, Elisa. 2010. La lotta al file sharing illegale e la "dottrina Sarkozy" nel quadro comparato: quali prospettive per libertà di espressione e privacy nella rete globale? *Diritto pubblico comparato ed europeo* 1: 74–106.

———. 2013. La nozione di indirizzo IP nel quadro della tutela della proprietà intellettuale in rete. In *Desafíos para los derechos de la persona ante el siglo XXI: Internet y nuevas tecnologías/Sfide per i diritti della persona dinanzi al XXI secolo: Internet e nuove tecnologie/Challenges of individual rights in the XXI century: The Internet and new technologies*, ed. Antonio Pérez Miras, Germán M. Teruel Lozano, Edoardo C. Raffiotta, 329–338. Pamplona: Editorial Aranzadi/Thomson Reuters.

Bocchini, Roberto. 2003. *La responsabilità civile degli intermediari del commercio elettronico*. Naples: Esi.

Bugiolacchi, Leonardo. 2005. La responsabilità dell'host provider alla luce del d.lgs. n. 70/2003: esegesi di una disciplina dimezzata. *Responsabilita Civile e Previdenza* 1: 188–209.

Camera, Guido, and Oreste Pollicino. 2010. *La legge è uguale anche sul web. Dietro le quinte del caso Google – Vivi Down*. Milan: Egea.

Caso, Roberto. 2007. Il conflitto tra copyright e privacy nelle reti Peer to Peer: in margine al caso Peppermint. Profili di diritto comparato. *Diritto dell'Internet* 5: 471–482.

Cassano, Giuseppe, and Iacopo Pietro Cimino. 2004a. Il nuovo regime di responsabilità dei providers: verso la creazione di un novello «consenso telematico»? Un primo commento agli artt. 14–17 del d.lgs. 70/2003. *Giur. It.:* 671–675.

———. 2004b. Il nuovo regime di responsabilità dei providers: verso la creazione di un novello «censore» telematico. *Contratti* 1: 88–96.

Costanzo, Pasquale. 2012. *Miti e realtà dell'accesso ad internet (una prospettiva costituzionalistica)*. www.giurcost.org. Accessed 12 June 2013.

———. 2013. La governance di Internet in Italia. In *Internet: regole e tutela dei diritti fondamentali*, ed. Elisa Bertolini, Valerio Lubello, and Oreste Pollicino, 41–58. Rome: Aracne.

Cuomo, Luigi. 2011. La Cassazione affonda la Baia dei Pirati. *Cassazione Penale* 3: 1102–1112.

De Cata, Marcello. 2008. Il caso «Peppermint». Ulteriori riflessioni anche alla luce del caso «Promusicae». *Rivista di Diritto Industriale* 2: 328–448.

———. 2010. *La responsabilità civile dell'internet service provider*. Milan: Giuffrè.

De Marco, Eugenio. 2008. *Accesso alla rete e uguaglianza digitale*. Milan: Giuffrè.

De Minico, Giovanna. 2010. I nuovi diritti e le reti. Verso nuove disuguaglianze? *Forum di Quaderni Costituzionali*. www.forumcostituzionale.it/site/images/stories/pdf/documenti_forum/paper/0251_deminico.pdf. Accessed 20 June 2013.

———. 2011. *Diritti Regole Internet*.www.costituzionalismo.it. Accessed 20 June 2013.

Di Ciommo, Francesco. 2003. *Evoluzione tecnologica e regole di responsabilità civile.* Naples: Esi.
Di Majo, Alessandro. 2012. La responsabilità del provider tra prevenzione e rimozione. *Corr. giur* 4: 553–560.
Duni, Giovanni. 2007. Voce. Amministrazione digitale. *Enc. dir. Annali I*: 13–49.
———. 2008. *L'amministrazione digitale, Il diritto amministrativo nella evoluzione telematica.* Milan: Giuffrè.
Flor, Roberto. 2007. La rilevanza penale dell'immissione abusiva in un sistema di reti telematiche di un'opera dell'ingegno: bene iudicat qui bene distinguit? *Dir. Informaz. e Informatica* 3: 557–578.
Franceschelli, Vincenzo. 2007. Musica in rete tra pirateria e uso personale (la libera circolazione delle idee in internet è cosa troppo seria per lasciarla al diritto penale). *Rivista di Diritto Industriale* 2: 77–92.
———. 2010. Sul controllo preventivo del contenuto dei video immessi in rete e i provider. A proposito del Caso Google/Vivi Down. *Rivista di Diritto Industriale* 2: 347–354.
———. 2012. Digital platforms in a competition law context. *Rivista di Diritto Industriale* 6: 289–300.
Frosini, Tommaso Edoardo. 2011. Il diritto costituzionale di accesso a Internet. *Munus. Rivista giuridica dei servizi pubblici* 1: 121–142.
———. 2013. L'accesso a Internet come diritto fondamentale. In *Internet: regole e tutela dei diritti fondamentali*, ed. Elisa Bertolini, Valerio Lubello, and Oreste Pollicino, 69–80. Rome: Aracne.
Ianni, Vincenzo. 2011. *La responsabilità in sede penale dell'internet service provider alla luce dei più recenti decisa giurisprudenziali.* www.neldiritto.it. Accessed 17 June 2013.
Ingrassia, Alex. 2012. Il ruolo dell'ISP nel ciberspazio: cittadino, controllore o tutore dell'ordine? *Diritto Penale Contemporaneo.* http://www.penalecontemporaneo.it/upload/1351711435II%20 ruolo%20del%20ISP%20nel%20cyberspazio%20DPC.pdf. Accessed 20 June 2013.
Gambuli, Marco. 2005. *La responsabilità penale del provider per i reati commessi in internet.* www.altalex.it. Accessed 2 May 2013.
Merla, Flaminia. 2010. Diffusione abusiva di opere in internet e sequestro preventivo del sito web: il caso "The Pirate Bay". *Dir. Informaz. e Informatica* 3: 448–462.
Musso, Alberto. 2010. La proprietà intellettuale nel futuro della responsabilità sulla rete: un regime speciale? *Dir. Informaz. e Informatica* 6: 795–828.
Nivarra, Luca. 2003. Voce: Responsabilità del provider. *Digesto IV - Discipline privatistiche, II, tomo 2*: 1195–1199.
Ortaglio, Eleonora, and Nicolò Zingales. 2011. Italy. http://www.ligue.org/uploads/documents/ rapportBitalie.pdf. Accessed 21 July 2013.
Osbourne, Dawn. 2008. User generated content (UGC): Trademark and copyright infringement issues. *Journal of Intellectual Property Law and Practice* 3: 555–562. doi:10.1093/jiplp/ jpn120.
Passaglia, Paolo. 2011. *Diritto di accesso ad internet e giustizia costituzionale. una (preliminare) indagine comparata.* www.giurcost.org. Accessed 20 July 2013.
Pisa, Roberto. 2010. *L'accesso a Internet: un nuovo diritto fondamentale?.* www.treccani.it/ Portale/sito/diritto/approfondimenti/2_Pisa_internet.html. Accessed 20 Dec 2011.
Pizzetti, Franco, ed. 2011. *I diritti nella "rete" della rete.* Turin: Giappichelli.
Riccio, Giovanni Maria. 2002. *La responsabilità degli internet provider.* Turin: Giappichelli.
———. 2003. La responsabilità degli internet providers nel d.lgs. n. 70/2003. *Danno e resp.* 12: 1157–1169.
Ricolfi, Marco. 2013. Contraffazione di marchio e responsabilità degli internet service providers. *Il Diritto Industriale* 3: 237–250.
Sammarco, Pieremilio. 2006. I diritti televisivi su manifestazioni sportive: natura giuridica e loro tutela dallo sfruttamento non autorizzato agevolato dalle tecniche informatiche. *Dir. Informaz. e Informatica* 6: 746–760.

————. 2011. *La posizione dell'intermediario tra l'estraneità ai contenuti trasmessi e l'effettiva conoscenza dell'illecito: un'analisi comparata tra Spagna, Francia e regolamentazione comunitaria*. Dir. Informaz. e Informatica 2: 285–292.

Scorza, Guido. 2007. Il conflitto tra copyright e privacy nelle reti Peer to Peer: in margine al caso Peppermint. Profili di diritto interno. *Diritto dell'Internet* 5: 461–470.

Siano, Manuela. 2011. La sentenza Scarlett della Corte di Giustizia: punti fermi e problemi aperti. In *I diritti nella "rete" della rete*, ed. Franco Pizzetti, 81–96. Turin: Giappichelli.

Smith, Emerald. 2011. Lord of the Files: International Secondary Liability for Internet Service Providers. *Washington and Lee Law Review* 68: 1555–1588.

Tosi, Emilio. 2003. *I problemi giuridici di Internet. Dall'E-Commerce all'E-Business*. Milan: Giuffrè.

————. 2012. La responsabilità civile per fatto illecito degli Internet Service Provider e dei motori di ricerca a margine dei recenti casi "Google Suggest" per errata programmazione del software di ricerca e "Yahoo! Italia" per "link" illecito in violazione dei diritti di proprietà intellettuale. *Rivista di Diritto Industriale* 1: 17–44.

Van Eecke, Patrick. 2011. Online service providers and liability: A plea for a balanced approach. *CML Review* 48: 1455–1502.

www.agcom.it
www.federalismi.it

Chapter 7
Secondary Liability of Service Providers in Brazil: The Effect of the Civil Rights Framework

Caitlin Sampaio Mulholland

Introduction

The Brazilian Civil Rights Framework (Law number 12.965, dated 23 April 2014) 'establishes principles, guarantees, rights and duties related to the use of the Internet in Brazil'. The project arose in 2009 and was approved in the House of Representatives on 25 March 2014 and in the Federal Senate on 23 April 2014, and immediately afterwards approved by President Dilma Rousseff. The legislation deals with themes such as network neutrality, protection of privacy, retention of data, and the social functions that the network must fulfill, in particular ensuring freedom of expression, preventing censorship and transmitting knowledge, in addition to imposing obligations on users and providers with regard to civil liability.

The Brazilian legislation expressly imposes secondary liability on an Internet Service Provider (ISP) for content generated by third parties in cases of non-compliance with a judicial notice requiring the ISP to take down unauthorised, abusive or illicit material. This measure was adopted after taking into account the general criteria for indirect liability in Brazil, as well as the approaches taken in other legal systems that share the same origins as the Brazilian legal system—that is, civil law jurisdictions. It operates on the premise that the provision of internet service should not include any guarantees regarding the content generated by third parties and that strict liability for damage resulting from such provision of services therefore could not be seen as a normal risk of doing business as an ISP.

C.S. Mulholland (✉)
Pontifical Catholic University (PUC-Rio), Rua Pereira da Silva, 121/602, Laranjeiras, Rio de Janeiro, Brazil
e-mail: caitlinsm@puc-rio.br

© Springer International Publishing AG 2017   171
G.B. Dinwoodie (ed.), *Secondary Liability of Internet Service Providers*,
Ius Comparatum – Global Studies in Comparative Law 25,
DOI 10.1007/978-3-319-55030-5_7

Nonetheless, the legislation does not ignore the damage that can result from the provision of Internet services. It also effectively makes ISPs responsible for the conduct of third parties in defined circumstances, as set out below.

In that sense, the Civil Rights Framework appears to be a law intended to establish procedures for the protection of Internet users in Brazil. According to Marcelo Thompson, the Civil Rights Framework 'inspires many of the fundamentals that it acknowledges for the Internet in Brazil, and is especially innovative in its use of a vast platform of collective deliberation to draw up its final text. Above all as regards its aspirations to guarantee what is understood as the rights of Brazilian citizens, it can be said that Civil Rights Framework is a fundamental charter, indeed a real Constitution for the Internet in Brazil' (Thompson 2012, 205).

This Chapter examines four of the elements of the Civil Rights Framework, namely, articles 18, 19, 20 and 21, and considers how those provisions affect the secondary liability of ISPs in Brazil.

The Concept of the Internet Service Provider

It is important to identify the situations where the Civil Rights Framework is applicable, or rather, to whom the law is addressed, or even more precisely, who will eventually be called to answer for the damages caused directly or indirectly to users of the Internet. Thus, the first issue to be addressed in this Chapter is the meaning of the term 'ISP'.

As Marcelo Leonardi sees it, 'the provider of Internet services is the *genus* of which all the other categories (backbone provider, Internet access provider, e-mail provider, hosting provider and content provider) are *species*. The Internet-services provider is the individual or corporate person who provides services related to the functioning of, or through, the Internet. The confusion is common because many of the chief providers of Internet services function as providers of information, content, hosting, access and e-mail' (Leonardi 2005, 21). Despite this notion of providing Internet services being widely accepted by doctrine and jurisprudence, the Civil Rights Framework makes no mention of it.

The Civil Rights Framework does contain the notion of a so-called 'autonomous-system administrator', to whom certain of its provisions are applicable.[1] Such a

---

[1] Despite the fact that there is no definition in the law that differentiates access and content providers, the courts have ruled that there is a fundamental difference between them in terms of civil liability. The former does not hold any responsibility for damages caused by third parties to others by using its provision. Mainly, this provider has the only purpose of giving others the architecture to access the Internet. On the other hand, content providers are those who manage different kinds of content, such as Facebook, Google, and Yahoo. They may be held responsible for damages caused by third parties if, after judicial notification, the ISP does not comply with a judicial order to take down potentially harmful content generated by third parties. This difference of concept can

person is defined as 'a physical or corporate person that administers specific Internet Protocol (IP) address blocks and the respective autonomous routing system, duly registered in the national body responsible for the registration and distribution of IP addresses geographically pertinent to the country' (article 5, IV).

Article 5 also does refer to two other more technically appropriate concepts, which perhaps are of greater relevance: (i) the concept of connection to the Internet, involving the setting-up of a terminal for sending and receiving packages of data over the Internet by means of attributing or authenticating an IP address (article 5, V); and (ii) the concept of Internet applications, being the set of functionalities that can be accessed by means of a terminal connected to the Internet (article 5, VII). The first concept has to do with the activity of providing a connection, the second with that of providing applications (content, search, hosting and email, for instance). That is to say, the rules of the Civil Rights Framework highlighted below apply (with some variation) in situations of civil liability for those whom the law calls *access* providers and *application* providers. The former is addressed in Article 18, while the latter is governed by Article 19.

Article 18: Exclusion of Liability of Access Providers for Content Generated by Third Parties

Article 18 of the Civil Rights Framework states that 'the party that provides connection to the Internet [an 'Internet access provider'] will not be made civilly responsible for damage caused by content generated by third parties'. An Internet access provider is the one who connects to a backbone provider through a good-quality line and supplies connectivity in his area of activity to other (usually smaller) providers, institutions and especially individual users through dedicated or even dialed lines ... being a retailer of Internet connectivity' (Leonardi 2005, 24).

The access provider allows an individual to have access to the Internet, but this is a service that is merely instrumental; in other words, a service that enables the use of other services that, for instance, offers applications whereby contents can be posted or generated.

One might say, then, that the risk assumed in the business of providing access to the Internet does not include the consequences of control over the contents generated by third parties. The fact that a third party posts improper content on the network does not—a priori and in the abstract—amount to negligent conduct on the part of the party that provides access (*culpa in vigilando*).

---

be identified in the way the Civil Rights Framework deals with civil liability of one or the other in its articles 18 and 19, as seen in sections "Article 18: Exclusion of Liability of Access Providers for Content Generated by Third Parties" and "Article 19: Secondary and Conditional Liability of the Application Provider" of this Chapter.

Article 19: Secondary and Conditional Liability of the Application Provider

Article 19 of the Civil Rights Framework states that 'in order to assure freedom of expression and impede censorship, the provider of Internet applications can only be held civilly responsible for damage resulting from content generated by third parties if, following a specific legal notice, the provider fails to take measures, within the scope and technical limitations of the service and the timeframe established [in such a court order], to render unavailable the content identified as infringing, except where otherwise established by legal provisions'.

This article emphasises the right to freedom of expression while mentioning at the same time impeding censorship, these matters being seen as rights to be guaranteed by the law.[2] Second, this norm derives from a contemporary juridical conception in terms of which the provider of applications should be held responsible only in exceptional circumstances—that is, when the conditions stipulated in article 19 are met.

The legislation undeniably takes a position in favor of free expression of ideas (and against censorship) by guaranteeing that ISPs will not be held responsible for merely including third-party content in their application, even though such content is a posteriori held to be illicit, abusive and an infringement of rights. This means that in the Brazilian legal system those who provide applications are not obliged to check beforehand and prevent content being posted by third parties (which would constitute censorship) because they will not subsequently be held responsible for any damage caused by this content. That is to say, liability for content generated, posted and/or disseminated over the Internet falls first and as a rule on the person who directly engages in the damaging conduct. This rule, nonetheless, makes a specific exception: providers will be held responsible, jointly with the direct perpetrator, if, after being judicially notified of improper content posted by third parties, they fail to take this down within the timeframe set by the court.

Thus, the following elements must be established for an ISP to be held legally responsible for content generated by a third party: (1) the existence of a request for legal notification made by a person who alleges that his or her rights have been violated; (2) judicial determination, albeit preliminary and provisional, of the potential harmfulness of the conduct of the person who has posted the content; (3) a preliminary decision notifying the application provider indicating the improper

---

[2]Article 5 of the Constitution of the Federative Republic of Brazil defines as fundamental rights, among others, freedom of expression of thought and intellectual, artistic, scientific and communications activities (article 5, IV and IX). The article further provides for inviolability of privacy, private life, and personal honor and image, as well as secrecy of correspondence and data via telegraphic and telephonic communications (article 5, X and XII). Also, article 220 provides that 'freedom of thought, creation, expression and information, in any form, process or vehicle will suffer no restriction, as provided for in this Constitution'; the second paragraph adds that 'any and all censorship is forbidden, be it of a political, ideological or artistic nature'.

content to be taken down and the lapse of time to do so; and (4) non-compliance by the ISP with the judicial order to take down content.

Only if these four requirements are satisfied can the ISP be held responsible. Even in these cases, it is possible to hold the person who post the improper content on the network directly responsible. This conclusion also leads us to the first major question as regards the application of article 19—that is, whether the ISP's civil liability is of a secondary nature, or shared with that of the person who directly caused the damage.

Secondary or Joint Liability?

Under Brazilian law, secondary liability applies under civil law whenever a person is responsible for damage caused directly by another person, resulting from a prior legal relationship of attribution of liability. Examples include employer-employee and parent-child relationships (art. 932).[3] The Brazilian legal system protects the victim of damage and allows him/her rapid reparation of the damage caused, by providing for joint rather than secondary liability[4], in accordance with the general rule provided in article 933 of the Civil Code, which expressly provides for a principle of solidarity.[5]

Not only can the person immediately responsible for the damage, causally speaking—that is, a minor-age child, a protected person, a ward, employee, a servant, etc.—be directly sued by the victim for compensation, but so can those indirectly responsible for the damage—that is, the parent, guardian, employer, etc. The legal suit can be brought against either or both parties, since joint tortfeasorship imposes joint and several liability. But where the indirectly liable party is sued, he or she has a right of recourse against the party directly responsible for the damage.

Further, under articles 932 and 933 of the Civil Code, guarantor and guardian liability is no-fault (so-called 'objective') liability, meaning he or she is obliged to

---

[3] Article 932 of the Brazilian Civil Code, which deals with indirect or secondary liability, states: 'the following are equally liable for civil reparation: I – parents, for under-age children under their authority and in their company; II – guardians and protectors, for the wards and protected ones in the same conditions; III – employers or proprietors, for their employees, servants and functionaries, while performing the work assigned to them or because of such; IV – the owners of hotels, hostels, houses or establishments which charge their guests, residents and pupils for accommodation, even if for the purpose of education; V – those who have at no charge taken part in the produce of the crime, up to the stated amount'.

[4] In Brazilian law, the term "secondary liability" means that the victim has to primarily seek reparation from the direct perpetrator of the damage. Where unable to obtain such restitution, he or she can then seek reparation from the person indirectly responsible for the direct perpetrator, such as the employer in the case of a damage caused by the employee.

[5] Article 933 provides that 'the persons indicated in sub-items I to V of the preceding article, even if there is no fault on their part, will be responsible for acts practiced by third parties referred to therein.'

compensate for the damage regardless of any fault on his part, even in cases where he or she took all reasonable procedures to prevent harm.

Similarly, the Brazilian Copyright Act (Law number 9.610/98) embraces joint and several liability as it establishes in article 104 that "whoever sells, exposes to selling, conceals, acquires, distributes, has in deposit or uses artistic works or sound recordings that are reproduced fraudulently, with the finality to sell, make a profit or advantage, benefit, direct or indirect gain, for him or others, will be held jointly liable with the counterfeiter". In the same way, article 110 establishes that "for copyright infringement in spectacles and public hearings, conducted in places or establishments referred to in art. 68 [theatrical works, musical or literary-musical compositions and sound recordings], its owners, directors, managers, business owners and renters shall be jointly liable with the organizers of the shows".

But how would this theory of indirect civil liability apply—if at all—in the case of ISPs, in respect of the content generated by a third party? Generally speaking, in order for there to be civil liability on the part of ISPs—regardless of the sort of legal relation established—there has to be damage, an attribution factor (fault or risk) and causality. Indirect or accessory liability imposes liability to compensate for the damage on the party who controls the activity, medium or instrument that caused direct damage to the person. And courts have extended liability to those who profit from infringing activity when an enterprise has the right and ability to prevent the infringement. A common example is when an employer answers for the damage caused directly by the employee since he has the control of the means used by the latter while the work is being carried out. Similarly, the owner of an animal is liable for the damage caused directly by it, since he has control over the animal and is therefore able to prevent it from causing and harm.

From the indirect nature of the provider's responsibility, one may conclude that his obligation to compensate would be joint with that of the person who is the direct cause of the damage. That is how the Brazilian courts proceeded before the enactment of the Civil Rights Framework. In 2012, the Superior Court of Justice held that 'when notified that a certain text or image possesses illicit content, the provider must act energetically to take the material down from the air immediately, under penalty of *answering jointly* with the direct author of the damage, by virtue of the omission practiced' (Supreme Court of Justice, Special Appeal 1,308,830/Rio Grande do Sul, Rapporteur Minister Nancy Andrighi, Third Panel, decided on 08/05/2012).[6]

This approach is advanced by Marcel Leonardi, for whom 'civil liability for the acts of users and third parties is based on a system that attributes secondary liability to providers in the case of willful misconduct or negligence, when they fail to fulfill their duties (thereby making it impossible to identify the person responsible for the illicit act) or else when they collaborate in the practice or neglect to block access to

---

[6] See also Supreme Court of Justice, Special Appeal 1.186.616/Minas Gerais, Third Panel, Rapporteur Minister Nancy Andrighi, decided on 31.08.2011. Similarly, see Special Appeal 1.193.764/SP, Third Panel, Rapporteur Minister Nancy Andrighi, decided on 08.08.2011.

illegal information after being notified of the existence of same' (Leonardi 2005, 49-50).

Under the Civil Rights Framework, the ISP will be liable for damage caused by his omission to take down the material following a legal notice requiring it to do so, even when the illicit or abusive content generated is not causally connected to the direct conduct of the ISP. Thus, it seems that the Framework has attempted to place secondary and joint responsibility on the provider for a misdeed committed by a third party. This means that, unlike article 933 of the Civil Code, which establishes objective indirect liability of the guarantor/guardian, the Civil Rights Framework embraces indirect civil liability based on fault—negligent omission to remove infringing content generated by a third party after legal notification—which also entails joint liability of the Internet provider for damage caused directly to the victim by a third party.

Liability Based on the Risk of the Provider's Activity or on the Basis of Presumed Fault?

ISPs provide the means for third parties to act in a way that causes damage to others. If the ISP can control the means and so prevent any damage, it must assume the responsibility for any damage that ensues. The nature of this responsibility is in principle objective, based on the theory of risk—that is, on the concept that whoever has an instrument available that could potentially infringe upon the rights of others should be responsible for any resulting harm. In other words, those who have greater capacity to prevent damage should assume the responsibility for the consequences. The ISP, however, has a right to reimbursement from the person who directly caused the damage.

The Brazilian system considers as co-existing grounds for civil liability on the part of the service provider contributory acts such as risk control (risk theory) and presumed fault in neglecting to take precautionary measures to lower the risk of infringement. This will depend on the type of control that the Internet-services provider has over the content previously made available, or the legal relationship that exists between provider and third party.

Accordingly, the provider of the connection answers civilly for the damage caused to the user by poor provision or undue interruption of service. Likewise, the provider of content is liable for damage caused to a person for defamatory material published on his site, if he acts as editor of the site, with previous control of the material to be posted.

On the other hand, the Internet service provider is responsible for fault based upon content generated by third parties when the provider is legally notified about the illegal content and and fails to take it down within the appointed timeframe; we refer to this hypothesis herein.

Thus, the higher courts have held that where an ISP is unable to foresee or control the risks of the activity, the theory of assumed risk does not provide a basis for the imposition of civil liability for content that is generated by third parties. This is based on the fact that ISPs do not possess the capacity to predict (and consequently prevent) the risks of damage caused by third parties—because there is no prior monitoring or selection of what is posted on the network—and because it is important to protect freedom of expression and avoid private or public censorship.

Erica B. Barbagalo argues that 'the provider of hosting services is not responsible for the content of the sites he hosts, since he does not intervene in their content, not having the editorial control of the electronic pages. Nor can the hosting provider be expected to carry out inspection activities: in most cases, the host has no access to the content of the site, which is only allowed to the owner, who can change the content of his pages as often as he wants. Furthermore, there are many pages and sites hosted in each server, which makes it impossible for the hosting provider to inspect the content' (Barbagalo 2003, 347).

Barbagalo also says that adopting the theory that the ISP has assumed the risk of infringing content would be a mistake. She argues that 'the activities carried out by the provider of Internet services are not by nature [risky] activities, they entail no greater risks to the rights of third parties than the risks of any commercial activity. And to interpret the norm as meaning that any damage must be compensated, regardless of the fault element, by the mere fact that an activity is carried out, would definitively burden those who regularly engage in productive activities, and consequently hamper development' (Barbagalo 2003, 360).

Thus, the basis of the civil liability of ISPs in respect of content generated by third parties is fault, with specific regard to the omission factor and by presumption, when the provider, after being legally notified, fails to take the necessary measures to take down inappropriate material from his network.

Legal Notification as a Formal Requirement to Establish Responsibility of the Internet Provider

Article 19 settled the controversy concerning the nature of the notice necessary to compel an ISP to remove infringing content. Prior to the introduction of the Civil Rights Framework, the courts had considered that an extra-judicial notification informing the ISP of an infringement of rights and requiring takedown of the material before a certain date, would suffice to define the provider's secondary liability if such notification were ignored.

The 'notice and take-down' remedy usually operated without judicial oversight, via extra-judicial notification; courts did not carry out an analysis of reasonableness of the remedy or even consider the existence of a right having been violated. The following two extracts are representative of many Superior Court of Justice decisions, insofar as their approach to notice and take-down remedy prior to the Civil Rights Framework was concerned:

> In this hypothesis, the decision made expressly states that the provider of Internet services was *notified extra-judicially* as to the creation of a false, defaming profile of the alleged holder, did not take the appropriate measures but rather opted to remain inert, which is the reason that he has been held jointly liable for the moral damage inflicted on the prosecuting party, thus configuring the subjective liability of the accused' (Supreme Court of Justice, Special Appeal under Specific Court Regulations 1402104/Rio de Janeiro, Rapporteur Minister Raul Araujo, Fourth Panel, decided on 27/05/2014).

> The decision reached expressly provides that the provider was served extra-judicial notice by means of an instrument that he himself makes available for denouncing abuses—for instance, creating a defamatory false profile of the supposed holder, this proving offensive to third parties—and failed to take the proper steps, preferring rather to remain inert, for which reason he became jointly liable for the moral damage inflicted on the plaintiff, thereby configuring subjective responsibility of the accused (Special Appeal under Specific Court Regulations no. 1396963/Rio Grande do Sul, Rapporteur, Minister Raul Araapp, Fourth Panel, decided on 08/05/2014).

The 'notice and take-down' system, however, is flawed because it allows arbitrary removal of content based on a simple complaint made by the interested person, without the necessary due process of law. Furthermore, it is a system that condones censorship, temporary or permanent, or else intimidates or restricts freedom of expression. The absence of judicial oversight may lead to an abusive exercise of rights. First, it might permit, by means of a simple notification, a restriction of freedom of expression. Second, it could result in undue or unjustified censorship (Leonardi, http://leonardi.adv.br/2010/04/o-problema-do-sistema-de-notificacao-e-retirada-na-web/).[7]

Imagine the hypothesis of a person who, in a blog maintained by an ISP (*blogspot*, for instance), writes a theatre review of a play that is currently being shown, using somewhat harsh words against the main actress, or even draws a not very favourable cartoon of the actress. The actress, perceiving the text or drawing as a violation of her dignity, could legally request the provider to remove the allegedly defamatory material, which in accordance with the above approach should be carried out under penalty of holding the provider jointly responsible. This example illustrates the dangers of censorship and the possible restriction of the freedom of expression (Bezerra 2013, 9).[8]

---

[7] As Marcel Leonardi puts it, 'the need for judicial analysis and specific notification to take down content cannot be ignored, since deciding on the legality or illegality of the material—in all its possible forms—is something necessarily subjective, besides being the exclusive prerogative of the Judiciary, not of the users or the providers. Jurisprudence is actually moving in this direction, with different decisions emphasizing that this is a role reserved for the State, and it cannot be usurped by the intermediaries or the users'. (Leonardi 2010)

[8] Bezerra offers the following opinion: 'It is undeniable that for the actual provider of content to take down contents from the air – as the STJ imposes – entails a discretionary judgment of the nature of the divulged data. Moreover, by suspending certain content, the provider will be making a comparison between the fundamental right to privacy on the part of the person who feels offended and the fundamental right to freedom of expression on the part of the person who spread

With the introduction by the Civil Rights Framework of a requirement of judicial notice before an ISP can be held legally responsible, the ISP will feel more confident that it will not be held responsible without prior judicial oversight. In order to grant the order, the judge must consider the following questions: (1) the existence of *fumus boni iuris* ('likelihood of success on the merits'); (2) *periculum in mora* ('danger in delay'), that is, whether maintaining the alleged violation will create a situation difficult to correct or else make the damage all the worse. Should these two requirements be satisfied, the judge will order the ISP to remove the relevant content.

In accordance with the first paragraph of article 19, the legal notice must contain, under penalty of invalidity, clear and specific identification of the content alleged to be infringing in order to enable unmistakable location of the material. The best interpretation to be given to this paragraph is that the order to take down the content must indicate by URL (Uniform Resource Locator) the exact rights-infringing material so that entire web-sites are not taken down or applications blocked, which of course would impede other users from accessing the platform.[9]

Legal Procedure

As regards legal procedure, the Civil Rights Framework provides that lawsuits concerning compensation for damages, resulting from content made available over the Internet, that are related to honour, reputation or personality and privacy rights, as well as such content being made unavailable by ISPs, can be presented before special courts. This is because there is often a need to speed up the process, since lawsuits that begin in special courts are usually quicker than those that begin in common civil courts. However, that there are two possible difficulties with this approach. First, technical proof by a skilled professional is not allowed in these special courts. Second, there is a limit to the value of lawsuits in such courts. These two restrictions may mean that the ordinary process is preferable. Since the plaintiff has the right to

---

the information. It is therefore no exaggeration to state that the understanding of the STJ eventually transfers to the providers competence and responsibility that can only be accorded to the Judiciary Power itself'. (Bezerra 2013).

[9] Following a preliminary judicial decision in the Daniela Cicarelli case, YouTube was blocked in Brazil in January 2007. The actress had been filmed having sexual relations on a beach in Spain with her boyfriend at the time, Tato Malzoni. The film was distributed over the Internet and the persons filmed filed a successful suit in the Brazilian courts, asking that all the sites that carried the video should take down the images. Sites whose content is controlled by editors complied with the notice. But in the case of YouTube, every time Google removed the video, some user posted the film again. In January 2007, Judge Ênio Santarelli Zuliani demanded that the telephone companies block access to YouTube in Brazil, supposedly for not obeying the legal requirement. The video site remained off the air for 48 h. Realizing the repercussions of the case and under pressure from critics claiming that he had acted like a censor, Ênio changed his decision and freed up access to YouTube. Source: http://info.abril.com.br/noticias/Internet/google-vence-no-caso-cicarelli-10052012-57.shl.

choose which route to follow, there is no disadvantage for the plaintiff in adopting the procedural strategy considered to be the most convenient.

Exclusion of Application of Article 19 to Copyright

Article 19 does not apply to cases concerning infringement of copyright or related rights, as provided in the second paragraph, which expressly states that 'application of the provision in this article to infractions of copyright or related rights depends on a specific legal provision, which must respect freedom of expression and other guarantees set forth in article 5 of the Federal Constitution'. This means that the liability of ISPs for copyright infringement perpetrated by their users will continue to be governed by case law.

The National Congress is discussing a Bill to change substantially the Copyright Law in force (L. 9.610/98) and to reform secondary liability of ISPs as regards the violation of copyright. As far as protecting copyright is concerned, Brazilian jurists, and more specifically those with government appointments (the Commission for Culture in the House of Representatives), are engaged in debating the adoption of two instruments designed to control content generated by third parties on the Internet: (i) 'notice and notice', in terms of which the holder of the violated copyright uses the prerogative of notifying (extra-judicially or judicially) the ISP to alert it to the fact that a third party is using its medium to make available content of which it is not the right-holder; following this notification, the ISP must notify the third party about the allegation; and (ii) equal remuneration to be granted to the holder of the violated copyright for each improper use that is made.

The second paragraph of article 19 of the Civil Rights Framework states that the civil liability of the ISP for infringement of copyright by a third party should be addressed in another specific law. Thus, until a proper legislative reform is made to adopt new measures to protect copyright online, what is currently available is the use of extra-judicial notification to remove infringing material.

The Brazilian Federal Supreme Court has proposed a public consultation on the collective management of copyrights involving Brazilian artists that are available on the Internet. According to Marcos Souza, director of Copyright Rights for the Ministry of Culture, the guarantee of copyright on the Internet is still a very obscure question and there is a growing pressure on governments around the world, for this to be regulated. As the Civil Rights Framework is already in place in Brazil, the idea of the Ministry is to discuss the digital environment as a whole.[10]

---

[10] In: http://www.ebc.com.br/cultura/2015/07/governo-fara-consulta-publica-sobre-cobranca-de-direitos-autorais-na-internet (09/25/2015).

Article 20 and the Obligation of Users Responsible for Offending Content to Inform the Reasons for Its Unavailability

Article 20 of the Civil Rights Framework states that 'whenever the provider of Internet applications has information concerning the user directly responsible for content referred to in article 19, he must inform the user of the reasons and provide information as to the content being made unavailable, together with information that allows justification and full legal defense, except where there is an express adversary legal provision or well-founded express adversary judicial decision'.

Once the ISP has been served legal notice about the infringing content, and has complied with the notice by removing from the network the material unduly posted by a third party, he is obliged to inform the infringing user—if he can identify him/her—of the reasons for, and other information on, making the material unavailable. This procedure has two main purposes: (a) to promote due legal process by providing the user of the network with information that allows him to defend himself in the event of any subsequent lawsuit; and (b) to exonerate the ISP from any liability vis-à-vis the user.

The user who distributed the removed material may, on the other hand, demand that the ISP, who exercises this activity professionally in an organised fashion and for economic reasons, substitute the content removed with the reason and the court order that led to its being made unavailable (Article 20, sole paragraph). This is to assure the user of the guarantee of free expression of ideas and at the same time to inhibit censorship, showing as it does that the material was removed by legal order and not by private censorship on the part of the provider.

Article 21 and the Secondary Civil Liability of the Internet Provider for Infringement of Privacy

Article 21 of the Civil Rights Framework sets a different procedure from article 19, inasmuch as it refers to a notification to remove infringing material generated by a third party. This article concerns the protection of the privacy of people portrayed in images, videos or other material containing scenes of nudity or sexual acts of a private nature, shown without permission on the network. The legislator understood that the unauthorised distribution of this potentially seriously damaging content dispenses with the need for legal notification to oblige the provider to remove the content from the network. To be more agile, an extrajudicial notification should be sufficient to make the content unavailable. Furthermore, the notification should contain 'under penalty of invalidity', 'elements that enable specific identification of the material alleged to infringe the privacy of the party concerned and verification of the legitimacy to present the petition' (Article 21, sole paragraph).

Exclusionary Elements of Liability: Prerequisites for the Provider's Non-compliance with a Court Notice to Remove Content

Article 21 requires that with extrajudicial notice the alleged victim should specifically identify the material held to be offensive so that the ISP can remove it. Article 19 also states that the court notice should offer clear and specific identification of the content alleged to be of an infringing nature so as to allow for unmistakable location of the material on the network.

Nevertheless, in article 19, unlike article 21, there is a provision for an exclusionary element that permits non-compliance with a court notice that calls for removal of illicit material based upon the technical limits of the ISP. If, given the state of technical development at the time, the ISP does not possess the technical capacity and the objective conditions to identify clearly and specifically the content held to be offensive, then the ISP can evade any liability for the undue content to continue on the network.

This article has faced much criticism, on the basis that it conflicts with the fact that the consumer (user or victim) is himself likely to be unable to identify the infringing content; and it is said also to conflict with the principle of enterprise risk, which would permit holding responsible the person who exercises the providing service and has the greater technical capacity to control it.

Conclusion

In conclusion, the above can be summarized as follows:

1. The Civil Rights Framework represents an important legislative step forward in Brazil, integrating as it does in a single body of law norms related to the juridical regime of the Internet and bringing more security as to how the norms are applied.
2. ISPs are in principle exempt from liability for offensive content generated by third parties who use their services.
3. The basis for excluding the ISP from responsibility is the non-existence of the duty to previously monitor the content to be generated or posted by third parties who use the service provided.
4. ISPs are not guarantors of the contents posted by third parties (as, for example, editors are), precisely because they are under no obligation to previously control or monitor such content.

5. The liability of ISPs for the abusive or illicit content generated by third parties can only be admitted in the circumstance of the provider being legally notified as to the abuse or illicit nature of the material and remaining inert and refusing to remove the content.

6. Legal notification is a formal requirement for holding the ISP responsible. Extra-judicial notification can only be admitted in the circumstance provided in article 21, that is, content that violates the privacy of a person portrayed without his or her permission.

7. The civil liability of the Internet-services provider is subjective in nature: the provider is responsible for culpable omission when, after being legally notified, he refuses to take down the infringing content posted by third parties.

8. The provider is also jointly responsible with the direct causer of the damage, based on the principle of social solidarity and non-compliance with a legal order (or extra-judicial notice, in the case of violation of privacy).

Acknowledgments The author would like to thank Eduardo Magrani for his work in the revision of this paper.

References

Barbagalo, Erica Brandini. 2003. Aspectos da Responsabilidade Civil dos Provedores de Serviços na Internet. In *Conflitos sobre Nomes de Domínio*, eds, Ronaldo Lemos e Ivo Waisberg. São Paulo, Revista dos Tribunais, p 341 and 363.

Bezerra, Márcia Fernandes. Apontamentos sobre o Marco Civil da Internet. In JICEX – Revista da Jornada de Iniciação Científica e de Extensão Universitária do Curso de Direito das Faculdades Integradas Santa Cruz de Curitiba, 1(1), 2013.

Leonardi, Marcel. 2005. *Responsabilidade civil dos provedores de serviços de internet*. São Paulo: Editora Juarez de Oliveira.

————. 2010. O problema do sistema de notificação e retirada na web. Conjur. Abril.

Rodrigues Junior, Otavio Luiz. Marco Civil e opção do legislador pelas liberdades comunicativas, In   http://www.conjur.com.br/2014-mai-14/direito-comparado-marco-civil-opcao-pelas-liberdades-comunicativas.

Thompson, Marcelo. 2012. Marco Civil ou Demarcação de Direitos? Democracia, Razoabilidade e as Fendas na Internet do Brasil (Civil Rights Framework or Demarcation of Rights? Democracy, Reasonableness and the Cracks on the Brazilian Internet) (May 28, 2012). *Revista de Direito Administrativo* (Journal of Administrative Law), 261, 2012, 203–251.

Chapter 8
Internet Commerce and Law

Katja Lindroos

Introduction

A lively debate has taken place in Finland regarding online copyright piracy and copyright infringement in the digital setting. Since 2010, Parliament has seen two bills, one withdrawn draft and one Citizen's Initiative, regarding liability of internet service providers (ISPs) for illegal content. The opposing views expressed have led to an increased awareness of the constitutional aspects of copyright law. Copyright owners still have a strong right, but it is not unlimited, and the rights of others, especially ISPs, may outweigh the interests of effective enforcement. The *Finnreactor*-case in 2010 and *Pirate Bay*-cases from 2013, both involving ISP liability, have further fuelled the debate, as the courts have now weighed in. The cases and proposals are discussed chronologically to show how the duties placed on ISPs are formed and understood in Finland.

Finnish Internet Law

Legislation for Telecommunications, Electronic Networks and ECHR Principles

An ambitious legislative reform, concluded in 2014, collected the existing regulation regarding the Finnish communications infrastructure, ISPs and the provision of various internet services into one law, the Information Society Code.[1]

---

[1] Law 917/2014.

K. Lindroos (✉)
UEF Law School, University of Eastern Finland, Joensuu, Finland
e-mail: katja.lindroos@uef.fi

© Springer International Publishing AG 2017
G.B. Dinwoodie (ed.), *Secondary Liability of Internet Service Providers*,
Ius Comparatum – Global Studies in Comparative Law 25,
DOI 10.1007/978-3-319-55030-5_8

The aim was to update the legislative landscape of electronic communications in its entirety. The Code includes guarantees of citizen access to communications services in terms of coverage, availability and quality of services. It includes 45 Chapters and over 300 provisions, which are divided into 13 parts. This massive legislative package was built on existing legislation, such as the Television and Radio Services Act, the E-Commerce Act,[2] the Communications Market Act[3] and the Electronic Communications Data Protection Act. Entire acts were included as chapters or as parts of the Information Society Code. The Information Society Code entered into force on 1 January 2015, thus repealing and replacing the old legislation. The sections relating to online copyright infringement were delayed in Parliament, but entered into force on 1 June 2015.[4] In addition to issues of ISP liability, the revised Copyright Act includes a provision releasing TV cloud storage services from copyright liability or compensation requirements.[5]

The former E-Commerce Act[6] was incorporated in full, as provisions 173–194 in Chapter 22 of the Information Society Code. Chapter 22 regulates ISP liability as a general matter. However, the INFOSOC Directive, which only slightly predated the E-Commerce Directive, has specific provisions relating to copyright. Both these directives were implemented in Finland by way of the E-Commerce Act and with changes to the various intellectual property acts.

Chapter 22 of the Information Society Code defines when *information society services* are exempt from liability for mere conduit, caching and hosting, regardless of cause of action.[7] It also lays down a separate notification procedure applicable to copyright. Chapter 22 of the Information Society Code goes further than the E-Commerce Directive in respect of two areas. A hosting service provider that knows that there is illegal material on its service must remove the content to avail itself of the liability exemption.[8] When the material includes pornographic images of children, violence or mixing with animals or hate speech, the service provider will be exempt from liability if it removes the material on its own initiative upon obtaining actual knowledge of this content on its platform. Chapter 22 of the Information Society Code does not apply to issues relating to privacy and data

---

[2] Law 607/2015.

[3] Law 393/2003.

[4] Law 607/2015.

[5] Section 25 l of the Copyright Act.

[6] Law 458/2002.

[7] Government Bill HE 194/2001.

[8] Government Bill HE 194/2001 states, in relation to the hosting safe harbor, that the requisite knowledge is specific, i.e. the provision requires knowledge of the infringing material, its location and its clearly illegal nature and the knowledge must be received through a notice as specified in the Act or court order.

protection. These are separately regulated in Chapters 17–20 of the Information Society Code.

Chapter 1 of the Information Society Code contains the same definition for information society services as the E-Commerce Directive. Information society services are services provided:

1. as distance services, i.e. without parties being present at the same time;
2. electronically, i.e. by sending and receiving services via devices handling information electronically or via storage of information so that only cables, a radio connection, optical equipment or other electro-magnetic equipment are used for sending, transmitting and receiving service;
3. as data transfers requested personally by recipients of services; and
4. usually against a payment.[9]

A service provider is a natural or legal person that provides an information society service. The preparatory legislative material to the Finnish legislation mentions search engines, information services, electronic publications and entertainment services as falling within the definition of information society services, despite not ordinarily being provided against payment. These services are not, however, specifically mentioned in relation to the safe harbours for mere conduit, caching and hosting. This does not necessarily preclude application of those safe harbours, but is merely indicative of the issue not having been specifically discussed, as in other countries.[10]

The Information Society Code addresses the specific obligations of service providers to cooperate with authorities. It also sets down the procedure and requirements for authorisation when the trade in question requires prior authorisation.[11] Authorization may be required to establish operations in Finland, since the number of actors on the market may be regulated. Depending on the activity an internet service provider may need a permit or simply file a notification to authorities. Once authorization is issued transactions do not require authorization. Any natural person resident in the European Economic Area, as well as a Finnish company or foundation, or foreign company or foundation established in the European Economic Area, can engage in commerce in Finland, provided that it is legal and in accordance with honest practices. Services requiring a permit include engaging in teleoperations in the communications market. This includes internet services, communications services and markets for the provision of services related to these services.

---

[9] Information Society Code Ch. 1, Sec. 3.1; 29)

[10] Commission Report on E-Commerce Directive at 74–75.

[11] Information Society Code Ch. 2–6.

The general provisions of the Information Society Code do not regulate the content of messages that are being transmitted. Thus, while more than 30 categories of services are included by definition,[12] that definition does not define the area of application of the rules. Instead, this section of the Code regulates the communications market in which the different services operate.

Intellectual property rights legislation regulates the removal of content that has been deemed infringing by a court.[13] That legislation does not use the definitions from the E-Commerce Directive or Information Society Code. Instead, it applies to "any maintainer of a transmitter, server or other device, or any other service provider acting as an intermediary" in the provisions empowering courts to order ISPs to give information or stop large scale infringement. These provisions were introduced when implementing the Enforcement Directive.[14] Enforcement of intellectual property rights may be a civil or a criminal matter. These cases are dealt with by general courts in summary (civil) proceedings or in criminal proceedings (if criminal charges are brought against the service provider). Examples of both are discussed below. It should be noted that Sections 174 and 175 of the Information Society Code, which specifically provide for freedom to provide information society services, do not apply to copyrights and related rights, industrial property rights and rights relating to semiconductor chips.[15]

The standards and procedures for establishing specific liability have been created by sector-specific statutes. Obligations to provide information to customers or police/government related to the regulation of services are enforced by the respective authority (i.e. the Competition and Consumer Authority or the Communications Authority). The administrative agency may issue a specific order on pain of fine, if the company does not voluntarily correct behavior that has been deemed contrary to its obligations under statute. An extreme sanction is prohibiting an ISP, such as a telecompany, from partially or completely engaging in commerce and/or withdrawing its license to operate. This sanction is available only as a last resort and for gross violations. The Authority can investigate cases on its own accord or on the basis of a complaint for violation of specific obligations under the statute filed by an aggrieved party. Cases are primarily settled after mutual correspondence and rarely go as far as formal proceedings to impose sanctions.

---

[12] Information Society Code Ch. 1 Sec. 3.

[13] Section 60c of the Copyright Act, Section § 48a of the Trademark Act, Section 57b of the Patent Act, Section 18a of the Trade Name Act, Section 34 of the Plant Breeder's Rights Act, Section 35a of the Design Act, Section 36a of the Utility Model Act and Section 37b of the Act on Integrated Circuits.

[14] Directive 2004/48/EC.

[15] See Directive 2001/29/EC of the European Parliament and of the Council of 22 May 2001 on the harmonisation of certain aspects of copyright and related rights in the information society, Official Journal L 167, 22/06/2001 P. 0010–0019. (InfoSoc Directive), Recital 16.

Safe Harbours and Liability in Practice

As a general rule, liability for damage to third parties is based on statutory liability or on a fault-based standard of negligence. Consequently, there seems to be a presumption against general liability, even in instances where some liability exists. The Finnish Supreme Court routinely starts its reasoning by stating that the risk for damage lies with the owner. In internet cases the language used by courts is different, even the complete opposite, with courts stating that the intensity of the harm establishes or extends liability, regardless of fault or other interests.[16] The right-holder's interest in preventing large scale copyright piracy is equated with society's interest. This may be because other interests, such as that of users, consumers or ISPs, are rarely addressed in legislation. The Information Society Code is an attempt at remedying this imbalance.

Greater remedies are available against direct infringers of intellectual property rights than against no-fault intermediaries. If ISPs engage in criminal activity they will be treated as direct infringers.[17] Under law ISPs can be subject to a duty to act in the society's interest without themselves having done anything wrong.[18] Since ISPs are public companies they are also not likely to avoid enforcement, like many direct infringers that can remain anonymous. Non-infringing intermediaries can be ordered to provide contact information of the content provider or to discontinue making available infringing content temporarily or permanently. However, the duty to act, which can be imposed on an intermediary ISP, is in practice more effective in curtailing large scale infringement. Thus, rights-holders have put pressure on courts to interpret the obligations in Section 60 a–c of the Copyright Act broadly. Section 60 a–c of the Copyright Act regulates when a copyright owner can bring suit against or retain a preliminary or permanent injunction against a direct infringer or an intermediary (ISP or other actor) for copyright infringement.

The available remedies against direct infringers and ISP intermediaries are separate and not tied to each other. However, there may be statutory provisions requiring the right-holder to initiate proceedings against the direct infringer within a certain period of time in order to bring a case against an intermediary or to maintain the injunction in force.[19]

Finnish law is more detailed than the E-Commerce Directive when it comes to obligations and safe harbours for ISPs. This is due to a tradition of public control and strongly regulated markets that are only slowly being relaxed through de-regulation. The Information Society Code continues de-regulation and seeks to provide more freedom of competition and provision of services than was previously the case, with there now being less government interference with market forces.

---

[16] As in the Finnreactor and Pirate Bay, which are discussed below.

[17] KKO 2010:47.

[18] HE 181/2014 at 32.

[19] The Pirate Bay case discussed below.

The safe harbour for an ISP that is a *mere conduit* applies when the service provider does not initiate the transaction, selects the recipient of the transaction, or selects or modifies the transferred data. The safe harbour applies to automatic, intermediate and transient storage insofar as this takes place for the sole purpose of carrying out the transmission in the communication network and provided that the information is not stored for any period longer than is reasonably necessary for the transmission.

The safe harbour for *caching* applies to automatic, intermediate and temporary storage of information for the sole purpose of making more efficient the information's onward transmission, provided the service provider:

1. does not modify the information;
2. complies with the conditions on access to the information;
3. complies with rules regarding the updating of the information, specified in a manner widely recognised and used by industry;
4. does not interfere with the lawful use of technology, widely recognised and used by industry, to obtain data on the use of the information; and
5. acts expeditiously to remove or to disable access to the information it has stored, upon obtaining actual knowledge of the fact that the information at the initial source of the transmission has been removed from the network, or access to it has been disabled, or that a court or an administrative authority has ordered such removal or disablement.

The safe harbour for *hosting* applies to the storage of information provided by a recipient of the service, who is responsible for the content. The ISP is not liable for the information stored or transmitted at the request of a recipient of the service, if it acts expeditiously to disable access to the information stored: (1) upon obtaining knowledge of an order concerning it by a court or if it concerns a violation of copyright upon obtaining the notification; (2) upon otherwise obtaining actual knowledge of the fact that the stored information is clearly hate speech,[20] or a pornographic image depicting a child, violence or mixing with animals.[21] The liability safe harbour does not apply if the content producer is acting under the authority or the control of the ISP.[22]

Finnish legislation imposes an obligation on the hosting service provider to provide a notification procedure that applies only to copyrighted material. In particular, Section 192 of the Information Society Code requires the right-holder to contact the content provider (direct infringer) before it can target the hosting service provider, unless the content provider cannot be identified or does not remove the material without delay after being asked to do so. The hosting service provider must provide information about where the notice may be sent, and this information must be available readily and continuously on a website or in the phone book.

---

[20] Contrary to Section 8 of Chapter 10 of the Penal Code.

[21] Contrary to Section 18 of Chapter 17 of the Penal Code.

[22] KKO 2010:47 and 48.

Section 191 of the Information Society Code stipulates the requirements of the form and content of the notice. The notice must be in writing, either in paper or electronic form, so that its content may not be changed, and it must include the name and contact information of the right-holder, as well as a signed sworn statement that:

1. the notifier is the legal right-holder or representative;
2. it has tried to reach the content provider without success; and
3. the notified content is to the best of its knowledge being made available online illegally.

Lastly, the notice must include specific information about the infringing material that is sought to be removed, including an explanation of its location. The information must be presented in a way that allows the service provider to locate the offending data easily. Even if the notice does not meet the requirements, the service provider must still make reasonable efforts to contact the notifier using the contact information provided.

Upon receipt of the notice, the hosting service provider must immediately inform the content provider of the removal of the content as well as provide a copy of the original notice. The content provider may have the content restored by filing a counterclaim (writing, contact, grounds and signature) for the relevant item(s). When the counterclaim is filed within 14 days, the hosting service provider may not prevent the content from being displayed, even if the right-holder files a new take-down notice. If a counterclaim is filed, the right-holder may initiate court proceedings for the sake of obtaining a court order mandating the removal of the content. If the content is in breach of the contract between the content provider and the hosting service, such that the contract is terminated by the hosting service provider, this process is not necessary.

Section 194 imposes strict liability on the notifier in the event of wrongful notice; that is, the right-holder is liable for all damages resulting from wrongful information, regardless of fault. The court may adjust the compensation partially or completely, if the faulty information was minor in relation to the notice as a whole or in relation to the grounds stated in the counterclaim.

A service provider is not liable for the infringement of copyright perpetrated by its users if the right-holder does not notify it of the infringing material in accordance with the specific requirements set for the notification in Section 191 of the Information Society Code. However, according to the *Finnreactor*–case, a service provider cannot avail itself of the safe harbour for hosting if its own activities amount to copyright infringement or if it works together with the content provider to promote illegal activity.[23]

Under Section 185 of the Information Society Code, courts can issue to a public prosecutor, the police or a right-holder a temporary or permanent injunction ordering the hosting service provider to prevent access to content, if it is *evident* that the content is illegal or illegally distributed. In such instances, the content provider has

---

[23] KKO 2010:47 and 48.

a right to be heard, unless this is not possible in the circumstances of the case. For the order to remain in force, the content provider must be heard, unless it cannot be identified. The courts would likely grant the police or other public authorities such orders when requested if the hosting service provider has not already removed the content on its own accord. This section also applies to copyright and related rights and constitutes a narrower alternative to Section 60 c of the Copyright Act. This section indicates—dealing here with an issue that was prevalent at the time of implementation—that the hosting safe harbour does not cover *mere conduit* service providers, such as teleoperators, but e.g. discussion forums or auction sites. In practice, this provision is secondary to the notification procedure.

Section 315 of the Information Society Code provides the Ministry of Transport and Communications, the Finnish Communications Regulatory Authority and the Data Protection Ombudsman and the Consumer Ombudsman a right of access to certain information from service providers, provided this is consistent with the citizen's right to privacy under data protection law. The right is limited to information that is necessary for the carrying out of their duties and does not apply to information dealing with confidential messages, identification data or location data.

As an exception to the above, the Finnish Communications Regulatory Authority may receive identification and location data that is necessary for investigating fault or disruption in the service or clarification of issues relating to billing the service. The same Authority and the Data Protection Ombudsman also have a right of access to such data when investigating violations of (or threats to) information security as well as compliance with data protection and privacy law. The types of offences (12 in total) that may give rise to this right are listed in Section 316 of the Information Society Code.[24] The right to use the information that has been accessed is limited and requires that the data be destroyed and kept confidential by the receiving authority. The ISP must provide the subscriber and, in case of a minor, the legal guardian of the subscriber, access to information regarding the location data of a mobile, smart phone or tablet at any given time.

The Act on the Exercise of Freedom of Expression in the Mass Media[25] provides that an official with the power of arrest (usually police), a public prosecutor or an injured party may obtain a court order obliging the keeper of a transmitter, server or

---

[24] (1) a breach of data protection in electronic communications under section349 of this Act; (2) unauthorised use as referred to in Chapter 28(7) of the Criminal Code; (3) endangering data processing as referred to in Chapter 34(9a) of the Criminal Code; (4) possession of a data network offence device as referred to in Chapter 34(9)(b) of the Criminal Code; (5) criminal damage as referred to in Chapter 35(1)(2) of the Criminal Code; (6) secrecy offence as referred to in Chapter 38(1) of the Criminal Code; (7) message interception as referred to in Chapter 38(3) of the Criminal Code; 120 (8) interference with communications as referred to in Chapter 38(5) of the Criminal Code; (9) petty interference with communications as referred to in Chapter 38(7 a) of the Criminal Code; (10) computer break-in as referred to in Chapter 38(8) of the Criminal Code; (11) offence involving an illicit device for accessing protected services as referred to in Chapter 38(8b) of the Criminal Code; or (12) data protection offence as referred to in Chapter 38(9) of the Criminal Code.

[25] Section 17. Laki sananvapauden käyttämisestä joukkoviestinnässä 13.6.2003/460.

other similar device, to release the information required for the identification of the sender of a network message to the requester, provided that there are *probable reasons to believe* that the contents of the message are such that its distribution to the public is a criminal offence.

The injured party is only entitled to such a court order if it has the right to bring a private prosecution for the offence.[26] Lesser offences are not necessarily subject to public prosecution. However, all private injuries do not necessarily amount to a criminal offence; hence, the limitation in the text of the statute. The same actors may apply for a court order to cease the distribution of a published network message if it is evident on the basis of the content that the provision of the message to the public is a criminal offence. The order is subject to appeal and is not enforceable until appeals are exhausted, unless the court provides otherwise. The safeguards applicable (hearing, notice, lapse, security) are the same as those that will be discussed below. An order to destroy the message may be issued even in cases where the perpetrator cannot be identified.

Copyright Liability Cases

Direct or Indirect Infringement (Finnreactor-Case)

The Supreme Court has issued three rulings in which it addressed criminal liability for intermediaries that facilitate copyright infringement. All verdicts were voted on. In Finland, only Supreme Court decisions may serve as precedent. The precedential value of a split decision is limited. In the *Mailbox*-case,[27] the defendant maintained an e-mail mailbox that he allowed members to use if they shared a copy of a computer program. After gaining access the members could download the computer programs that other members had shared in the mailbox. At issue was whether the defendant could be held criminally liable (as a commercial copyright offence)[28] for making available copyrighted works by setting up the system (that is, not only for the copies he had made and shared with others). While the system was predominantly free of charge, the court found the defendant guilty of the aggravated offence based on evidence: (1) that the defendant had gained access to several valuable computer programs; (2) of trading for payment; (3) of the mailbox containing copies of 443 software programs; (4) that the act was intentional and planned.

---

[26] Most criminal offences are under public prosecution, which means that a prosecutor or police will make the request. In practice, the prosecutor or police may not prosecute minor offences. In that case, the injured party may only make a request, if the injury amounts to a criminal offence that could be independently prosecuted as private prosecution (asianomistajarikos). This rule precludes requests based on civil liability.

[27] KKO 1999:115.

[28] Under the law then in force, the defendant was prosecuted for the copyright offence, which required commercial activity.

The defendant's sophisticated computers were forfeited to the government, since they were essential for the execution of the crime.[29] When calculating compensation (not damages), the Supreme Court rejected the right-holders' requests for compensation based on the purchase price of the software. The law provides for reasonable compensation, which mandates consideration of the circumstances of the case. The court found that the intentionality of the acts of the defendant required that the compensation be significant, weighed it against the hobbyist nature of the activity, and concluded that it would be reasonable to calculate the compensation based on half of the purchase price (app. 680,000 FM[30]). The court rejected the supplemental claim for damages, because no proof of actual damages had been presented. Of the dissenting judges, one would have determined compensation at a rate of app. 100% of the purchase price (1,320,000 FM) and the other would have accepted compensation based on 50% but would also have ordered damages of 132,000 FM.

The *Finnreactor*-cases involved a P2P-network and the Supreme Court assessed the criminal liability of the administrators of the network (as complicit main offenders or abettors) as well as the compensation/damages payable by them to the right-holders.[31] A private person that had shared files in the network was not held responsible as an administrator, but instead was guilty of the petty offence of making available copyright infringing material.[32] The network divided its users in seven categories based on their individual "ratio number" that indicated the frequency of shares to the network. The higher the ratio number of the user, the greater the rights, duties and perks they received. Users with a low ratio number could receive a warning, get demoted to a lower category or have their accounts closed, if they were not actively sharing. Starting from level 4 the users also had administrator duties. Level 4 administrators could promote or demote users between categories and delete files from the network.[33] Level 5 administrators could also delete or add users. These administrators would also advise users and monitor their sharing habits.[34] The highest level 6 administrators were also in charge of the technological functioning of the network.[35]

Each defendant was charged with the petty offence as direct infringers/accomplices or alternatively as abettors. The level 5 and 6 administrators were found guilty as main offenders/accomplices, while the level 4 administrators were found guilty for aiding and abetting. Two of the abettors were juveniles. Although the Copyright

---

[29] KKO 1999:115.
[30] The amount of compensation is roughly equivalent to the price of three family homes, or 10 annual salaries.
[31] KKO 2010:47.
[32] KKO 2010:48.
[33] 4 defendants.
[34] 5 defendants.
[35] 2 defendants.

Act had been toughened prior to the Supreme Court verdict the defendants were convicted based on the old law.[36]

All defendants argued that their actions did not constitute copyright infringement, since they had not part taken in the actual copying or sharing of copyrighted works, nor had they possessed the copyrighted works at any time. The acts of the direct infringers were not direct consequences of the defendants' acts as administrators of the network. Their administrative acts did not target single copyrighted works, nor had they any knowledge of the specific works that were available in the network.[37]

The Supreme Court referred to the *Mailbox*-case in determining that it is not required that the perpetrator actually make physical copies available to the public, although this has been the main rule. Thus, the act of giving physical copies to someone else that makes them available to the public has not been considered copyright infringement.[38] However, the act of selling the licensing documents that had been removed from the software packaging did constitute the making available of copyrighted works, although the buyer did the actual copying of the software.[39] These previous cases had concerned an individual offender and not, as in this case, a group. The Court thus had to consider whether the defendants' acts constituted copyright infringement individually or jointly.

According to Finnish (criminal) law,[40] two or more people that intentionally commit a crime together are each punished as main offenders. The prerequisite for complicity is knowledge that one's own acts in accord with the acts of others will satisfy the essential elements of the offence X. If so, the offender is responsible for the result of the acts taken together as X, not merely for his own acts—provided the offender's acts constitute a significant contribution to the crime. A lesser contribution will be assessed as aiding and abetting in the commission of a criminal offence.[41]

The Supreme Court took as its (copyright law) starting point the position that electronic copies constitute "the making of copies" that require the copyright holder's consent. This decided the issue of whether copyright law applies to new technology, or whether peer-to-peer technology was within the private use exception.[42] This is so whether or not these copies arose from copying actual works or transfers between computers or other media equipment. Because the network required sharing of files for the actual copying of works, it did not fall within the exception of

---

[36] According to general doctrines of criminal law, like the principle of legality and *in dubio pro reo*, an act is punishable only to the extent it was criminal at the time of commission of the alleged offense.

[37] KKO 2010:47 at 7.

[38] KKO 1999:8.

[39] KKO 2003:88.

[40] Chapter 5:3 of the Penal Code.

[41] Chapter 5:6.1 of the Penal Code.

[42] Sharing copies of music between private persons has traditionally been fair use in Finland.

"making a few copies for private use[43]".[44] The acts of all the users of the network nonetheless constituted "making available to the public" of copyrighted works, without the copyright holder's consent.

The Supreme Court found that the torrent-files that the network made available were essential for the actual copying by users, as well as for the users to know what copyrighted works were available on the network. The network was also specifically designed for the efficient distribution of copyrighted works. Thus, while the acts of each administrator were not essential for the actual sharing and copying of works, it did not preclude complicity or viewing the acts together as one single act. On this basis it had to be assessed whether the acts of each defendant constituted a significant contribution to or aiding and abetting of the crime.[45]

The fact that the activity was planned, based on a clear distribution of work, and the duties of the administrators were specifically designed to maintain a high rate of availability of new material and increasingly efficient distribution of all available material (improving torrent files), together was sufficient to satisfy the court that the acts of the level 5 and 6 administrators were essential for the continued copyright infringements occurring on the network.[46] They had access to and knowledge of which works were popular and newly made available and they were aware of their own role in the efficient operation of the network. Like the Court of Appeal, the Supreme Court found that the acts of the level 4 administrators did not constitute a significant contribution, but satisfied the elements of aiding and abetting.

This case was not about criminal punishment, since the petty offence carries a maximum penalty of a fine. The main issue on appeal was the amount of compensation for each of the defendants. The punitive elements built into the concept of compensation are unique to copyright[47] law in Finland. Finnish tort law caps damages to actual damages, which can be adjusted. The concept of reasonableness in tort law is thus severely strained to master the gap between tort law and intellectual property law. In essence compensation amounts to statutory damages, also a concept that is foreign to Finnish law. Statutory damages are normally used to cap (rather than enhance) actual damages in Finland.

The right to compensation for the unlawful use of a copyrighted work does not require proof of intent or criminal negligence, nor of the defendant having received economic gain from the crime.[48] The right-holder is entitled to reasonable compensation for the use of copyrighted works. Since the level 4 administrators were not criminally liable *for using* the works, only aiding in the use by others, they were not required to pay compensation. Even if the level 4 administrators were also ordinary users of the network, such use did not incur criminal liability nor could it influence

---

[43] Under section 56a § 2 of the Copyright Act, such copying is not punishable as a crime.

[44] KKO 2010:47 at 16.

[45] KKO 2010:47 at 17 and 18.

[46] KKO 2010:47 at 19.

[47] The same provision is found in all intellectual property laws, but has been applied only in copyright cases.

[48] KKO 2010:47 at 24.

the assessment of whether compensation should be paid.[49] Thus, the Supreme Court relieved these defendants of the duty to pay compensation that the lower courts had issued.

The level 5 and 6 administrators were jointly liable to pay compensation.[50] Their individual share of the compensation was however, capped at 10% of the total amount of compensation. One defendant that had been an administrator for a shorter period than the others was responsible for a smaller share.[51] The starting point for calculating the compensation is the licensing fee for the lawful use of the work. The purchase price is problematic as a basis for calculations, since it includes value added tax and other unrelated costs, not attributable to the use of the work.[52] According to settled case law it is often not reasonable to order the payment of compensation in the amount of normal licensing fees.[53] Thus, the punitive element built into the concept of compensation was in effect limited by the compensation culture in Finnish law.[54] The amount of compensation must be based on the court's assessment, as well as the special circumstances of the case, not matters such as the number of copies made, the purchase price of a legal copy or the normal license fee for a legal use.

The intent and premeditation of the defendants, as well as the efficiency of the network in affecting actual sales and harming the right-holder, had a bearing on the court's assessment that the compensation must be substantial. However, the structure of the network did tempt users to copy works that may have remained unused, and to which access would not have occurred had they not been freely available. The fact that the number of copies made of any individual work was largely outside the control of the defendants also precluded using the full number of works as a basis for the calculation. Thus, the court awarded compensation in the amount of 15% of the claimed purchase price for non-musical works, 25% of the wholesale price for musical works and 50% of the licensing fee for use of musical works (because it is based on what is paid to artists).[55] Because the Tort Damages Act does not apply to this type of compensation the amount could not be adjusted based on (general) reasonableness or other grounds.[56]

---

[49] KKO 2010:48 and KKO 2010:47 at 27.

[50] KKO 2010:47 at 25.

[51] KKO 2010:47 at 41.

[52] KKO 2010:47 at 34 rejecting earlier case law KKO 1998:91 and KKO 2001:42.

[53] KKO 2010:47 and KKO 1989:151, KKO 1999:115 and KKO 2002:101.

[54] KKO 2010:47 at 35.

[55] The dissenting judge would have ordered higher compensation (20%, 30% and 100% respectively, based on the number of actual copies).

[56] KKO 2010:47 at 42.

Preliminary Injunction (Pirate Bay Cases)

Finnish law separately addresses general injunctions against intermediaries (injunction to discontinue), where the direct infringer can be heard and the situations in which the direct infringer is unknown or cannot be reached (blocking order).[57] In the *IFPI v Elisa*–case the Finnish branch of International Federation of the Phonographic Industry[58] filed an application[59] for an injunction to discontinue against the tele-operator Elisa[60] under Section 60 c of the Copyright Act, requesting the court to place a duty on Elisa to block its users from accessing 101 copyrighted works on the Pirate Bay–site.[61]

Section 60 c of the Finnish Copyright Act (21.7.2006/679) states, in relevant part[62]:

---

[57] An "injunction to discontinue" is unique to Section 60 c of the Copyright Act and targets intermediaries to copyright infringement. Preliminary injunctions are quite rare in Finland and hardly ever used against others than the main perpetrator, in contested cases or when the law is unsettled. Keskeyttämismääräys. An injunction against direct infringers in absentia or blocking order (estomääräys) refers to a stronger remedy with roots in Article 11 of the Enforcement Directive.

[58] IFPI Finland representing EMI Finland Oy Ab, Sony Music Entertainment Finland Oy, Universal Music Oy and Warner Music Finland Oy.

[59] Precautionary measures follow a separate summary procedure under Chapter 7:3 of the Code of Procedure. The court has reduced competence to accept evidence or assess questions of substance in summary proceedings. There is no plaintiff or defendant, but an applicant and respondent. The loser pays-principle does not apply to the issue of attorney costs.

[60] Elisa is the second largest teleoperator in Finland. IFPI has since filed suit against the other major teleoperators as well.

[61] For an in depth comment of the case, see Weckström 2013.

[62] The provisions regarding injunctions in the Copyright Act were amended by Parliament after the Supreme Court denied hearing of the *IFPI Finland v. Elisa*-case on the issuance of an interim preliminary injunction. The amendment entered into force on 1 September 2013. Changes in wording were not intended to change the state of the law (Proposed Bill HE 124/2012 vp at 94 and 119). Sections affected are italicised in the text below. The complete wording of Section 60 c of the Copyright Act before amendment by Law 118/2013 as unofficially translated by Finlex:

(1) In trying a case referred to in section 60b, the court of justice may, upon the request of the author or his representative, order the maintainer of the transmitter, server or other device or any other service provider acting as an intermediary to discontinue, on pain of fine, the making of the allegedly copyright-infringing material available to the public (injunction to discontinue), unless this can be regarded as unreasonable in view of the rights of the person making the material available to the public, the intermediary and the author. (2) Before legal action referred to in section 60b is taken, the court of justice referred to in said section may, upon the request of the author or his representative, issue an injunction to discontinue, if the conditions mentioned in subsection 1 for its issue are met and if it is apparent that the author's rights would otherwise be seriously prejudiced. The court of justice shall reserve an opportunity to be heard both for the person against whom the injunction is sought and for the person making the allegedly copyright-infringing material available to the public. A service of a notice to the person against whom the injunction is sought may be delivered by posting it or by using fax or electronic mail. *The handling of the matter shall otherwise come under the provisions of Chapter 8 of the Code of Judicial Procedure.* (3) Upon the request the court may issue an interim injunction to discontinue referred to

1. In trying a case [...], the court [...] may, upon the request of the author [...], order the maintainer of the transmitter, server or other device or any other service provider acting as an intermediary to discontinue, on pain of fine, the making of the allegedly copyright-infringing material available to the public (injunction to discontinue), unless this can be regarded as unreasonable in view of the rights of the person making the material available to the public, the intermediary and the author.

2. Before legal action [...] is taken, the court [...] may, upon the request of the author [...], issue an injunction to discontinue, if the conditions mentioned in subsection 1 for its issue are met and if it is apparent that the author's rights would otherwise be seriously prejudiced. The court [...] shall reserve an opportunity to be heard both for the person against whom the injunction is sought and for the person making the allegedly copyright-infringing material available to the public. [...]

3. Upon the request the court may issue an interim injunction to discontinue referred to subsection 2 without hearing the alleged infringer, if deemed necessary for the urgency of the case.

4. An injunction to discontinue [...] shall not prejudice the right of a third person to send and receive messages. The injunction to discontinue shall enter into force when the applicant provides the security [...] to the execution officer, unless otherwise provided in [...] Code of Judicial Procedure. The injunction to discontinue issued by virtue of subsection 2 or 3 [...] shall expire, if a legal action has not been taken within 1 month from the issuing of the injunction.

5. If the [main] legal action [...] is dismissed or ruled inadmissible or the case is discontinued due to that plaintiff has cancelled his legal action or failed to appear [...], the person requesting the injunction to discontinue must recompense the person against whom the injunction is issued, as well as alleged infringer for damage caused by the enforcement of the injunction and for the costs incurring

---

subsection 2 without hearing the alleged infringer, if deemed necessary for the urgency of the case. The injunction shall remain in force until further notice. After the injunction has been issued, the alleged infringer shall be reserved an opportunity to be heard without delay. After hearing the alleged infringer, the court shall decide without delay whether it retains the injunction in force or cancels it. (4) An injunction to discontinue issued pursuant to this section shall not prejudice the right of a third person to send and receive messages. *The injunction to discontinue shall enter into force when the applicant provides the security referred to in [section16 of Chapter 7 of the Enforcement Act (37/1895)] to the execution officer, unless otherwise provided in section 7 of Chapter 7 of the Code of Judicial Procedure. The injunction to discontinue issued by virtue of subsection 2 or 3 of this section shall expire,* if a legal action has not been taken within one month from the issuing of the injunction. (5) If the legal action referred to in section 60b is dismissed or ruled inadmissible or the case is discontinued due to that plaintiff has cancelled his legal action or failed to appear in court of justice, the person requesting the injunction to discontinue must recompense the person against whom the injunction is issued, as well as alleged infringer for damage caused by the enforcement of the injunction and for the costs incurring in the matter. The same shall apply when the injunction to discontinue is cancelled by virtue of subsection 3 or expires by virtue of subsection 4. The taking of legal action for the compensation of damages and costs shall be governed by the provisions of section 12 of Chapter 7 of the Code of Judicial Procedure.

in the matter. The same shall apply when the injunction to discontinue is cancelled by virtue of subsection 3 or expires by virtue of subsection 4.

The District Court discussed two issues: (1) whether the requirement for the making of the allegedly copyright-infringing material available to the public was met; and (2) whether an injunction to discontinue can be regarded as unreasonable in view of the rights of the person making the material available to the public, the intermediary and the author. The District Court emphasised that the requirement of reasonableness relates not to the request itself, but to the effect of the requested measures. Further, it is noted that the constitutional requirement of proportionality prevents an injunction that interferes with the public distribution of material to a greater extent than is necessary for the protection of copyright holder's right. The District Court held that the requirement of making available copyright-infringing material to the public had been met; it referred to the Swedish judgments convicting the administrators of The Pirate Bay-site. While the District Court recognised that there likely is not a technical means available that would totally prevent access, nor a means that would prevent access to only plaintiff's copyrighted works, this did not render the requested injunction unreasonable or disproportionate. Neither did the possible ineffectiveness or circumvention of any or all measures prevent the issuance of an injunction. The District Court rejected the claim that Elisa could not honor the injunction by implementation of technically available measures. The court issued a temporary injunction against Elisa ordering it to discontinue making available to the public the 101 copyrighted works, on pain of a fine of 100,000 euros.

Elisa appealed, alleging that the District Court unlawfully granted more than the applicant had requested by supplementing instead of rejecting the applicant's application that had failed to specify in the required way the measures that were requested. The main issue on appeal was whether the applicant should have itemised the domain names and IP addresses targeted by the injunction and, if so, whether they should have done so in the application, or could legitimately do so later, when applying for enforcement of the judgment.[63] It should be noted that the Court of Appeal was asked and agreed to consider many aspects, evidence and material that related to events (enforcement proceedings) after the judgment of the District Court had issued. While the Appellate Court upheld the judgment of the District Court, it seems clear that the actual scope and content of the measures required by the injunction had narrowed considerably, thus rendering moot many of Elisa's initial objections. The Appellate Court upheld the general wording of the injunction and dismissed all claims of over-inclusiveness and harm to Elisa or third parties.

This case is discussed further below in the context of comparing the 2013 Draft bill and 2014 Bill, which sought to address many concerns raised by the case.

---

[63] In subsequent applications against teleoperators DNA Oy and TeliaSonera Finland Oyj, IFPI Finland mentioned that it would specify the domain names and IP addresses in enforcement proceedings, which would indicate that the Execution Officer required that the injunction be specified as a condition for enforcement. It would seem that the District Court viewed the itemisation issue as strictly procedural, which according to standard rules of procedure legitimately shifts detailing issues to enforcement proceedings.

The Copyright Battles

The 2010 Bill and Citizen's Initiative for Common Sense into the Copyright Act

In 2010, a bill was proposed suggesting that the Finnish legislature adopt rules based on the French system of graduated response.[64] This system would have obliged ISPs to send notices (on behalf of right-holders) to the owner of the internet connection through which illegal downloading of copyrighted material had occurred. Under the bill, ISPs could also be ordered by a court to collect personal data of infringing connections and transfer the data to the court (which could then release the data to the right-holder). The bill was followed by massive debate and resistance in Parliament, owing to constitutional concerns and expired before the 2012 elections.

Since 2012, the Citizens' Initiative Act has mandated Parliament to process citizen's initiatives which have received at least 50,000 signatures. On 26 of November 2013 the 'Common Sense into the Copyright Act' initiative (The Initiative) was submitted to Parliament. The Initiative targeted the enforcement of copyright on the internet. Digital downloading would remain illegal, but punishable only as a petty copyright offence, if the perpetrator is a private person. Reportedly 71% of offenders are minors. The copyright offence would be reserved for large scale commercial copyright infringement. Remedies such as search, seizure and damages would only be available for the copyright offence.

The Initiative also sought to repeal the possibility of compensation calculated on the basis of the purchase price of the item, which mirrors punitive damages. Punitive damages are not recognized in Finnish tort law. The Initiative also sought to reintroduce broad exceptions for research and educational purposes, especially relating to audiovisual content, which until the 1980s had been free for use in Finland. The Initiative sought to have Sections 60a to 60c of the Copyright Act, relating to injunctions requiring release of personal data of the content provider and blocking access to illegal sites, repealed. These provisions allow for injunctions against non-fault intermediaries. The Initiative suggested that the measures necessary to implement the Enforcement Directive were already in force in Finland. The Initiative was rejected by Parliament. However, the Initiative did affect the contents of the Information Society Code and a Proposed Bill for the amendment of the Copyright Act, especially regarding obligations of ISPs for internet content.

In December 2013 a Draft Bill was circulated for comments, which included the option of ordering a teleoperator to slow down the internet connection to IP-addresses of known sites that include illegal material. This option was viewed as a milder alternative to an injunction to discontinue in Section 60c of the Copyright Act, which was already available to right-holders. However, the Bill that was presented to Parliament in the Fall of 2014 no longer included this proposal due to massive

---

[64] HE 235/2010.

criticism regarding concerns about technological utility and protection of fundamental rights. The 2014 Bill addressed all of the issues raised by the *IFPI Elisa*
case. It was adopted with only minor revisions in May 2015, and passed into law 1
June 2015.

The 2013 Proposal to Slow Down Internet Connections

The 2013 proposal for a government bill focused exclusively on digital piracy and
proposed amendments to section 60c of the Copyright Act.[65] The main amendments
were: (1) the ability to obtain a new order slowing down internet connection; (2)
changed (alleviated) requirements for orders without bringing suit against the direct
infringer; and (3) changes in requirements for obtaining and maintaining any order
in force. Several features of these amendments are worth noting.

First, an order to slow down the internet connection would have been a global
novelty. It would have served as a more lenient alternative to the injunction to discontinue already available under Section 60c of the Copyright Act. The Ministry of
Education and Culture had evaluated the following options for remedies against
digital piracy: (1) letter procedure (the so-called "graduated response")[66]; (2)
restricting access; (3) slowing down internet traffic; and (4) closing of the domain
name. Alternative (3) was selected as the most viable alternative, in addition to the
specific (DNS/IP) order to discontinue already in force. The technical measures
would target specific IP addresses and would not affect domain names or require the
use of filtering software, such as DPI-technology.[67]

Teleoperators raised doubts about whether this technology would actually differ
in effect from an order to block access to specific IP addresses. However, it was
proposed that this measure limits the rights of users less, since they would still
retain access (although slow) to lawful material on the site. It was also suggested
that this measure may be more effective in redirecting users to lawful sources for the
same material.[68] The effect of the measure would be seen as a delay (several seconds
to 30 s for a 5 MB download) in accessing protected works or the delay could be
triggered by reaching a certain number of users that simultaneously access the link.
However, it would not affect traffic between users, nor access to the same links
through search engines. This measure could be targeted at the IP addresses of individual users of P2P networks, an illegal streaming service or P2P-service providers.

---

[65] The draft was circulated for comments (by 1 January 2014) and was delivered to Parliament in
the Spring of 2014.

[66] This was included in the failed bill of 2010 and the proposal cites the constitutional problems
involved. Since the *Finnreactor*-case, maintaining illegal services in Finland was already criminal,
rendering the fourth option without effect, since Finnish authorities do not control domain names
outside the .fi domain.

[67] Draft Bill at 12.

[68] Draft Bill at 13.

This measure should not be confused with sanctions related to graduated response, under which recurring infringers could be subject to a general slowing down of the internet connection. The target is the IP address of the source of the illegal material, not the user's access to the internet. The right-holder already has the right to information about the individual user (from service providers) under Section 60a of the Copyright Act. In addition to proof of the illegality of the site, the applicant would be required to present evidence of the number of users of a service, since the measure could primarily only be used against services that impact the copyright holders' rights. The number of users is not essential if unpublished copyrighted material is made available to the public. Making available unauthorized recordings of live performances or new releases prior to their release, is not copying existing material, but usurping the market of the copyright holder.[69] If there is a significant amount of legal material on the service the court should not be able to order an injunction to discontinue, but instead would have ordered the slowing of the connection.[70]

Second, the amendment would have allowed both orders to be issued in cases where the direct infringer/service provider is located abroad or otherwise conceals its contact information. It was emphasised that the rule that right-holders primarily target the direct infringer/service provider should remain strong. Likewise, direct infringers should also be heard, which is a prerequisite for all injunctions under 60c. However, it was stressed that the option of action when the infringer cannot reasonably be served or reached with a subpoena should be made clear in the statutory text. The right-holder would have to show that reasonable attempts have been made to reach the direct infringer; or that the direct infringers could not be identified. It was proposed that this option could only be used in the most serious cases of copyright infringement and that a requirement of special circumstances should be added to the statutory language.[71] If the order is cancelled due to the measure being unfounded the applicant should cover all costs incurred by the operator.[72]

Third, with regard to changes relating to ordering and maintaining the injunction in force, there were several proposals that at first glance seemed reasonable but

---

[69] Draft Bill at 14.

[70] Draft bill at 22.

[71] Arguments for strong protection (copyright) advance remedies against ISPs that are "effective, proportionate and dissuasive" preventive measures against copyright infringement according to Article 11 of the Enforcement Directive Directive 2004/48/EC of the European Parliament and of the Council of 29 April 2004 on the enforcement of intellectual property rights (OJ L 157, 30.4.2004) and Article 8 of the INFOSOC Directive (Directive 2001/29/EC of the European Parliament and of the Council of 22 May 2001 on the harmonisation of certain aspects of copyright and related rights in the information society OJ L 167, 22/06/2001 P. 10 – 19).

Arguments against strong protection (copyleft) advance narrowly-tailored or no remedies against ISPs relying on Article 9 of the Enforcement Directive, fundamental rights of ISPs (freedom of enterprise) and users (freedom of expression) and the prohibition against imposing an obligation to monitor the internet contrary to Article 15 of the E-Commerce Directive Directive 2000/31/EC of the European Parliament and of the Council of 8 June 2000 on certain legal aspects of information society services, in particular electronic commerce, in the Internal Market OJ L 178, 17/07/2000 P. 1 – 16.

[72] Draft bill at 24.

which may have been problematic. One proposal related to the ability of the right-holder to extend a granted injunction to other teleoperators without separate proceedings. While the largest teleoperators were in favor of this option, local teleoperators (of which there are 24 in Finland) were opposed to injunctions that would be measured against the capabilities and resources of large teleoperators. Yet, the draft bill proposed that such a possibility would be added to Section 60c. An effort to transfer the authority to grant such extensions to the Communications Agency was resisted by both the Agency itself and the Ministry of Justice, since such duties are reserved to courts.[73]

The Draft bill also contained a proposal that the requirement of initiating procedures against the direct infringer within 1 month (60c§ 4) be extended to 6 months. This proposal seems contrary to the goal of alleviating the requirements of attempting to bring proceedings first against the direct infringer. If the applicant is required to initiate proceedings and attempt to locate the direct infringers, why is there a need to extend the deadline for maintaining the order against the ISP in force without having filed a claim against the direct infringer? If the simultaneously advanced proposal, which would allow a blocking order against the direct infringer by presenting evidence that reasonable attempts have been made was accepted that provision would have allowed the court to consider, whether the order against an ISP could issue. It seems backwards to start from the exception of an interim injunction to discontinue without hearing the direct infringer having issued (and 1 month has passed), instead of starting from the position of infringement occurring and determining whether measures should be sought against the direct infringer or an intermediary. In the *IFPI v. Elisa*–case, it took 5 months from the filing of the application against the intermediary for the court to issue the order and an additional month for IFPI to file against the direct infringers. It is difficult to imagine a case where the applicant would not be able to *provide proof of having made reasonable efforts to reach/serve the direct infringer* within that time, so as to render the lapse of the injunction unreasonable against the applicant. The Draft bill did not address when the requirement of urgency would be met, so as to require that an injunction be issued against an ISP without hearing the direct infringer, rather than having to proceed through the normal channels of hearing the affected parties before issuing an interim or permanent injunction.[74]

Several proposals went to the heart of the controversy in the *IFPI v. Elisa*–case. Under the proposal, the applicant was required to list the IP-addresses in the application in order so that for the order and its enforcement can be sufficiently tailored.[75] If the application is too vague it could amount to an unlawful monitoring obligation under Article 15 of the E-Commerce Directive, which is prohibited by EU law.[76] However, the Draft Bill also proposes that an exception be made to the general rules

---

[73] Draft Bill at 13.

[74] For a more detailed account, see the discussion on the Pirate Bay-case above, the 2014 Bill below, and Weckström 2013.

[75] Draft Bill at 13 and 21.

[76] Draft Bill at 21.

of procedure regarding who is responsible for the costs of the measure. According to the general rule in Finnish law, the applicant is responsible for the costs of the measure.[77] It is further proposed that the court should have discretion on this matter based on a standard of reasonableness.[78] It is not clear why it would be reasonable that a disinterested party carry the costs of preventing copyright infringement.[79] The general rule would, however, apply to attorney's fees, for which the parties would bear their own costs.[80]

It was also deemed necessary to extend the reasonableness inquiry to include the interests of the users of the affected service. This proposed amendment of Section 60c was intended to strengthen the constitutional right of third parties to send and receive messages under section 60c 4.[81] The court should balance the right to property, the freedom to provide services and the freedom of expression of the affected actors. Whether the material is legally available in other channels would affect the inquiry as well as the amount of material that is legally available on the service with the consent of the copyright holder or because in the public domain.[82] When assessing the purpose of the service, regard must be had to: (1) the importance of illegal material in the overall operations of the service; (2) the degree of legal material and the importance of the availability of that material in the overall operation of the service; (3) the proportion of legal vs. illegal material; (4) the absolute number of illegal works distributed via the service and; (5) the rate of use of the copyrighted works in relation to the legal material on the service.

It was proposed that the order would as a rule be set to expire at a specific date (*määräaikainen*). The order could be renewed, amended, extended, cancelled or allowed to lapse. If the order is cancelled the applicant would not have to bear the costs, unless the order was made without hearing the service provider. The proposal does not specifically mention Section 60c 5; however, this proposal would seem to

---

[77] For measures under Ch. 7:3 of the Code of Judicial Procedure.

[78] The bill proposed that the court assess evidence of the effect of the infringement and compare those effects to the costs of enforcement and the appropriateness of the request for the measure. Draft Bill at 22.

[79] It was mentioned elsewhere in the Draft bill that the enforcement costs in the Pirate Bay case amounted to 12,500 euros/order, which the operators paid as part of their general cost of operation. It was noted that the new order could involve greater cost because of the new technology. Draft Bill at 17. It is questionable whether it is necessary to change the statute to prevent recovery of legitimate costs, because of one instance of the operators forfeiting their right to have their cost covered. Nothing in the current statute prevents the (large) operators from doing so in the future. Increased discretion in summary proceedings may be problematic.

[80] Precautionary measures follow separate summary procedure under Chapter 7:3 of the Code of Procedure. The court has reduced competence to accept evidence or assess questions of substance in summary proceedings. There is no plaintiff or defendant, but an applicant and respondent. The loser pays-principle, which is standard in civil cases, does not apply to the issue of attorney costs. Instead, the general rule is that the applicant bear the cost of the measure and that the parties bear their own cost for proceedings (attorney fees).

[81] Draft Bill at 21.

[82] Draft bill at 15.

suggest an amendment to the rule that the applicant is responsible for all damages arising from the measure having been issued without cause.

Lastly, it was proposed that the applicant be released from the obligation to provide a security when the court has stated that the applicant has not had an opportunity to bring proceedings against the direct infringer.[83] The proposal seems to take as a starting point that the purpose of the security is mainly to secure enforcement of interim injunctions (i.e. cover costs of enforcement before a final order has issued). After "the order is independently in force" the applicant's security should be released.[84] The section 60c 5 that existed at the time, which connects the security to securing compensation of the respondent, was not mentioned.[85]

The 2014 Bill and the Revised Copyright Act

The Draft Bill and the proposal to slow down the speed of internet connections were severely criticised in the round of comments. The final Bill (HE 181/2014) that reached Parliament in 2014 had excised the "slow-down" proposal, citing several technical, constitutional and enforcement related concerns.[86] Instead, a revision of former Section 60c was proposed, which distinguishes between injunctions to discontinue, interim injunctions to discontinue, and a blocking order.[87] This distinction separately addresses general injunctions against intermediaries (injunction to discontinue), where the direct infringer can be heard and the situations in which the direct infringer is unknown or cannot be reached (blocking order). In this way the constitutional interests of the intermediary are better addressed in the procedure.

The 2014 Bill included extensive discussion on the constitutional balancing of competing interests.[88] The new Sections 60d - 60f addressed procedural concerns separately and guidance is given to courts on how to weigh competing interests in

---

[83] Draft bill at 17.

[84] Draft bill at 17.

[85] It should be noted that said section refers to the general rules of procedure in the Code of Judicial Procedure and the Enforcement Act, where the general rule of the applicant carrying the cost is cemented. The proposal included two exceptions to the general rules of procedure for precautionary measures; shifting the cost from the applicant to the respondent; and allowing orders on pain of fine without a security.

[86] HE 181/2014 at 31.

[87] Estomääräys. The term has not been officially translated. The term "blocking order" is not meant to refer to general injunctions.

[88] This is in line with a general reform that aims at making Finnish legislation more transparent, understandable and clear. The rigid legal language is replaced with clear language stating criteria for application, which competing interests should be weighed and how. In a way the statutory legal language has been translated to ordinary Finnish without thereby changing the state of the law. The preparatory works are a significant legal source for courts, and they are unlikely to depart from the interpretations offered.

the different circumstances.[89] While the meaning of the actual provision was not changed significantly, the tone and reasoning of the 2014 Bill was more balanced and took account of the interests of all of the direct infringer, users of internet services, the intermediary and the right-holder. While the goal of Sections 60a–60g of the Copyright Act is to secure the right of the copyright owner to efficient remedies against online infringement, this right visibly competes with the rights of others in the new Section 60c–60f.[90] Sections 60c–60f address both minimum requirements for when an injunction may issue, as well as statutory guidance on the division of cost.

The proposal in the 2013 Draft bill relating to the right-holder extending an already-granted injunction to other teleoperators without separate proceedings was rejected. The 2014 Bill discusses the opportunity for the court to consider the same evidence of copyright infringement; the constitutional rights of intermediaries to a fair trial, however, took precedence over efficiency concerns.[91] The 2014 Bill stresses the peculiarity of the procedural setting, since the intermediary is not an infringer, nor is it responsible for the infringement. While it is accepted that intermediaries take some part in the fight against rampant copyright infringement, courts are advised to give more weight to third party interests in all stages of proceedings.[92]

The 2013 proposal that the requirement of initiating procedures against the direct infringer within 1 month be extended to 6 months was partially approved. Although the period was extended to 2 months for practical reasons, it is now clear that proceedings must be initiated within 2 months in order to maintain in force an interim injunction.[93] The position entertained in the *IFPI v. Elisa*–case, that right-holders would not have to seek enforcement of the order (and post a security) or continue proceedings against the direct infringer after obtaining an interim injunction to discontinue against the teleoperator, was rejected.[94] A new blocking order (Section 60e) is reserved for instances when the direct infringer is unknown or cannot be sued in Finland. The right-holder must show reasonable efforts to locate the direct infringer and, if possible, must pursue regular means of filing suit against absentee

---

[89] There is no English translation of Sections 60a–60g of the Copyright Act. Sections 60d–60f address the procedural concerns previously imbedded in Section 60c, separately. The new Section 60c simply addresses (permanent) injunctions to discontinue. The new Section 60d addresses temporary injunctions to discontinue. The new Section 60e addresses blocking orders and the new Section 60f addresses costs and attorney fees. Section 60 g extends all measures to concern also infringement of related, as in neighboring, rights. It is now clear that different rights and interests are at stake, depending on the procedural setting. What was muddled and unclear, is now clearly stated in the statutory text and preparatory works. We discuss the outcomes of the debate below. For a detailed analysis of the relevant procedural concerns in English, see Weckström 2013.

[90] The courts in the Pirate Bay –cases were corrected. They should have balanced competing rights, instead of merely considering the copyright owners' interest in efficient remedies.

[91] HE 181/2014 at 33.

[92] HE 181/2014 at 32 and 35.

[93] HE 181/2014 at 36.

[94] Section 60 e of the Copyright Act.

defendants. The blocking order requires widespread copyright infringement, which requirement severely prejudices the right-holder's capacity to secure its right.[95]

All injunctions are to set to expire and cannot remain in force indefinitely.[96] Even if the injunction has not expired, an interested party can petition the court to have it cancelled. Thus, the intermediary and infringers are afforded a procedural means to relieve themselves from liability if the right-holder presses for enforcement of an outdated injunction.[97] This is a novelty in Finnish law.[98] If necessary, the injunction can be renewed and revisions made. Minor amendments, such as removing IP addresses, can be made by the execution officer. A broader order, however, requires new court proceedings. This settles the question at the heart of the *IFPI Elisa*–case that the injunction must be clearly and narrowly tailored, so as to be enforceable, at the time of application to the court.[99]

The new Section 60f clarifies the issue of enforcement and attorneys costs. The 2014 Bill proposed that intermediaries bear the enforcement costs as a rule, since these costs were often minor. However, it was proposed that this rule be reversed if the measure required developing new technological means or techniques that would raise such costs.[100] This distinction was not maintained in the final law. Instead, the intermediary as a rule bears the cost of enforcement.[101] This is logical, since the broad and continuous orders envisioned by the right holders in *IFPI v. Elisa* were rejected by Parliament. Naturally, both the right-holder and intermediary may claim costs against the direct infringer. The issue of the costs of attorneys was also settled, and the standard rule of procedure now prevails, with the parties having to bear their own costs.[102] Here, the position of the 2013 proposal is reversed, since Section 60f must be interpreted to allow courts to order that all costs be covered by the right-holder if the injunction is unfounded. There is extensive guidance to courts and execution officers relating to when and how the right-holder may reclaim the security.[103] Here, the exception to general rules of procedure was rejected and right-holders must, according to the general rules of procedure, post a security when the infringer is unknown.

The Bill emphasizes that the purpose of the security is to provide protection to the intermediary or alleged infringer in the event that the injunction is unfounded.

---

[95] Section 60 e of the Copyright Act.

[96] HE 181/2014 at 25.

[97] Section 60 c 5, 60 d 4 and 60 e 4.

[98] Finnish law has not traditionally recognised declaratory judgments or allowed private causes of action against disinterested parties (intermediary liability). As intellectual property law has created such a cause of action, it is a natural consequence to allow the intermediary to relieve itself from liability. Any other situation would raise a conflict with constitutional rights (namely the right to a fair trial).

[99] HE 181/2014 at 32.

[100] HE 181/2014 at 37.

[101] Section 60e of the Copyright Act.

[102] Section 60f of the Copyright Act.

[103] HE 181/2014 at 37–38.

When proceedings are concluded and an infringement established, the right-holder may petition the court to have the security released. It is recognised that courts may establish infringement without hearing the direct infringer and sometimes do so in summary proceedings. As a rule, right-holders may not secure an interim injunction against an intermediary without posting a security for the proceedings.[104] This is prompted in part by the *IFPI v. Elisa*-case, where the District court ordered that an (untailored) interim injunction be enforced whilst Elisa's appeal was pending, but without requiring security be posted. The special requirements of the new Sections 60d and 60e are designed to assess the validity of the right-holders claim. Courts are advised that the special constitutional interests involved in these circumstances require a reasonableness inquiry that is uncustomary in summary proceedings.[105] Thus, right-holder claims to have their security released or diminished are referred to executive proceedings, and if necessary, revisited through (administrative) appeal.[106] This decision is in line with the more balanced tone of the 2014 Bill that emphasises that the constitutional interests of third parties must be protected against possible abuses of expedited summary proceedings by right-holders.

Conclusion

Many proposals have been written, from the "copyleft" to the "copyright". The rich debates have brought intellectual property enforcement closer to the general rules of procedure, constitutional law and criminal law. It is noteworthy that the final bill and statutory language includes the rights of the infringer, the users, the intermediary and the right holder as "stakeholders", when the issue of ISP liability is decided. It recognises that summary proceedings are not necessarily suited for reasonableness review, but nonetheless must be addressed by courts. Naturally, the Finnish debate was influenced by international developments. The EU Court of Justice has recognized the need to balance competing fundamental rights in *Promusicae*[107] and *UPC Telekabel*[108] decisions, when interpreting directives. Yet, it has left to national law to determine how fundamental rights should be balanced in national proceedings. The Finnish legislator has answered this call with clearer statutory provisions to aid national courts in striking a fair balance between competing interests relating to online copyright infringement.

---

[104] The Draft bill mentions that the enforcement costs in the Pirate Bay case amounted to 12,500 euros/order, which the operators paid as part of their general cost of operation.

[105] HE 181/2014 at 36.

[106] HE 181/2014 at 35–36

[107] Case C-275/06 Promusicae [2008] ECR I-271.

[108] Case C-314/12, UPC Telekabel Wien GMbH v. Constantin Film Verleih, 27.3.2014.

References

Legislation

Communications Market Act 393/2003 (repealed 31.12.2014).
Copyright Act (404/1961).
Directive 2000/31/EC of the European Parliament and of the Council of June 2000 on certain legal aspects of information society services, in particular electronic commerce, in the Internal Market.
Directive 2004/48/EC of the European Parliament and of the Council of 29 April 2004 on the enforcement of intellectual property rights.
E-Commerce Act 458/2002 (repealed 31.12.2014).
Information Society Code (917/2014).
Law 607/2015 changing the Copyright Act.

Bills

HE 181/2014 vp Hallituksen esitys eduskunnalle laiksi tekijänoikeuslain muuttamisesta.
HE 221/2013 vp Hallituksen esitys eduskunnalle tietoyhteiskuntakaareksi sekä laeiksi maankäyttö- ja rakennuslain 161 §:n ja rikoslain 38 luvun 8b §:n muuttamisesta.
HE 124/2012 vp Hallituksen esitys eduskunnalle markkinaoikeutta ja oikeudenkäyntiä markkinaoikeudessa koskevaksi lainsäädännöksi.
HE 235/2010 vp Hallituksen esitys Eduskunnalle laeiksi tekijänoikeuslain 60 a §:n ja sähköisen viestinnän tietosuojalain muuttamisesta (expired).
HE 194/2001 vpHallituksen esitys Eduskunnalle laiksi tietoyhteiskunnan palvelujen tarjoamisesta ja eräiksi siihen liittyviksi laeiksi.

Draft Bills

HE-luonnos_luvaton verkkojakelu. The draft has been circulated for comments (by 17.1.2014) and was set for delivery to Parliament in the Spring of 2014.

Cases

KKO 2010:48.
KKO 2010:47.
KKO 2003:88.
KKO 2002:101.
KKO 2001:42.
KKO 1999:115.
KKO 1999:8.
KKO 1998:91.
KKO 1989:15.

Other

First Report on the Application of Directive 2000/31/EC of the European Parliament and of the Council of June 2000 on certain legal aspects of information society services, in particular electronic commerce, in the Internal Market (Directive on electronic commerce); Brussels, 21.11.2003; COM (2003) 702 final (Commission Report on E-Commerce Directive)

Weckström, Katja (2013): preliminary injunctions against intermediaries for trademark and copyright infringement, p. 213–238 in Weckström, Katja (Ed.): governing innovation and expression: new regimes, strategies and techniques, Publications of the Faculty of Law at the University of Turku, Private Law Series A: 132.

Chapter 9
Common Law Pragmatism: New Zealand's Approach to Secondary Liability of Internet Service Providers

Graeme W. Austin

Introduction

New Zealand is a small common law country. Its laws are significantly influenced by the law of England and Wales, along with those of other cognate common law jurisdictions. The approach of its legislature, courts and regulators is typically piecemeal and pragmatic. No overarching standards have emerged, either in decisional or statutory law, on the liability of internet service providers (ISPs) in relation to the conduct of others using their services. In part, this may be due to the lack of consensus at the international law level on secondary liability issues, especially in intellectual property law. As a trading nation, New Zealand tends to be very responsive to public international law obligations. Without a clear consensus on the issue of secondary liability at the international level,[1] an important catalyst for law reform is absent (New Zealand has, however, emulated key aspects of the safe harbour provisions to be found in the US Digital Millennium Copyright Act (DMCA)[2] and the EU Directive on Electronic Commerce[3]). In addition, the small size of the New

Chair of Private Law, Victoria University of Wellington, New Zealand & Professor of Law, Melbourne University Law School, Australia. The author has served as a Member of the New Zealand Copyright Tribunal; the views expressed in this chapter are his own, and should not be attributed to the Tribunal. The author thanks Professors Bill Atkin and Graeme B. Dinwoodie for their comments on an earlier draft.

[1] See Graeme B. Dinwoodie, Chap. 1.

[2] Digital Millennium Copyright Act, Pub. L. No. 105-304, sec. 202, §512, 112 Stat. 2860, 2877 (1998).

[3] Council Directive 2000/31, 2000 O.J. (L 178) 1 (EC), arts. 12–15.

G.W. Austin (✉)
Victoria University of Wellington, Wellington, New Zealand

Melbourne University Law School, Melbourne, Victoria, Australia
e-mail: Graeme.Austin@vuw.ac.nz

© Springer International Publishing AG 2017   213
G.B. Dinwoodie (ed.), *Secondary Liability of Internet Service Providers*,
Ius Comparatum – Global Studies in Comparative Law 25,
DOI 10.1007/978-3-319-55030-5_9

Zealand economy has inhibited the development of a large body of case law in which detailed secondary liability principles might have evolved.

There are no generally applicable definitions of ISPs. In the defamation context, for example, the New Zealand Court of Appeal has recently said that the popular social media service "Facebook" is not an ISP.[4] In the Copyright Act 1994, however, the definition of ISP includes a party that: "(b) hosts material on websites or other electronic retrieval systems that can be accessed by a user," a definition that certainly catches Web 2.0 businesses, such as Facebook.[5]

Accordingly, New Zealand law has not yet engaged with the relative merits of horizontal standards or context-specific approaches in areas such as trademarks, copyright, invasion of privacy, and defamation.[6] Nor have New Zealand courts engaged with questions such as the relevance to primary liability of the possibility of imposing secondary liability on ISPs. And, for the most part, there has been no opportunity to determine, as a matter of decisional law, whether liability standards for ISPs are, or should be, different from secondary liability standards adopted in other contexts.

In 1999, the New Zealand Law Commission[7] published a study of Electronic Commerce, which advocated adoption of a general standard for liability of ISPs.[8] In broad outline, the Commission recommended enactment of a new statutory provision confirming that ISPs would generally be immune from liability unless they have actual knowledge of the existence of information on a website which would be actionable at civil law or constitute a criminal offence, and fail to remove promptly any offending information of which they have knowledge.[9] To date, the Commission's recommendation has not been adopted.

In the copyright context, there is potential for ISPs to be liable under s. 16(1)(i) of the Copyright Act 1994 (NZ), which provides that the exclusive rights of the copyright owner include the right "to authorise another person" to do an act that would infringe copyright. In New Zealand, there has been no case comparable to

---

[4] *Murray v. Wishart* [2014] 3 NZLR 722, at [143] (NZ Court of Appeal).

[5] Copyright Act 1994 (NZ), ss. 2(1), 92(4)(b) (safe harbour provisions for ISPs, discussed *infra*).

[6] Moreover, unlike in Australia, New Zealand's superior courts have not articulated a principle of "coherence" in private law that might drive common law development toward the development of standards that are consistent with statutory regimes. Cf. *Miller v. Miller* [2011] HCA 9; (2011) 242 CLR 446 (High Court of Australia).

[7] The New Zealand Law Commission is an independent Crown Entity under the Crown Entities Act 2004. Funded by the government, it reviews areas of the law that need updating, reforming or developing. Under the New Zealand Law Commission Act 1985, s. 5, the New Zealand Law Commission has the following functions: (a) to take and keep under review in a systematic way the law of New Zealand; (b) to make recommendations for the reform and development of the law of New Zealand; (c) to advise on the review of any aspect of the law of New Zealand conducted by any government department or organisation […] and on proposals made as a result of the review; (d) to advise the Minister of Justice and the responsible Minister on ways in which the law of New Zealand can be made as understandable and accessible as is practicable.

[8] New Zealand Law Commission, *Electronic Commerce: Part Two* (Wellington, New Zealand 1999), p. 137, para [333].

[9] Id.

Roadshow Films Pty v. iiNet Ltd.,[10] in which the High Court of Australia declined to hold an ISP liable for authorising copyright infringement when subscribers used file-sharing software. Moreover, as is discussed in more detail below, because of the recent adoption of a "graduated response" regime that applies to the activities of *primary* infringers, copyright owners will be unlikely to spend resources on establishing principles of secondary liability under which ISPs can be held responsible for facilitating the infringement of others.

The Copyright Act 1994 does include detailed safe harbour provisions that limit the remedies that can be ordered against ISPs.[11] For this purpose, the Act defines "Internet service provider" as follows:

[...] a person who does either or both of the following things:

(a) offers the transmission, routing, or providing of connections for digital online communications, between or among points specified by a user, of material of the user's choosing;

(b) hosts material on websites or other electronic retrieval systems that can be accessed by a user.[12]

The second part of the definition—"hosts material on websites or other electronic retrieval systems that can be accessed by a user"—extends the safe harbour provisions beyond firms that provide the technological "backbone" for the Internet. The definition would extend the concept of "Internet service providers" to "Web 2.0" platforms, bulletin boards, blogs, or even websites operated by firms, public entities, and private parties.

With the graduated response regime, however, Parliament did not deploy the concept of "service provider" or "Internet service provider". Instead, the relevant label term is "Internet protocol address provider" or "IPAP,"[13] defined as follows:

IPAP, or Internet protocol address provider, means a person that operates a business that, other than as an incidental feature of its main business activities,—

(a) offers the transmission, routing, and providing of connections for digital online communications, between or among points specified by a user, of material of the user's choosing; and

(b) allocates IP addresses to its account holders; and

(c) charges its account holders for its services; and

(d) is not primarily operated to cater for transient users.

Some of the firms that are the beneficiaries of the safe harbour provisions will also be "IPAPs" for the purposes of the "graduated response" regime. The drafting reflects the prevailing pragmatic attitude discussed above. In the safe habour context, the definition reaches people and firms on whose websites infringing material

---

[10] [2012] HCA 16.

[11] Copyright Act 1994 (NZ), s. 92B, *et seq.*

[12] Section 2(1).

[13] Section 122A.

has been posted. The graduated response regime, in contrast, is concerned only with "file sharing", and enforcing the cooperation of firms responsible for assigning IP addresses. Accordingly, in the latter context there is no need for any wider definition.

The pragmatic, "issue-by-issue", approach to ISP liability is illustrated by a decision on a summary judgment motion in *A v. Google New Zealand Limited*,[14] in which a medical practitioner brought a claim against Google New Zealand Ltd., alleging that the latter's search engine results displayed defamatory material as well as links to a third party website containing defamatory material. The defendant cross-applied to strike out the claim. The judge noted that this was the first time the issue of responsibility of a "search engine service provider" for the content of information on third party websites had come before the courts.[15] The focus of the court's analysis was confined to the defendant's search engine service, *not* any possible broader function the defendant might perform as a "service provider." Moreover, the analysis of the liability issue was strictly *obiter*, as the court had earlier held that the wrong defendant had been sued: Google Inc., which was incorporated in the State of Delaware in the United States, was responsible for the operation of the search engine, not the New Zealand-based named defendant, Google New Zealand Ltd.

In this, and in another first instance defamation case,[16] the court did opine that it was "reasonably arguable" that a search engine is a publisher in respect of the words and URLs displayed. Had the correct defendant been sued, it would have been inappropriate to strike out the claim on a summary judgment motion. Even so, on the current state of New Zealand law, the defendant would have been primarily liable as a "publisher" of the defamatory material. Secondary liability principles were therefore not apposite.

The latter case recently went on appeal to the New Zealand Court of Appeal,[17] New Zealand's second-highest court. It was a claim in defamation against the operator of a Facebook page in respect of third party postings. The Court of Appeal held that the operator of such a website could be liable only if he actually knew of the defamatory postings and failed to remove them within a reasonable time.[18] While the Court noted that a website operator was different from an ISP,[19] it did not explore the significance, if any,[20] of this distinction. Moreover, the analysis was directed at the scope of the defendant's *primary* liability (as a publisher), and only in that context was it concerned with the degree of the defendant's responsibility for and control over the content of the Facebook page. *Sensu stricto*, liability of a publisher in

---

[14] [2012] NZHC 2352.

[15] Ibid, [1].

[16] *Wishart v. Murray* [2013] 3 NZLR 246, 262.

[17] *Murray v. Wishart* [2014] 3 NZLR 722.

[18] Id, [144].

[19] Id, [143].

[20] Cf. *Trkulja v. Google LLC (No 5)* [2012] VSC 533, rejecting, for the purposes of the law of the State of Victoria, the proposition that an ISP performing a passive role cannot be a service provider.

the defamation context does not raise secondary liability issues. In passing, the Court did ask whether it might be appropriate to introduce, presumably by way of legislative reform, a new liability theory of "inciting defamation," genuinely a secondary liability matter, but then noted that this issue was not relevant on the facts before it.[21]

Immunity from Secondary Liability

The safe harbour provisions in the New Zealand Copyright Act 1994[22] provide the only comprehensive set of immunities from secondary liability for ISPs. No safe habours have been created in the trademark context, either by legislation or in case law. And, as noted above, in the defamation context any safe harbours that might develop are likely to concern primary liability for "publication" of the defamatory statements—unless the Court of Appeal's suggestion of adopting a liability theory based on inciting defamation is explored further. As a matter of common law principle, we might see safe harbours develop in context such as passing off and privacy.[23] As yet, this has not occurred.[24]

The copyright safe harbours were created by statute. Their scope remains untested. The safe harbour provides a set of general immunities:

92B Internet service provider liability if user infringes copyright

(1) This section applies if a person (A) infringes the copyright in a work by using 1 or more of the Internet services of an Internet service provider to do a restricted act without the consent of the copyright owner.

(2) Merely because A uses the Internet services of the Internet service provider in infringing the copyright, the Internet service provider, without more,—

   (a) does not infringe the copyright in the work:
   (b) must not be taken to have authorised A's infringement of copyright in the work:
   (c) subject to subsection (3), must not be subject to any civil remedy or criminal sanction. [...]

(3) However, nothing in this section limits the right of the copyright owner to injunctive relief in relation to A's infringement or any infringement by the Internet service provider.

---

[21] *Murray v. Wishart* [2014] 3 NZLR 722, [135].

[22] Copyright Act 1994, s. 92B *et seq.*

[23] New Zealand courts have recognized a common law tort of invasion of privacy. *See Hosking v. Runting* [2005] 1 NZLR 1.

[24] The importance of non-pecuniary remedies in the defamation and privacy contexts may account for this.

(4) In subsections (1) and (2), Internet services means the services referred to in the definition of Internet service provider in section 2(1).

The provisions were influenced by the United States regime, even if they were drafted somewhat differently. It is unlikely, however, that the drafting changes will result in any major substantive distinctions. The United States statute provides that the service provider "shall not be liable for monetary relief",[25] while specifying the circumstances in which injunctive relief *will* be available.[26] In contrast, and perhaps somewhat awkwardly, the New Zealand statute provides in s. 92B(2) that the Internet service provider "does not infringe" when within the safe harbours. In the subsection immediately following, however, the statute specifically preserves the availability of injunctive relief in relation to the primary infringer's activity "or any infringement by the Internet service provider."[27] Given that an injunction cannot issue *unless* there is infringement, courts will be likely to interpret s. 92B(2A) to mean that the provisions are only triggered when an infringement has occurred (based on another ground), and that an ISP that complies with the conditions does not attract liability, *except* for the purposes of an application for injunctive relief. Substantively, therefore, the interpretation of the New Zealand provisions is likely to be in line with the equivalent US law.

Specific safe harbours apply where the Internet service provider "stores material provided by a user" that infringes copyright in a work,[28] and where the ISP caches material.[29] Both of these safe harbours concern the services referred to in the definition of Internet service provider provided in the Copyright Act 1994 (extracted above). The safe harbour is conditioned on the service provider having "store[d] material provided by a user of the service"; however, the description of relevant "services" that brings a service provider within the ambit of the provision refers to the "host[ing]" of material. The relevance of the terminological distinction drawn in the definition and the safe harbour provision itself is not obvious. No case law has addressed the significance (if any) of this drafting choice.

The safe harbour for storing infringing material is described as follows:

92C Internet service provider liability for storing infringing material

(1) This section applies if—

    (a) an Internet service provider stores material provided by a user of the service; and

    (b) the material infringes copyright in a work (other than as a result of any modification by the Internet service provider).

---

[25] Copyright Act 1976 (US), s. 512(a).

[26] Copyright Act 1976 (US), s. 512(j).

[27] Copyright Act 1994 (NZ), s. 92B(3).

[28] Copyright Act 1994 (NZ), s. 92C(1)(a) and (b).

[29] Copyright Act 1994 (NZ), s. 92C.

(2) The Internet service provider does not infringe copyright in the work by storing the material unless—

    (a) the Internet service provider—

        (i) knows or has reason to believe that the material infringes copyright in the work; and

        (ii) does not, as soon as possible after becoming aware of the infringing material, delete the material or prevent access to it; or

    (b) the user of the service who provided the material is acting on behalf of, or at the direction of, the Internet service provider.

(3) A court, in determining whether, for the purposes of subsection (2), an Internet service provider knows or has reason to believe that material infringes copyright in a work, must take account of all relevant matters, including whether the Internet service provider has received a notice of infringement in relation to the infringement.

(4) An Internet service provider who deletes a user's material or prevents access to it because the Internet service provider knows or has reason to believe that it infringes copyright in a work must, as soon as possible, give notice to the user that the material has been deleted or access to it prevented.

(5) Nothing in this section limits the right of the copyright owner to injunctive relief in relation to a user's infringement or any infringement by the Internet service provider.

While there is an obligation to notify the third party whose content has been blocked or removed, there is no obligation to put back the material, following submissions by that party. No "counter notice" procedure is stipulated, in contrast to the DMCA.[30] Moreover, there are no statutory protections for ISPs that act on a notice and take down material in good faith.[31]

As originally enacted, the safe harbours applied only to ISPs that had "adopted and reasonably implemented a policy that provides for termination, in appropriate circumstances, of the accounts of repeat infringers."[32] The prospect of cutting off Internet access attracted significant public controversy, and the provision has never come into force.[33]

Perhaps the most interesting aspect of the drafting concerns the question of knowledge. An ISP loses the benefit of the safe harbour where it knows or has reason to believe that the material is infringing and does not take steps to delete or remove access to it. To determine whether the ISP had the requisite knowledge, the court is required to take into account all relevant factors, including whether the ISP received a notice of the infringement. This implies that knowledge is not necessarily

---

[30] Copyright Act 1976 (US), s. 512(g)(2) and (3).

[31] Cf. Copyright Act 1976 (US), s. 512(g)(1).

[32] Copyright Act 1994, s. 92A (repealed).

[33] Copyright (Infringing File Sharing) Amendment Act 2011, s. 4.

to be equated with receipt of an infringement notice—suggesting that other factors might be relevant in particular cases, including the quality of the notice and the information it provides. Knowledge from other sources might also be relevant. That said, in most cases, the information that would be relevant to establishing knowledge of the infringement is prescribed by the regulations setting out the form of the required notice:

5B Notice to Internet service provider of infringement

A notice to an Internet service provider … must be in the form of a notice that—

    (a) contains the following information:

        (i) the name of the person signing the notice and the name of the copyright owner or the copyright owner's duly authorised agent (if different from the person signing the notice):

        (ii) the contact details of the persons referred to in subparagraph (i), which must include telephone number, postal address, email address, and fax number (if a fax number is available):

        (iii) the date and time when the alleged infringement was discovered:

        (iv) a description of the specific material that is alleged to be infringing:

        (v) the nature of the alleged infringement:

        (vi) the online location where the allegedly infringing material is found; and

    (b) is signed by—

        (i) the copyright owner; or

        (ii) the copyright owner's duly authorised agent.[34]

Presumably, there is also scope for other matters to be included in the notice, or drawn otherwise to the attention of the ISP—which may be relevant to a court's decision as to whether the knowledge requirement is satisfied. For instance, an analysis of why a permitted use or other defence did not apply to the storage of the particular work might be relevant to establishing the requisite knowledge (There are no statutory provisions penalising parties for serving false or misleading notices[35]). The quality of notices, including notices generated automatically, has been discussed in leading New Zealand media.[36]

---

[34] Regulation 5B of the Copyright (General Matters) Regulations 1995, as amended by regulation 4 of the Copyright (General Matters) Amendment Regulations 2008 (SR 2008/352).

[35] Cf. Copyright Act 1976(US), s. 512(f). In New Zealand, these may in some circumstances be addressed by prohibitions in the Fair Trading Act 1986, s. 9, against misleading conduct in trade. The Copyright Act 1994, s. 302, includes a provision creating civil liability for unjustified proceedings. Unlike in some jurisdictions, this does not reach unjustified allegations. Section 112 of the Act provides a remedy against false claims of as to copyright ownership and/or licensing arrangements.

[36] Pat Pilcher, *HBO Issues Takedown on VLC*, New Zealand Herald, 16 July 2013, available at: http://www.nzherald.co.nz/business/news/article.cfm?c_id=3&objectid=10898892. Courts have not engaged with this question.

Finally, from the drafting of the regulations that stipulate the requirements for notices, it appears that the power to issue notices is vested in the copyright owner or the copyright owner's authorized agent. Licensees do not appear to have standing to issue notices.

File Sharing: The Graduated Response Regime

A detailed "graduated response" regime that is focused on responsibility for file sharing was created by the Copyright (Infringing File Sharing) Amendment Act 2011. While the graduated response regime does not directly concern the secondary liability of ISPs, it is closely related to the issues explored in this volume. For example, the regime imposes obligations on ISPs to assist with the operation of the graduated response process. Most significantly, however, this regime, together with the safe harbour provisions, is likely to account for the lack of jurisprudence on the secondary liability of ISPs. It is likely that the existence of an efficient, relatively low cost, system for securing relief against *primary* infringers will mean that there will be little need to develop secondary liability principles trained on the activities of ISPs themselves. That said, the regime applies only to "file sharing", and not copyright infringement more generally.

The New Zealand Copyright Tribunal is the body principally responsible for enforcing the new regime.[37] Appeals from the Tribunal's decisions may be made to the High Court,[38] and the Tribunal may of its own volition state a case to that Court.[39] To date, there have been no High Court decisions under the new regime.

For the purposes of the regime, "file sharing" is defined as follows: "file sharing is where—(a) material is uploaded via, or downloaded from, the Internet using an application or network that enables the simultaneous sharing of material between multiple users; and (b) uploading and downloading may, but need not, occur at the same time."[40] "Infringement" means an incidence of file sharing "that involves the infringement of copyright." Accordingly, the new regime is concerned only with *infringement by file sharing*. The definition of "file sharing" appears to be quite wide. An email program, for example, could allow "simultaneous sharing of material between multiple users". So far, however, no cases have examined the scope of this concept. The term "file sharing" is not used in any other context in New Zealand's copyright legislation.

The New Zealand scheme requires IPAPs, at the instigation of rights owners, to issue three infringement notices to subscribers: a detection notice; a warning notice; and an enforcement notice. After an enforcement noticed is issued to an alleged

---

[37] The New Zealand Copyright Tribunal was constituted by s. 30 of the Copyright Act 1962; Copyright Act 1994, s. 205 provides that the Tribunal "shall continue in being."

[38] Copyright Act 1994, s. 224.

[39] Id, s. 223.

[40] Id, s. 122A.

infringer, the rights owner may take enforcement action by seeking an order from the Copyright Tribunal for an award of sum of up to NZ$15,000. The new legislation also establishes "quarantine periods" at each stage of the process—affording alleged infringers the opportunity to stop infringing copyright by file sharing once serviced with an initial notice. The Parliament has also provided a section empowering a District Court to require an IPAP to suspend the subscriber's Internet account for 6 months.[41] The relevant provision has been enacted in the new legislation, but no commencement date has been set.

The definition of "rights owner" is important. It includes a copyright owner as well as a person acting as agent for one or more copyright owners.[42] It is also specified that:

If a rights owner acts as agent for 1 or more copyright owners,—

(a) a reference to the copyright of a rights owner is to be taken as a reference to the copyright of any of the copyright owners for whom the rights owner acts as agent; and
(b) a reference to infringement against a rights owner is to be taken as a reference to infringement against the copyright of any of the copyright owners for whom the rights owner acts as agent.[43]

In practice, the agent for rights owners in proceedings under the new regimes has been the collective rights management agency, the Recording Industry Association of New Zealand (now Recorded Music NZ). In this respect, the system reflects the typical collective management of rights in commercial recordings. These provisions address a potential coordination problem: it would be quite likely that the three instances of infringement by file sharing occurring at any individual's Internet account were in respect of works with different copyright owners. Without delegation to an agent, rights owners would need to share potentially quite significant amounts of information about detection. Moreover, the use of the detection technology can be centralised under the control of the designated agent.

Where a subscriber challenges an infringement notice (which must be done within 14 days of the infringement notice to which it relates[44]), the IPAP must forward the challenge notice to the rights holder. The rights holder has the option of accepting or rejecting the challenge notice. A challenge is deemed to be accepted if it is not rejected within 28 days after the date of the infringement notice to which it relates.[45]

If a rights owner provides an IPAP with information that identifies an IP address at which an infringement of its copyright is alleged to have occurred as a result of file sharing, the IPAP is obliged to match the IP address with the account holder to whom it related at the time of the infringement, and issue the appropriate notice

---

[41] Copyright Act 1994, s. 122P.
[42] Id, s. 122A(1), defining "rights owner" for the purposes of the new regime.
[43] Id, s. 122A(2).
[44] Id, s. 122G.
[45] Id, s. 122H. (If the challenge notice is rejected, the rights holder must provide reasons.)

within 7 days of receiving the information.[46] The obligation to match IP addresses expires if the alleged infringement occurred more than 21 days before the IPAP received the relevant information.[47] The new regime also imposes a number of record keeping obligations on IPAPs. An IPAP must retain, for a minimum of 40 days, information on the allocation of IP addresses to each of its account holders. In addition, every IPAP must retain, for a minimum of 12 months, information about: infringements that is sent by rights owners to the IPAP for the purpose of matching infringements to subscribers and information about challenge notices; and which infringement notices (if any) have been cancelled or have expired.

Regulations[48] made under the authority of the new legislation provide further detail as to the notice requirements under the regime, strictly prescribing the information that must be sent to subscribers before the IPAP issues any infringement notices. Regulation 4, which prescribes the information that rights owners must provide to the IPAPs, is typical:

(1) Every rights owner notice provided to an IPAP must include the following information about the rights owner who provides the notice:

   (a) the rights owner's name:
   (b) the rights owner's contact details, which must include an email address, telephone number, and physical address:
   (c) if the rights owner does not have a physical address in New Zealand, an address in New Zealand to which the IPAP can send documents for the rights owner:
   (d) if the rights owner is acting as an agent for a person whose copyright is alleged to have been infringed, evidence of the rights owner's authority to act as agent for that person.

(2) Every rights owner notice provided to an IPAP must—

   (a) identify the IP address at which the infringements are alleged to have occurred; and
   (b) state the date on which the infringements are alleged to have occurred at that IP address; and
   (c) in relation to each copyright work in which copyright is alleged to have been infringed,—

      (i) give the name of the owner of copyright in the work; and
      (ii) give the name of the work, along with any unique identifiers by which it can be identified; and
      (iii) describe the type of work it is (in terms of section 14(1) of the Act); and
      (iv) describe the restricted act or acts (in terms of section 16(1) of the Act) by which copyright in the work is alleged to have been infringed; and

---

[46] Id, s. 122C.
[47] Id, s. 122C.
[48] Copyright (Infringing File Sharing) Regulations 2011 (SR 2011/252).

      (v) give the New Zealand date and time when the alleged infringement occurred or commenced, which must specify the hour, minute, and second; and

     (vi) identify the file sharing application or network used in the alleged infringement.

(3) Every rights owner notice must include a statement that, to the best of the rights owner's knowledge, the information provided in the rights owner notice is true and correct; and that statement must be verified by a signature (physical or digital) of the rights owner or a person authorised to sign on behalf of the rights owner.

(4) If a rights owner sends a series of rights owner notices to an IPAP that relate to alleged infringements occurring at different IP addresses, the rights owner need only make the statement referred to in subclause (3) once in relation to all those notices, as long as it is clear which notices the statement relates to.

(5) If an IPAP specifies a process by which rights owners can provide rights owner notices to the IPAP, then as long as that process is consistent with the Act and these regulations, rights owners must comply with the process specified by the IPAP, unless otherwise agreed by the IPAP.

The regulations also address the issue of cost, prescribing the maximum amount that an IPAP may charge a rights owner for performing the functions required of the IPAP. Currently this is set at a maximum of $25 per notice sent to the IPAP.[49]

The third enforcement notice triggers the copyright owner's right to apply to the Copyright Tribunal for compensation. The Tribunal is required to issue an order against the subscriber if: (1) the three infringement notices were properly issued; (2) there was an infringement of the right owner's copyright; and (3) the infringement occurred "at an IP address of the [subscriber]."[50] Most cases will be determined "on the papers" without the need for a hearing. If a hearing does occur, parties will not normally be represented by legal counsel.[51] All parties have a right to be heard if a proceeding before the Tribunal is conducted.[52]

In proceedings before the Copyright Tribunal, the rights owner has the advantage of a number of important evidentiary presumptions. Section 122 N of the Copyright Act 1994 provides:

(1) In proceedings before the Tribunal, in relation to an infringement notice, it is presumed—

    (a) that each incidence of file sharing identified in the notice constituted an infringement of the rights owner's copyright in the work identified; and

    (b) that the information recorded in the infringement notice is correct; and

    (c) that the infringement notice was issued in accordance with this Act.

---

[49] Id, reg. 7.
[50] Copyright Act 1994, s. 122O.
[51] Id, s. 122M.
[52] Id, s. 122C.

(2) An account holder may submit evidence that, or give reasons why, any 1 or more of the presumptions in subsection (1) do not apply with respect to any particular infringement identified in an infringement notice.

(3) If an account holder submits evidence or gives reasons as referred to in subsection (2), the rights owner must satisfy the Tribunal that, in relation to the relevant infringement or notice, the particular presumption or presumptions are correct.[53]

No clear view has emerged as to the evidential threshold either party must satisfy when it is not possible to rely on the statutory presumptions. Once the account holder "submit[s] evidence or give[s] reasons why" one or more of the presumptions do not apply, it is for the rights owner to satisfy the Tribunal that the relevant presumption or presumptions are correct. The quality of the evidence that will be sufficient to reverse this onus may raise difficult issues in future cases,[54] perhaps requiring a case to be stated to the High Court.

If the Tribunal is satisfied that the required three instances of file sharing occurred, the Tribunal is required to order the account holder to make a payment to the rights owner (In practice, the sum is usually payable to the rights owner's agent). In determining the amount of that payment, the accompanying regulations require the Tribunal to consider a number of factors, including: (1) the reasonable cost of purchasing the infringed work in electronic form if it was legally available for purchase at the time of the infringement; (2) if the was not available for purchase in electronic form, the reasonable cost of purchasing it in another format, if available; and (3) the amount claimed by the rights holder, or any other reasonable amount determined by the Tribunal. The Tribunal must also consider the application fee for initiating proceedings and any amount that the Tribunal considers appropriate to be a deterrent against further infringing. When assessing the "deterrence" figure, the Tribunal is required to consider: the flagrancy of the infringement, the possible effect of the infringing activity on the market for the work; and whether the sum calculated in the light of the reasonable marketplace cost of the work.[55] The Tribunal has taken the view that the issue of "flagrancy" in the graduated response regime should be approached differently from the way that a court might assess the flagrancy of infringement in ordinary copyright infringement proceedings. In the latter type of proceedings, a court is likely to have the opportunity to have heard and assessed witnesses giving *viva voce* evidence, which is a very different situation from the Tribunal's summary proceedings determined on the papers.[56]

The regulations specify that the Tribunal must consider the extent of the infringement of "works" that occurs at IP addresses, rather than the number of infringing copies that were "shared". In practice, the awards made by the Tribunal have been

---

[53] Id, s. 122N.

[54] This issue has been raised by the Tribunal in recent decisions. See, e.g., [2013] NZCOP 7, [18].

[55] Copyright (Infringing File Sharing) Regulations 2011 (SR 2011/252), reg 12(3).

[56] See, eg, *Recording Industry Association of New Zealand v TCLE[A] T7364885* [2013] NZCopyT 17; [2013] NZCOP 17.

significantly below the $15,000 threshold. A few awards have been close to $1000.[57] Most have been significantly lower.[58]

The IPAPs are themselves precluded from disclosing identifying information to rights holders. The legislation states: "On issuing an enforcement notice to an account holder, the IPAP must send a copy of the notice to the relevant rights owner, but must omit any information that discloses the name or contact details of the account holder."[59] The Tribunal itself "must make all reasonable efforts to ensure that, unless it orders otherwise or an order is made against the account holder, the identity and contact details of the account holder are not disclosed to the rights owner."[60] The IPAPs have the information that matches IP addresses to individual subscribers; the copyright owners might have only the information that matches IP addresses to alleged acts of infringements. Only the Copyright Tribunal will have access to both.

A significant aspect of the regime for the present study is the legislature's decision to make account holders the target of the claims for compensation, rather than individual infringers. In some decisions, account holders have claimed that their children engaged in the primary act of infringement.[61] By imposing liability on the account holder, regardless of the identity of the primary infringer, the new regime does not involve the Tribunal in resolving difficult evidentiary issues as to identifying the actual infringer. In this sense, the regime is akin to making the owner of a motor vehicle liable for parking infractions. The regime thus encourages supervision of internet access by account holders.

Conclusion

Consistent with its common law tradition, the New Zealand legal system has not adopted a coherent or comprehensive approach to the issue of the secondary liability of ISPs. Instead, the issue has evolved in a piecemeal fashion—responding in a contexualised way to issues as they have arisen from time to time. The prevailing character of that tradition may in part account for successive governments' lack of interest in adopting the Law Commission's recommendation in favour of a *general* immunity for ISPs based on their level of knowledge of the infringements. Also relevant in the New Zealand context is the small size of the jurisdiction and the lack of a large body of case law enabling a more comprehensive evolution of secondary

---

[57] [2013] NZCOP 8 ($914.35); [2013] NZCOP 9 ($803.62).

[58] Since the inauguration of the scheme, 17 decisions of the Copyright Tribunal have been posted to the website of the New Zealand Ministry of Justice: http://www.justice.govt.nz/tribunals/copyright-tribunal/decisions-1

[59] Copyright (Infringing File Sharing) Amendment Act 2011, s. 6, inserting a new s. 122F(5) into the principal Act.

[60] Copyright Act 1994, s. 122 L(4) into the principal Act.

[61] See, e.g., [2013] NZCOP 7; [2013] NZCOP 2.

liability issues. Finally, the lack of consensus at the public international law level is likely to account for the absence of law reform initiatives in this context. In intellectual property law, in particular, the New Zealand legislature is highly responsive to the exigencies of public international law. The absence of a clear international law approach to secondary liability is likely to mean that New Zealand's law on the topic of secondary liability is likely to continue to develop in a piecemeal fashion.

Chapter 10
Secondary Liability of Internet Intermediaries and Safe Harbours Under Croatian Law

Ivana Kunda and Jasmina Mutabžija

Introduction

The issue of civil liability of Internet intermediaries[1] for content placed on the Internet by users of their services is still largely unsettled in Croatian law. Lacking a specific provision on secondary liability, the Croatian courts are seeking ways to address this issue under existing legal rules. As will be seen below, from the described developments in scarce case law, this is a challenging process given that these legal rules were not drafted in view of recent technological advances and business models. This chapter attempts to clarify the legal status of Internet intermediaries by discussing various issues arising in searching for the legal basis of their secondary liability and the effort to define the scope of the safe harbour for intermediaries.

---

[1] Because the term "Internet service providers" is mostly used to denote companies which provide Internet access and sometimes additional services such as basic webhosting or e-mail account services, the authors of this Chapter opt for the term "Internet intermediaries", being a broader term which includes entities providing all sorts of newly-developed services facilitating the use of Internet, such as search engines, social networks, comment sections on blogs and websites.

I. Kunda (✉)
Faculty of Law, University of Rijeka, Rijeka, Croatia
e-mail: ikunda@pravri.hr

J. Mutabžija
POSLuH hosting d.o.o., Zagreb, Croatia
e-mail: jasmina@posluh.hr

© Springer International Publishing AG 2017
G.B. Dinwoodie (ed.), *Secondary Liability of Internet Service Providers*,
Ius Comparatum – Global Studies in Comparative Law 25,
DOI 10.1007/978-3-319-55030-5_10

Legal Basis for Secondary Liability of Internet Intermediaries

This section is aimed at discussing in detail the legal characterisation of Internet intermediaries' liability, constitutive elements for establishing liability, type of liability, liability standards according to the cause of action in question, types of remedies and criminal liability of Internet intermediaries related to the conduct of third parties using their services.

Legal Characterisation

When in 2005 the issue of secondary liability of Internet intermediaries first reached the Croatian courts,[2] they were confused as to the legal ground on which to decide the case. It was a classic Internet defamation case. An anonymous user published allegedly defamatory content on a blog. The blog was owned by a natural person from Vukovar. The blog used the domain name www.vukovarac.net and it was hosted by a Croatian company. Anyone could freely and anonymously post articles or comments on the blog's web page. The blog owner was sued by the victim for the damages caused by the comment of an anonymous user, which the blogger refused to take down despite a request of the victim. He was also sued for damages caused by defamatory content that he placed on the blog.

Deciding in the first instance, the Municipal Court in Vukovar upheld the claim for damages based on the primary, but not on the secondary, liability of the blog owner. In respect of the secondary liability, the Court relied on the Media Act (hereinafter, the MA),[3] holding that the blog owner is neither the publisher nor the editor within the meaning of that Act, and thus not liable. The Court also oddly concluded that the blog owner, although he was a "lessee" of the web page, did not have the opportunity to intervene in the articles and comments on the blog's web page.[4] On appeal, the plaintiffs claimed that the web page in question was a sort of a bulletin board in electronic form, owned by the defendant, a telecommunications expert by profession, possessing all necessary knowledge for making and maintaining the web page; and that he, as the owner of the web page, allowed its free use, including for the purpose of placing defamatory content at the expense of both plaintiffs, which was known to the defendant and not removed intentionally, although he had the opportunity to do so. The plaintiffs concluded that the defendant should be liable for the damage caused to the plaintiffs by such conduct.

---

[2] Since there is no all-encompassing database of court decisions in Croatia, it is not possible to claim with certainty that there has not been a decision on the issue which was not taken into account in this Chapter. This said, it is unlikely given that such disputes are as a rule discussed on the Internet.

[3] Media Act, Official Gazette of the Republic of Croatia 59/2004, 84/2011 and 81/2013.

[4] Municipal Court in Vukovar, P-737/05, judgement of 5 June 2006.

On appeal, the County Court in Vukovar affirmed the decision in relation to primary liability, but quashed it in relation to secondary liability, remitting the latter issue to the court of first instance to decide anew. In their reasoning, the judges held that the compensation of damages should have been decided under the Obligations Act (hereinafter, the OA),[5] rather than the MA.[6]

In the new proceedings, the Municipal Court in Vukovar ruled in favour of the plaintiffs and held that the blog owner was liable. The Court's judgment in part related to wrongfulness was based on Article 12 of the Criminal Procedure Act (hereinafter, the CPA),[7] which provides that the court hearing a civil case is, in respect of whether there is a criminal offence and criminal liability, bound by the guilty verdict of the criminal court. Therefore, this judgment helps little in solving the conundrum on the exact legal basis of Internet intermediary's liability in relation to wrongful acts of its users. Indeed the court invokes several provisions of the Obligations Act that were in force at the time the tort was committed:[8] the provision on damage, including damage to immaterial property (*nematerijalna šteta*), on fault (*krivnja*) and on fair compensation of damage (*pravična novčana naknada*).[9] All these provisions have their equivalents in the OA, which has been in force since the beginning of 2006.[10] Hence, this judgment is relevant for cases falling under the OA as well. The Court noted that the blog owner was the operator of the webpage on which the disputed content was published, and that he had individual access to the IP logs from which he could have determined who accessed the page at the material time and who wrote the defamatory comments. The Court concludes that "he could have had, but did not preclude the publishing of defamatory articles and comments on the page" and that he, "a natural person owning the respective webpage where he enabled defamation of the plaintiff", is liable for non-pecuniary damage. This judgment, however, is not final as the appeal proceedings are pending at the time this Chapter is submitted for publication.

Based on the decisions and reasoning in the aforementioned case, in which the final decision is still awaited, it may be concluded that the primary liability of Internet intermediaries is governed by the general provisions on non-contractual

---

[5] Official Gazette of the Republic of Croatia 35/2005, 41/2008, 125/2011 and 78/2015.

[6] County Court in Vukovar, Gž-241/07, judgment and decree of 20 November 2007.

[7] Official Gazette of the Republic of Croatia 53/1991, 91/1992, 112/1999, 129/2000, 88/2001, 117/2003, 88/2005, 2/2007, 96/2008, 84/2008, 123/2008, 57/2011, 25/2013 and 89/2014.

[8] Official Journal of the Socialist Federative Republic of Yugoslavia 29/1978, 39/1985, 46/1985, 57/1989 and Official Gazette of the Republic of Croatia, 53/1991, 73/1991, 3/1994, 111/1993, 107/1995, 7/1996, 91/1996, 112/1999, 88/2001 and 35/2005.

[9] Municipal Court in Vukovar, P-70/2013, judgment of 25 November 2013.

[10] See Articles 1046 (damage, including to immaterial property), 1049 (fault) and 110 (fair compensation of damage) of the OA.

obligations in the OA, and in case of personality rights, including defamation, also the specific provision of Article 19 of the OA.[11] Rightly so, the MA has been rejected as the basis for deciding on secondary liability of Internet intermediaries. However, it is impossible to conclude with certainty as to the court's interpretation of the basis for secondary liability of Internet intermediaries in Croatian law because the Municipal Court in Vukovar relied on criminal liability as the basis for civil liability in the case in question. Besides, this decision is being tested before the second instance, although it is unlikely that it will be annulled or overruled on these grounds.

There are two possible legal bases that may be relied on under the OA, in particular in Title IX of the OA—"Non-Contractual Obligations", for the purpose of non-contractual liability: (1) the liability for one's own wrongful acts/omissions (*vlastita odgovornost*); or (2) liability for the acts/omissions of another person (*odgovornost za drugoga*). Most commonly, the person is liable for damage caused by her or his own acts or omissions, except where there are reasons for exclusion of liability. However, it is possible that in certain instances a person will be liable for the damage caused by another person. These are the cases enumerated under Section 3 of the OA, titled "Liability for Another", and include: (1) liability for mentally ill or incapable persons; (2) parent's liability for their children; (3) other person's liability for a minor; (4) employer's liability for its employees; and (5) legal person's liability for its body.[12]

One possible interpretation could be that the blog owner's liability should be based on liability for another person's acts or omissions. This might seem logical taking into consideration that the legal notion of "liability for another person" in the OA is largely derived from the need to ensure the liability of a person in control of and/or responsible for the acts of the other person because there is some legally relevant relationship between them (such as filiation, guardianship, employment etc.). Similarly, it is possible to claim that an Internet intermediary is in control of its user's acts and hence should bear liability for them. On the other hand, one might claim that the Internet intermediary is simply providing means for communication between its users and as such should not be under the obligation to control their acts and be liable for them.[13] Whatever the case may be, from the system and wording of the provisions of Articles 1055–1062 of the OA relating to the liability for another person, it may be concluded that these are limited to the five abovementioned situations and cannot be extended by analogy or otherwise to other situations, including the situation of potential Internet intermediary liability discussed in this Chapter.

Assuming this legal basis is rejected, the basis for secondary liability of Internet intermediaries under Croatian law is limited to the liability for their own acts or omissions under the general provision of Article 1045 of the OA. Based on the usual

---

[11] The authors' own translation of the provision of Article 19(1) of the OA reads: "Every natural and legal person has the right to protect her or his personality rights under conditions defined by the statute."

[12] See Articles 1055–1062 of the OA; Gorenc 2014, 1734–48; Slakoper 2005, 750–64.

[13] Such argument is usually coupled with the reservation that an Internet intermediary should be liable if remaining idle although aware of the wrongdoing.

fact pattern, those acts or omissions would often entail enabling, aiding, helping, abetting, or inducing a person (the user), to commit a wrongful act against the third person (the victim). Thus, although factually related to an act of the user of Internet intermediaries' services, these acts may be legally characterised as acts or omissions of the Internet intermediaries. Therefore, Article 1045 of the OA may serve as the basis for secondary liability of Internet intermediaries for the acts of users of their services which result in violation of third party's rights, albeit that the term "secondary liability" or its equivalent is not known in Croatian law. It would be complemented by Article 19 of the OA, insofar as the matter relates to the violation of personality rights.

After a decade-long saga of vukovarac.net, it is still not possible to conclude with absolute certainty whether there is secondary liability of Internet intermediaries under Croatian law, and, if so, how that liability would be characterised. Nevertheless, it is the opinion of the authors that, taking account of the court's reasoning and for the additional reasons explained above, the general provision of Article 1045 of the OA may be applied to cases of service providers' liability, including that of the Internet (or online) intermediaries and that they should be characterised as their own wrongful acts (which complement the wrongful acts of their users).

Requirements for Non-contractual Obligations to Arise

In principle there is a horizontal standard of liability applicable to the majority of the causes of action detailing the liability of service providers. There are, however, special standards contained in intellectual property laws as discussed below.

The basic provision for establishing liability in tort contained in Article 1045 of the OA states that whoever causes damage to another person has a duty to compensate that person, unless she or he proves that she or he was not at fault. Five conditions must be satisfied for liability in tort to arise: (1) subjects (person liable and person suffering damage); (2) wrongful act or omission; (3) damage; (4) causal link between the wrongful act or omission and damage; and (5) wrongfulness of the act or omission.[14] In our view, Article 1045 may also be used as grounds for secondary liability of Internet intermediaries for the acts of users of their services.

When discussing the requirements for secondary liability of Internet intermediaries, some of the elements demand particular attention. One such element is the causal link between a wrongful act or omission and damage, as there may be some difficulty in establishing the causal link between the acts of the service provider and the damage suffered by the third person to whom the service provider had no direct connection.[15] This is one of the main arguments relied upon by the defendant, on appeal, in the

---

[14] Gorenc 2014, 1699. In respect of liability for intellectual property rights infringement, in particular see Henneberg 2001, 204–5.

[15] This is not as problematic in case of liability for the acts of another person, where the causal link has to exist between the act of the factual tortfeasor (for whom the presumptive tortfeasor is liable) and the damage. Gorenc 2014, 1705.

vukovarac.net litigation. However, if we are to characterise the Internet intermediary's wrongful acts or omissions as its own, the causal link is present. The damage to defamation victim arises (or does not end) if an Internet intermediary fails to remove a defamatory content hosted by the Internet intermediary. The fact that the user of the Internet intermediary's services is the primary tortfeasor and that Internet intermediaries' acts or omissions are of secondary importance, does not alter this contention. Regardless of whether a defendant has an assisting or main role, it is liable under law. In explaining the existence of a causal link between the articles and the related comments, on the one hand, and harm of the plaintiffs, on the other, the Municipal Court in Vukovar in its 2013 judgment,[16] reasoned that the articles and comments were blatantly defamatory and that "in that sense, the defendant had a duty to prevent defamation in the articles and comments which were published on the respective web page".

Besides, wrongfulness might be an important element in discussing secondary liability of Internet intermediaries. Wrongfulness means that a legal norm has been violated (objective aspect of wrongfulness) by a conduct not meeting the standards of due care (subjective aspect of wrongfulness). The liability standard of due care under this rule is *culpa levis* (*obična nepažnja*) and it is presumed. Presumed fault means that a default presumtion of fault applies once the victim has shown the existence of all other conditions of liability in tort listed above. This presumption is rebuttable and the tortfeasor wishing to be exonerated from liability has to prove that there was no fault on her or his part, i.e. that she or he acted according to the applicable standard of due care, and hence did not cause the harm in question.[17] The standard of due care varies according to the characteristics of the legal relationship at stake. It is described as the due care which is required in the particular type of a relationship, namely, that used by good trader (in case of commercial contracts) or good host (in case of non-commercial contracts)[18] or, generally, due care used by a particularly careful man.[19] Where there is an obligation attaching to the professional activity of the person, the standard of due care is elevated, and care consistent with the rules of the applicable art and custom is required, i.e. that used by the good professional.[20] However, there might also be a difference between the general standards of due care prescribed in the OA and those required under the special regime for intellectual property rights. Indeed, the standard of *culpa levis* is, in cases concerned with intellectual property rights, applicable to the majority of the remedies available under the respective laws, including permanent injunctive remedies and condemnatory remedy for compensation of damages. However, in situation of an infringement of the right of an author

---

[16] See supra n. 10.

[17] Different standards of liability are sometimes prescribed for special cases. For instance, causal (objective) liability which exists regardless of the fault is prescribed in cases of liability for damage caused by a hazardous thing or a hazardous activity in Articles 1963 et seq. of the OA. Furthermore, compensation of damage caused by the employee to a third person may as a rule be claimed from the employer, and exceptionally directly from the employee only if the employee intentionally caused the damage (*dolus*).

[18] Article 10, paragraph 1 of the OA.

[19] Supreme Court of the Republic of Croatia, Rev. 387/1998 of 19.03.1998, Izbor 1998/91.

[20] Article 10, paragraph 2 of the OA.

(other than moral rights), the so-called penalty is available under Article 183 of the Croatian Copyright and Related Rights Act (hereinafter, the CRRA).[21] This remedy entitles right-holder to ask for double the amount of the agreed compensation, or if nothing is agreed, double the amount of the usual compensation. Whereas the proof of damage is not required, the proof of infringer's acting intentionally or with gross fault (*culpa lata*) is necessary. One should keep in mind that certain notions in the CRRA are of EU law origin and subject to interpretation by the Court of Justice of the European Union. For instance, the issue of liability for the linked copyright-infringing content which is gaining increasing attention due to the recent CJEU case law.[22]

It has to be pointed out that in cases in which an obligation is created based on unjust enrichment (*conditio sine causae*), the requirements are different as this is not characterised as liability in tort but a quasi-delict, both of which fall under the common denominator of non-contractual obligations. The requirements for unjust enrichment are: (1) increase in property (enrichment) on the part of the defendant; (2) reduction in property (impoverishment) on the part of the plaintiff; (3) causal link between the defendant's increase and the plaintiff's reduction in property; and (4) the party's act must not be a wrongdoing (otherwise, there would be liability for damages).[23] Article 1111 of the OA provides that, where a part of one's property is transmitted into another's property without legal basis (such as legal transaction, decision by the court or other authority or statute), the latter must return it, or if that is not possible, compensate the former for the gained value. The former (debtor) may seek the recovery from the latter (creditor).

The CRRA, the Trademark Act (hereinafter: the TA),[24] the Industrial Design Act (hereinafter, the IDA)[25] and the Patent Act (hereinafter, the PA)[26] offer a remedy alternative to damages and in the form of payment of the fee which is usually charged for a particular type of use (or the fee which is set under the tariff of the collective society).[27] Although the legal nature of this remedy is somewhat unclear,[28] legal scholarship states that in reliance on these provisions, it is not necessary to prove any requirement for liability in tort. Instead, the plaintiff only needs to prove the fact that the subject matter protected by an intellectual property right was used without authorisation by the right-holder.[29]

---

[21] Official Gazette of the Republic of Croatia 167/2003, 79/2007, 80/2011, 125/2011, 141/2013 and 127/2014.

[22] See CJEU judgment of 8 September 2016, GS Media BV v Sanoma Media Netherlands BV and Others, C-160/15, EU:C:2016:644.

[23] Gorenc 2014, 1866. Wrongfulness and fault are not at all relevant in cases of unjust enrichment. Klarić and Vedriš 2008, 646; Bukovac Puvača, Slakoper and Belanić 2015, 257.

[24] Official Gazette of the Republic of Croatia 173/2003, 54/2005, 76/2007, 30/2009 and 49/2011.

[25] Official Gazette of the Republic of Croatia 173/2003, 54/2005, 76/2007, 30/2009, and 49/2011.

[26] Official Gazette of the Republic of Croatia 173/2003, 54/2005, 87/2005, 76/2007, 30/09, 128/2010, 49/2011 and 76/2013.

[27] Article 179, paragraph 1 of the CRRRA, Article 78, paragraphs 2–4 of the TA, Article 54, paragraphs 2–4 of the IDA and Article 95c, paragraph 2–4 of the PA. These two cases are alternative additional options to determining the amount of damages caused.

[28] See Matanovac 2006, 164.

[29] Blažević et al. 2004, 31.

Furthermore, the TA, the IDA and the PA contain provisions entitling the right holder to file a lawsuit against the persons providing services as a part of their professional activities where these services are used in acts infringing or threatening to infringe the protected intellectual property right.[30] Therefore, liability of the service providers is, in addition to the above provisions applicable in specific situations, dependent on establishing infringement or threat of infringement of a third party's intellectual property right by their users and on establishing that the services of the intermediary are used in the act of infringement or the threat thereof.

Type of Liability

Upon establishing that a service provider is potentially liable in relation to the act or omission of the user of its services for the damage caused to a third person, it is necessary to turn to the issue of the type of liability where there is more than one liable person. According to the OA, the liability of two or more parties for the same damage may be several liability (*podijeljena odgovornost*) or joint and several liability (*solidarna odgovornost*). Several liability means that each tortfeasor is liable for the proportion of damage he or she had caused, and if that proportion cannot be established they are severally liable in equal shares. On the other hand, joint and several liability exists where each tortfeasor is liable for the entire damage regardless of her or his contribution to it and claim may be addressed to any of them for the entire amount.[31]

There is a general rule in Article 1107, paragraph 1 of OA providing that, for the damage caused jointly by more than one person, all of them are jointly and severally liable. In paragraph 2 it is further stated that the inducer and the contributor are also jointly and severally liable along with them, as well as the party aiding in non-revealing the tortfeasors. Paragraph 3 additionally provides that the tortfeasors who did not act jointly are also jointly and severally liable where their respective contributions to the caused damage cannot be established.[32] Likewise, under paragraph 4 of the OA, where it is certain that the damage was caused by

---

[30] Article 76 of the TA, Article 54 of the IDA and Article 95c of the PA.

[31] Klarić and Vedriš 2008, 625; *Enciklopedija imovinskog prava i prava udruženog rada*, Vol. 2, 1978, 523 et seq.

[32] An exception from this provision is contained in Article 99, paragraphs 2 and 3 of the Labour Act (Official Gazette of the Republic of Croatia, 149/09, 61/11, 82/12 and 73/2013), which provides that workers who cause damage to their employer are severally liable for the proportion of the damage they have caused, except where the respective proportions cannot be established in which case they are severally liable for equal shares of damage. This exception has long-term history and has been justified by the fact that the employer has the possibility to establish the extent to which each of the employees causing the same damage contributed to the damage. Being the employees, the tortfeasors are accessible to the employer as the victim, there is no need for the joint and several liability, unlike in fields of law different from labour law. See *Enciklopedija imovinskog prava i prava udruženog rada*, Vol. 1, 1978, 454.

two or more persons which are connected in some manner, but is impossible to determine which one caused the damage, these persons are all jointly and severally liable as well.

It might be inferred from the non-final 2013 judgment of the Municipal Court in Vukovar,[33] that secondary liability of Internet intermediary is joint and several liability: the Court does not question whether the blog owner is only partially liable and to what extent, but holds it fully liable for the entire damage claimed by the plaintiffs. However, without a final and explicit judgment to that effect,[34] the authors of this Chapter may only assume that such future cases would be decided by analogy to those in other fields.

Instances of several and joint liability addressed in the case law include: (a) liability of the inducer in inflicting grievous bodily harm on a person by another person, the latter being criminally liable and ordered to pay civil compensation;[35] (b) liability of borrower and collateral provider for the damage consisting in the loss of profit on the part of the lender for not being able to lease an apartment;[36] (c) liability of employer of an employee injured in the course of work and hospital that negligently treated the injured employee so that the latter died, for the damage claimed by the employee's wife;[37] (d) liability of employer of an employee injured in the course of work and owner of the heavy working vehicle that hit the employee while working so that the employee suffered injuries;[38] (f) liability of employer of an employee injured in the course of work and the insurer of the motor vehicle which hit the structure that the employee was working on so that the employee fell off and was injured, because it was impossible to establish the proportions of damage each of them caused to the injured employee;[39] and (g) liability of producer of the infringing bottle, producer of the infringing package and producer of the infringing medallion attached to the package, the bottle, the package and the medallion being protected under trademark law or law on unfair competition, because it was impossible to establish the extent to which each of them contributed to the

---

[33] See supra n. 10.

[34] To the knowledge of the authors there is no Croatian court decision stating the type of secondary liability of Internet intermediaries.

[35] Municipal Court in Rijeka, P-2582/07 of 20 September 2007, confirmed by County Court in Rijeka Gž-366/08 of 16.9.2009.

[36] Regardless of the transfer of ownership over that apartment from the collateral provider to the lender for the purpose of securing the loan pursuant to the loan agreement between the lender and the borrower, inability to lease the apartment in question resulted from the fact that the collateral provider failed to transfer the possession of that apartment to the lender upon the borrower's failure to repay the loan. Supreme Court of the Republic of Croatia, Rev-867/2009-2 of 25 November 2010.

[37] Supreme Court of the Republic of Croatia, Rev-919/2008-2 of 8 July 2010.

[38] Supreme Court of the Republic of Croatia, Revr-30/2007-2 of 14 February 2007.

[39] Supreme Court of the Republic of Croatia, Revr-689/2008-2 of 1 April 2009.

damage.[40] There is also a special provision in Article 173, paragraph 2 of the CRRA stating that the joint activity of two or more persons, resulting in infringement of author's right or related right, gives rise to their joint and several liability.

The essential feature of joint and several liability is that the victim may claim compensation from all tortfeasors jointly and severally liable in the same proceedings or separately,[41] choosing that or those whom she or he believes would have the deepest pockets or who are the closest targets. As a consequence, the obligations of other jointly and severally liable debtors towards the victim cease to exist,[42] but the relationship among the debtors becomes very important. Article 1109, paragraph 1 of the OA, provides that the person who is jointly and severally liable and pays in excess of her or his share of the damage caused, may request other persons jointly and severally liable for the same damage to repay her or him what she or he paid for the other person so liable. Paragraph 2 provides that the court shall determine the share of each of the persons jointly and severally liable for the same damage taking account of the level of the fault and gravity of the consequences resulting from her or his act or failure to act. Under paragraph 3, if it is not possible to determine the shares of each of the persons jointly and severally liable for the same damage, they shall be liable in equal shares unless fairness in a specific case requires otherwise. Legal theory further explains what happens in case of insolvency of one of the jointly and severally liable debtors or other similar situations where the repayment of what the other debtor(s) paid in excess cannot be obtained. The insolvent debtor's share is then proportionately divided among all other severally and jointly liable debtors.[43]

Types of Remedies Granted by the Courts

If liability concerns the cases falling under the general rules in the OA, the primary remedy is reparation of damage that entails elimination, compensation or alleviation of the caused damage.[44] There are three ways to repair the damage: (1) natural restitution that is intended to restore the situation *in integrum*, which due to its nature

---

[40] In this last case, the three defendants, Croatian companies, had infringed the intellectual property rights of the plaintiff, a Swedish company, by producing and marketing the goods (bottle, package and medallion) because aspects of these goods were either protected by trademark law (trademark ABSOLUT COUNTRY OF SWEDEN for spirits) as there was likelihood of confusion due to identity/similarity between the signs and goods in questions as well as under unfair competition law as the goods contained signs and information which created or could have created confusion regarding their origin, the manner of production, quality or their other properties. High Commercial Court of the Republic of Croatia, Pž-6665/02-3 of 17 May 2006.

[41] Gorenc 2014, 94; Supreme Court of the Republic of Croatia, Gž-345/78 of 25 January 1979, PSP 15/26.; Supreme Court of the Republic of Croatia, Rev.-1789/95 of 30 November 1999; Supreme Court of the Republic of Croatia, Rev-1264/97 of 6 December 2000.

[42] Supreme Court of the Republic of Croatia, Rev-2659/91 of 5 March 1992., Izbor 94/100.

[43] *Encikolpedija imovinskog prava i prava udruženog rada*, Vol. 2, 1978, 529.

[44] Gorenc 2014, 1792 et seq.

is rarely claimed in practice;[45] (2) compensation of damage, i.e. the pecuniary equivalent to the sustained harm (which is paid to the person suffering the harm);[46] and (3) satisfaction, which is intended to satisfy subjectively the person who suffered some harm, usually the harm to personality rights, and which can be moral or pecuniary.[47]

In addition to the reparation of damage, there is another injunctive remedy—the claim to remove the threat of damage, intended to prevent the damage that has not yet occurred, as well as to stop the acts/omissions that would cause damage in the future.[48] Under the rules of civil procedure, it is also possible that a party suffering harm seeks declaratory relief, i.e. asks the court to establish the infringement (or non-infringement, although this is nearly never used in practice). Such a claim is permissible if the substantive law specifically provides for it or if the plaintiff proved that she or he has a legal interest in submitting such claim.[49]

The preliminary measures under the Enforcement Act (hereinafter, the EA)[50] are available regardless of the cause of action, but the conditions differ for measures related to pecuniary and those related to non-pecuniary claims. A person seeking a preliminary measure related to a pecuniary claim must prove: (a) the likelihood that the claim exists; and (b) the risk that in the absence of such measures the other party will prevent or make it significantly more difficult to collect on a claim by transferring, hiding or otherwise disposing of her property. A person seeking a preliminary measure related to a non-pecuniary claim must prove: (a) the likelihood that the claim exists; and (b) the risk that in the absence of such measures the other party will prevent or make it significantly more difficult to collect on a claim by altering the present state of the matter, or that the measure is necessary to prevent violence or the occurrence of irreparable harm.[51] However, the court may dispense with these conditions provided the person seeking the provisional measure provides a guarantee to indemnify any damage that might be suffered by the other party.[52]

In cases of intellectual property infringement under the TA, the IDA and the PA discussed elsewhere in this Chapter, a secondarily liable service provider may be faced with several different claims: a declaratory claim of infringement,[53] an injunctive claim ordering the infringer to stop the infringement and prohibiting such or similar infringement in the future,[54] and an injunctive claim to stop the act which is

---

[45] Article 1085, paragraph 1 of the OA.

[46] Article 1085, paragraphs 2–3 of the OA. Article 1046 of the OA states that damage consists of actual damage (*damnum emergens*) and lost profit (*lucrum cesans*).

[47] Article 1046 of the OA.

[48] Article 1047 of the OA.

[49] Article 187, paragraph 2 of the CPA.

[50] Official Gazette of the Republic of Croatia 112/2012, 25/2013 and 93/2014.

[51] Article 346, paragraph 1 of the EA.

[52] Article 349 of the EA.

[53] Article 76, paragraphs 1 and 4 of the TA, Article 54, paragraphs 1 and 4 of the IDA and Article 95c, paragraphs 1 and 4 of the PA.

[54] Article 76, paragraphs 2 and 4 of the TA, Article 54, paragraphs 2 and 4 of the IDA and Article 95c, paragraphs 2 and 4 of the PA.

a serious threat of infringement as well as to prohibit the infringement by such an act.[55] This is explicitly provided in the law and is related to the obligation of the EU Member States to ensure that right holders are in a position to apply for an injunction aimed at prohibiting the continuation of infringement against intermediaries whose services are used by a third party to infringe an intellectual property right under Article 11 of the Directive 2004/48/EC of the European Parliament and of the Council of 29 April 2004 on the enforcement of intellectual property rights.[56] Curiously, the CRRA does not contain such a specific provision entitling right holders to raise claims against service providers-intermediaries. However, due to the necessity to comply with the cited requirements of the Directive, the provisions of the CRRA which relate to injunctions (Article 177) should be interpreted to be available also against the service providers-intermediaries. This conclusion would probably not be valid for other remedies envisaged in the intellectual property laws referenced above due to the fact that they are not explicitly allowed against service providers.

Under the MA, in case of publication of information whereby the rights or interests of a person have been violated, that person is entitled to ask the general editor to publish the correction to that information, free of charge.[57] Due to the nature of this claim and the reluctance of the courts to characterise all Internet intermediaries as media, it appears that this remedy would not be available against the Internet intermediary whose services are used in publishing the information in question, unless it falls under the definition of media within the meaning of Article 2 of the MA.[58]

Based on the above, it is apparent that some of the remedies available against users as primary tortfeasors would not be available against Internet intermediaries as secondary tortfeasors. They would, of course, be available against Internet service providers when they act as primary infringers.

---

[55] Article 76, paragraphs 3 and 4 of the TA, Article 54, paragraphs 3 and 4 of the IDA and Article 95c, paragraphs 3 and 4 of the PA.

[56] Official Journal of the European Union, L 157 of 30 April 2004.

[57] Article 40 of the MA.

[58] The authors' translation of the Croatian text of Article 2 of the MA reads: "Media are: newspapers, and other prints, radio and television programmes, programs of new agencies, electronic publications, teletext and other forms of daily or periodic publishing of editorially shaped program contents by transmission of text, voice, sound or picture. Media are not: books, textbooks, bulletins, catalogues or other holders of published information intended exclusively for educational, scientific and cultural process, advertising, business communication, work of the companies, institutes and institutions, associations, political parties, religious and other organisations, school gazettes, Offical Gazette of the Republic of Croatia, official gazettes of the local self-government and regional units and other official notifications, posters, flyers, prospects and banners, as well as video pages without live picture and other information free of charge, unless otherwise provided for in this Act."

Criminal Liability of Internet Intermediaries

As a rule, according to the Criminal Act (hereinafter, the CRA),[59] Internet intermediaries might be criminally liable in relation to the conduct of third parties who use their services, to the extent they aid or abet third parties in committing criminal offences.

Abetting in committing criminal offences is defined in Article 37 of the CRA, which in paragraph 1 provides that a person who intentionally abets another in committing a criminal offence shall be punished, as if she or he committed the criminal offence herself or himself. Paragraph 2 of the same Article states that if a person intentionally abets another in committing a criminal offence for which an attempt is punishable and the offence is not even attempted, the abetter shall be punished as if she or he attempted to commit this criminal offence herself or himself. Similarly, Article 38, dealing with aiders, prescribes that a person who intentionally aids another in committing a criminal offence shall be punished as if she or he committed the criminal offence herself or himself, but he can also be punished more leniently.

The necessary precondition for their criminal liability is, therefore, the intention of the abetter or the aider. The definition of intention is contained in Article 28 of the CRA, which differentiates between direct intent and indirect intent. Where someone acts with direct intent she or he is aware of the characteristics of the criminal offence and wants or is sure of their realisation.[60] On the other hand, when someone acts with indirect intent, she or he is aware that with own actions she or he might facilitate a criminal offence and she or he accepts such consequences.[61] The aforestated applies not only in situations when the service provider is a natural person, but also when the service provider is a legal person, such as a company, provided that by acts of responsible persons within the legal person some duty of the legal person is breached, or the legal person obtained or might have obtained an illegal financial benefit for itself or another.[62]

In addition, some criminal acts already incorporate abetting or alike in the definition of the act, as in the case of criminal defamation sanctioned under Article 144 of the CRA. The provision of this Article prescribes a fine for anyone who "in relation to another, asserts or disseminates a falsehood which can damage her or his honour or reputation". In the situations of criminal defamation, abetting takes the form of dissemination. The abovementioned owner of the blog at the website www.vukovarac.net was also prosecuted for criminal defamation. Deciding in the first instance, the Municipal Court in Ilok found the blog owner guilty of the offence and reasoned that, although he did not make the defamatory statements himself, he dis-

---

[59] Official Gazette of the Republic of Croatia 125/11 and 144/12.

[60] See Article 28, paragraph 1 of the CRA.

[61] See Article 28, paragraph 2 of the CRA.

[62] See Articles 2, 3 and 4 of the Liability of Legal Persons for Criminal Offences Act, Official Gazette of the Republic of Croatia 151/03, 110/07, 45/11 and 143/12.

seminated them because they were published on the Internet web page which he had the power and responsibility to manage, service and administer, including by controlling and selecting its contents.[63] These judgements were quashed by the Supreme Court of the Republic of Croatia because of the sanctions, not the verdict.[64] In the retrial, the blog owner was convicted again for the same reasons.[65] This judgment was confirmed by the County Court in Vukovar and thus became final. The County Court stated that the blog owner actively participated in stating and disseminating the defamatory content and that he also enabled further defamation by publishing defamatory comments.[66]

An important issue in this respect is related to the bearing of criminal conviction over civil liability. Article 12 of the CPA provides that the court hearing a civil case is, in respect of whether there is a criminal offence and criminal liability, bound by the guilty verdict of the criminal court. Essentially, where the same event has been the cause of both civil and criminal proceedings, the civil court has to find the defendant liable where the criminal court reached a guilty verdict, but it may decide either way where there is a non-guilty verdict.[67]

Safe Harbours Protecting Internet Intermediaries from Secondary Liability

Internet intermediaries in Croatia enjoy a safe harbour against liability for the conduct of users of their services. The safe harbour provisions are part of the Electronic Commerce Act (hereinafter, the ECA).[68] The ECA was adopted in 2003, *inter alia*, to harmonise Croatian law with Directive 2000/31/EC of the European Parliament and of the Council of 8 June 2000 on certain legal aspects of information society services, in particular electronic commerce, in the Internal Market (hereinafter, the E-Commerce Directive).[69] Prior to the adoption of the ECA, there was

---

[63] There were two parallel proceedings against the blog owner: one for the defamation of the husband: Municipal Court in Ilok, K-177/06, judgment of 4 July 2007, and the other of the wife: Municipal Court in Ilok K-178/06, judgment of 28 February 2007. Both contained the same reasoning. They were both confirmed by the second instance courts, for the husband: County Court in Vukovar, Kž-458/07, judgment of 17 March 2009, and for the wife: County Court in Vukovar, Kž-223/07, judgment of 13 November 2007. This judgment was confirmed by the County Court in Vukovar, Kž-5/11, of 13 January 2011 which became final. The County court states that the blog owner actively participated in stating and disseminating the defamatory content and that he also enabled further defamation by publishing defamatory comments.

[64] Supreme Court of the Republic of Croatia, Kzz-34/09, judgment of 20 October 2009.

[65] Municipal Court in Vukovar K-568/09, judgment of 28 October 2010. This judgment concerns only husband's case.

[66] County Court in Vukovar, Kž-5/11, of 13 January 2011.

[67] For details, see Kaleb 2008.

[68] Official Gazette of the Republic of Croatia 173/2003, 67/2008, 130/2011, 36/2009 and 30/2014.

[69] See Republika Hrvatska, Ministarstvo Gospodarstva, *Nacrt prijedloga Zakona o elektroničkoj trgovini, sa konačnim prijedlogom zakona*, Zagreb, 2003., available at: http://hidra.srce.hr/web-

neither a statutory provision in Croatian law immunising Internet service providers against liability in relation to the conduct of users of their services, nor any judicial practice of which the authors are aware. The following sections of this Chapter provide insight into the scope of application of the safe harbour provision, distinguish four categories of Internet intermediaries and their respective safe harbours, set out the requirements which those intermediaries have to meet in order to benefit from a particular safe harbour, detail remedies available against internet Intermediaries regardless of their liability, and discuss the practice of Croatian hosting providers in relation to takedown requests.

Scope of Application

Article 2 of the ECA, which defines the general terms used in the ECA, states that a service provider is "any legal or natural person providing information society services."[70] The same Article, in subparagraph 2, prescribes that "an information society service" is "a service provided for remuneration by electronic means at the individual request of a recipient of services, in particular online sales of goods and services, online offering of information, online advertising, electronic search engines, and the possibility of searching information and services which are transmitted via an electronic network, acting as intermediaries in providing access to a network or in hosting information of a recipient of the service."[71] This subparagraph is clearly the result of the transposition of Recitals 17 and 18 of the Directive on electronic commerce into Croatian law, although some terms or parts of the sentences in the Recitals are poorly translated or omitted in the ECA. In any case, since information society services span a wide range of economic activities which take place online (and not offline),[72] Article 2 of the ECA limits the immunity conferred solely to Internet (or online) service providers.

The ECA does not specifically state whether immunity is available regardless of the cause of action or limited to certain causes of action; the authors of this Chapter are not aware of any court decision on the matter either. Failing any limitation in the

---

pac-hidra-rdrh-pregled1/?rm=results&last_PAGER_offset=0&show_full=1&v=sloboda%20 pru%BEanja%20usluga%20004713&f=SubjectIndexHR&path=hidra-rdrh%233983, last visit: 3 December 2013.

[70] This is the authors' own translation of the Croatian text of Article 2, subparagraph 4 of the ECA, which reads: "*davatelj usluga* – pravna ili fizička osoba koja pruža usluge informacijskog društva".

[71] This is the authors' translation of the Croatian text of Article 2, subparagraph 3 of the ECA which reads: "usluga informacijskog društva – usluga koja se uz naknadu pruža elektroničkim putem na individualni zahtjev korisnika, a posebno Internet prodaja robe i usluga, nuđenje podataka na Internetu, reklamiranje putem Interneta, elektronički pretraživači, te mogućnost traženja podataka i usluga koje se prenose elektroničkom mrežom, posreduju u pristupu mreži ili pohranjuju podatke korisnika".

[72] See Recital 18 of the E-Commerce Directive.

ECA, we conclude that service providers enjoy immunity against all forms of liability irrespective of the cause of action as long as it falls within any of the types of non-contractual obligations. It is important to add that the provisions of the ECA do not apply to the following: data protection, taxation, activities of public notaries, client representation and protection of their interests before courts, as well as games of chance with monetary stakes, including lottery games, casino games, betting games and games of chance on automatic machines. Other than that, the ECA confers immunity against liability without regard to the specific area of law in question exempting Internet intermediaries from liability in a horizontal way for any kind of unlawful content provided by their users. Below, we discuss the requirements that Internet intermediaries have to fulfil to benefit from the safe harbours, remedies against Internet intermediaries independent of their secondary liability, and the practice of Croatian hosting providers concerning takedown requests.

Categories of Safe Harbour

Chapter IV of the ECA, titled "Liability of Information Society Service Providers – Intermediaries", follows very closely the language of Section 4 of the E-Commerce Directive titled "Liability of Intermediary Service Providers", with only a few relatively minor differences in the order, title and contents of the provisions. Yet instead of three categories in the Directive, there are four categories of Internet intermediaries and their respective safe harbours in the ECA.

The safe harbour for caching was transposed into Croatian law and corresponds nearly verbatim to Article 13 of the E-Commerce Directive.[73] The same applies for the safe harbour concerning hosting providers in Article 14 of the Directive.[74] However, the knowledge standards laid down in Article 14 of the E-Commerce Directive, namely, "actual knowledge" and "awareness", have been transposed to the ECA without paying attention to the fact that these are legal terms and not everyday language. This is particularly so in the case of "awareness", for which different Croatian words (*biti upoznat*—to be acquainted with, to be informed of, to be familiar with—and *biti svjestan*—to be aware of, to be conscious of), which do not have an identical meaning, were used in the first and second paragraph of Article 18 of the ECA. Conversely, in the E-Commerce Directive, the notion of "awareness" is consistently used in both paragraphs. The official explanations to the ECA offer no justification for this inconsistency.[75] Most likely, it is a case of inappropriate translation that should not affect the interpretation of the provision in question.

---

[73] See Article 17 of the ECA.
[74] See Article 18 of the ECA.
[75] See Republika Hrvatska, Ministarstvo Gospodarstva, *Nacrt prijedloga Zakona o elektroničkoj trgovini, sa konačnim prijedlogom zakona*, Zagreb, 2003., available at: http://hidra.srce.hr/web-pac-hidra-rdrh-pregled1/?rm=results&last_PAGER_offset=0&show_full=1&v=sloboda%20

Certain differences between the E-Commerce Directive and the ECA exist also in relation to the mere conduit safe harbour. Article 16 of the ECA, entitled "Technical Enabling of Access and Transmission of Information on a Communication Network",[76] is in its first part a somewhat compressed version of Article 12 of the E-Commerce Directive, and it reads: "The provider of information society services which transmits electronic messages provided by a recipient of the service is not liable for the contents of the sent message and its sending, on condition that [...]". The remainder of the Article is almost exactly the same as its counterpart in the E-Commerce Directive.

Aside from these slight differences between the E-Commerce Directive and the ECA, the latter in its Article 19 contains a new exemption not prescribed by the Directive, which deals with hyperlinks and reads as follows:

> The service provider which by means of electronic linking provides access to third party information is not liable for that information:
>
> – if she or he does not have the knowledge nor she or he could have known of the illegal activities of the user or the information,
> – if she or he immediately upon obtaining knowledge or awareness of illegal activities or information removes or disables access to the information.[77]

As can be seen from its text, the provision is very similar to and it was obviously modelled on the safe harbour rule for hosting services. Important to note here is CJEU case law which, due to the principle of supremacy of EU law, affects the operation of the provision of Article 19 of the ECA. In *GS Media*[78] the CJEU provided interpretation of the concept of "communication to the public" within the meaning of Article 3(1) of Directive 2001/29/EC of the European Parliament and of the Council of 22 May 2001 on the harmonisation of certain aspects of copyright and related rights in the information society (hereinafter, Information Society

---

pru%BEanja%20usluga%20004713&f=SubjectIndexHR&path=hidra-rdrh%233983, last visit: 3 December 2013.

[76] Prior to the third amendments to the ECA in 2011, this Article had the title "General Provision". The title was changed to the current one following a comment of the European Commission. See VLADA REPUBLIKE HRVATSKE, Konačni prijedlog Zakona o izmjenama i dopunama Zakona o elektroničkoj trgovini, Zagreb, September 2011, available at: https://vlada.gov.hr/UserDocsImages//Sjednice/Arhiva//155_6.pdf, d.bGE, last visit: 8 January 2014.

[77] This is the authors' own translation of the Croatian text of Article 19 of the ECA, which reads: "Davatelj usluga koji putem elektroničkog upućivanja otvori pristup trećim podacima nije odgovoran za te informacije:

– ako nema saznanja niti je mogao znati o nedopuštenom djelovanju korisnika ili sadržaju podataka u tim informacijama,
– ako odmah čim je saznao da se radi o nedopuštenom djelovanju ili podatku ukloni ili onemogući pristup podacima."

[78] CJEU judgment of 8 September 2016, GS Media BV v Sanoma Media Netherlands BV and Others, C-160/15, EU:C:2016:644. See also CJEU judgment of 13 February 2014, Svensson and Others, C-466/12, EU:C:2014:76.

Directive).[79] It held that hyperlinking to a copyrighted work is "communication to the public" where two conditions are fulfilled: (1) no consent by the right-holder for placing the work online, and (2) no knowledge (did not know or could have reasonably known) on the part of the hyper-linker. The latter condition differs for hyper-linkers pursuing financial gain. Where a hyper-linker acts in pursuit of financial gain, it is presumed that she or he knew or could have reasonably known of the unlawful publication of the work. This presumption is rebuttable (*presumptio iuris*). Conversely, where a hyper-linker acts without pursuit of financial gain, there is no such presumption. The essential difference is thus in the burden of proof. Against this background, the safe harbour provisions in Article 19 of the ECA need to be applied with modification to cases of infringed copyrighted works. Thus, for internet service providers who hyperlink to unlawfully published copyrighted works and who act in pursuit of financial gain, knowledge will be presumed (although with the possibility of rebutting the presumption). For those who act without pursuing financial gain, the question may be posed as to whether the knowledge standard in Article 19 is the same as that under Article 3(1) of the Information Society Directive. Without speculating on the interpretation by the Croatian courts, it suffices to say that the interpretation of Article 3(1) of the Information Society Directive should prevail if copyright infringement is at issue.

Requirements to Benefit from a Safe Harbour

The conditions which an intermediary providing an Internet service must satisfy in order to be immunised against liability vary according to the type of the service: mere conduit, caching, hosting and linking.

In order to avail itself of the safe harbour under Article 16 of the ECA, the service provider technically enabling access and transfer of information on a communication network must satisfy the following conditions: the provider must not initiate the transmission; the provider must not select the data or documents which are subject matter of the transmission; the provider must not exclude or modify the data contained in the messages or documents; and the provider must not select the receiver of the transmission. This is prescribed by paragraph 1 of Article 16 of the ECA, while paragraph 2 of the same Article states that the transmission of messages and provision of access to messages referred to in paragraph 1 must be performed in such manner to enable automatic, intermediate and transient-temporary storage of the messages transmitted and the data contained therein, whereas they can be temporarily stored only for the period necessary for message transmission.

The conditions for immunity to be met by the service provider with regard to caching are prescribed by Article 17 of the ECA, and are almost a word-by-word translation of the conditions listed in Article 15 of the E-Commerce Directive, with only slight alterations. According to the said Article of the ECA, the service pro-

---

[79] OJ 2001 L 167, p. 10.

vider that transmits in a communication network information provided by a recipient of the service shall not be liable for the automatic, intermediate and temporary storage of the that information, performed for the sole purpose of making more efficient onward transmission of the information to other recipients of the service upon their request, on condition that: the provider does not modify the information; the provider complies with conditions on access to the information; the provider complies with rules regarding the updating of the information;[80] the provider acts in accordance with the lawful use of technology to obtain data;[81] and the provider instantaneously removes or disables access to the data it has stored immediately upon either obtaining actual knowledge of the fact that the data at the initial source has been removed from transmission via the network or access to it has been disabled, or receiving a court order or an order of other competent authority to remove the data or disable access to it.

Similarly to Article 14, paragraph 1 of the E-Commerce Directive, Article 18, paragraph 1 of the ECA states that the provider of the service of storage of information shall not be liable for the information stored at the request of the recipient of the service under two conditions: first, that the provider does not have the knowledge nor could have known of the illegal activity of the user or the contents of the stored data, as well as of the court proceedings related to claims for damages, which would arise from the illegal activity of the user or the contents of the stored data, and if she or he was not nor could have been acquainted with the facts or circumstances from which the illegal activity of the user would be apparent; and, second, that the provider, immediately upon obtaining the knowledge or becoming aware of the illegal activity or data, removes or disables access to the data.

However, modelled on Article 14, paragraph 2 of the E-Commerce Directive, Article 18, paragraph 2 denies immunity where the recipient of the service is acting under the authority or the control of the provider. As pointed out above, the conditions to be satisfied for the immunity of hosting providers in the ECA are substantially the same as the ones provided by the Directive, however, the knowledge standards to be applied under Croatian law might be subject to discussion due to lack of precision in the translation process. So far there is no judicial practice which would interpret the knowledge standards and provide guidance on the issue.

The only case known to the authors of this Chapter, in which the safe harbour for hosting providers, or any safe harbour prescribed by the ECA for that matter, was invoked before Croatian courts was the case in which a Croatian hosting provider was sued for damages by a university professor for an anonymous defamatory article posted on the then active Croatian blog platform MojBlog.com.[82] The defama-

---

[80] Compared to Article 13, paragraph 1, subparagraph (c) of the E-Commerce Directive, the proviso "specified in a manner widely recognised and used by industry" is left out of the ECA version of the caching safe harbour.

[81] Compared to Article 13, paragraph 1, subparagraph (d) of the E-Commerce Directive, the proviso "widely recognised and used by industry" is left out of the ECA version of the caching safe harbour.

[82] Municipal Court in Zagreb, Pn-4423/06.

tory article was, after its publication on the blog platform, picked up by a couple of very popular Croatian news portals and newspapers,[83] which are widely read in Croatia, and was allegedly read by tens of thousands of people. These follow-up articles specifically stated that the defamatory article was published on the blog platform MojBlog.com. The blog platform was hosted by the defendant hosting provider, which was not aware of the disputed blog article prior to receiving the plaintiff's statement of claim, since the plaintiff contacted neither the owners of the blog platform nor the hosting provider with a request for removal of the article prior to initiating the court proceedings. Immediately upon receiving the statement of claim and thus becoming aware of the defamatory article, the hosting provider removed the article in question from the blog platform.[84]

In reply to the plaintiff's statement of claim, the hosting provider invoked Article 21 of the ECA, stating that hosting providers are under no obligation to monitor the information which they store or to seek facts or circumstances indicating illegal activity. Later in the proceedings, the hosting provider invoked the immunity prescribed by Article 18, paragraph 1 of the ECA. Following these submissions, the plaintiff withdrew his claim against the hosting provider and the dispute ended without a judicial decision on the merits.

This is unfortunate because the court lost the opportunity to interpret the applicable ECA provisions, while, in the opinion of the authors of this Chapter, certain arguments could have been made by the plaintiff to remove the defendant's immunity. For instance, the question of the standard of knowledge could have been debated since the allegedly defamatory article was picked up by several popular Croatian news portals and newspapers mentioning the fact that it was published on MojBlog.com. Another interesting fact of the case, potentially relevant to removing immunity, was that the hosting provider did not charge the owners of MojBlog.com for the service of hosting the blog platform in return for free advertising on the platform. This was regulated by a sponsorship contract between the hosting provider and the blog platform owners, which also contained a provision entitling the hosting provider to receive a part of the sale price if the owners sold the blog platform to someone else. Therefore, the plaintiff might have argued under Article 18, paragraph 2 of the ECA, that the recipient of the service was acting under some sort of a control, if not the authority, of the hosting provider.

As pointed out above, the ECA introduced a safe harbour not existing in the E-Commerce Directive, namely, the one related to hyperlinks. According to Article 19 of the ECA, immunity is conferred upon the service provider which by means of electronic linking provides access to third party information. Such service provider shall not be liable for the content of third party information under two cumulative conditions: first, if it does not have the knowledge nor could have known of the illegal activities of the user or the information; and, second, if, immediately upon obtaining knowledge or awareness of illegal activities or information, it

---

[83] For example, Index.hr.

[84] Interestingly, the owner of the blog platform was never sued by the plaintiff, even though its identity was very well known to the plaintiff.

removes or disables access to the information. These conditions are very similar to those related to the hosting safe harbour and even incorporate the same weakness regarding the terminology for knowledge standard. The main difference is that in the hyperlinks safe harbour, the illegality relates to the linked content and not to the hosted content.

Remedies Against Internet Intermediaries Independent of Their Secondary Liability

Article 20 of the ECA prescribes that the provisions of the ECA, particularly those pertaining to the liability of Internet intermediary service providers, shall not affect the authority of courts and other competent bodies of the Republic of Croatia to order the information society service providers and their users, at the request of the authorised person and in accordance with Croatian laws, to eliminate or prevent infringements of law and regulations in force, and take other measures against them prescribed by law. Therefore, the possibility of remedies being awarded against service providers to help restrain wrongful conduct by others independent of the service providers being secondary liable is explicitly allowed by the ECA. The principal purpose of allowing such remedies is to enable prevention and swift termination of online violations of the law.

The Croatian laws related to the protection of intellectual property rights contain several provisions allowing remedies against service providers independently of them being secondarily liable. The CRRA, in Article 185, paragraph 1 states that the court can order any preliminary injunction against the intermediary, whose services are used by third parties to infringe the rights protected by the CRRA, with the aim of stopping or preventing the infringement. The court can grant such remedy at the request of the right-holder, provided that the latter shows the likelihood that her or his right is being infringed or that there is threat of infringement. Such preliminary injunction can be granted *ex parte* if the right-holder shows the likelihood that the injunction would otherwise be ineffective or that irreparable harm would occur. Additionally, such preliminary injunction can be granted even if the right-holder has not yet filed a lawsuit against the infringer. However, in that case, in the decision granting the preliminary injunction, the court will order the right-holder to file the lawsuit within a fixed period of time to justify the preliminary injunction.[85] If the lawsuit is not filed at all or is filed with delay, the preliminary injunction will lapse. Other matters concerning the procedure for granting the preliminary measure are governed by the Enforcement Act.[86] Provisions identical to those in the CRRA are

---

[85] See Article 185, paragraph 5 of the CRRA.

[86] Official Gazette of the Republic of Croatia 112/12. A person seeking the issuing of preliminary measures related to pecuniary claims has to prove: (a) the likelihood that the claim exists, and (b) the risk that in the absence of such measures the other party will prevent or make it significantly more difficult to collect on a claim by transferring, hiding or otherwise disposing of her property.

contained also in the TA,[87] the IDA[88] and the PA,[89] regarding the protection of trademarks, industrial designs and patents, respectively.

Furthermore, the right-holder that commenced the civil proceedings for protection of its rights against infringement can request the delivery of information on the origin and the distribution channels of goods or services by which its right is being infringed. This may be requested from, *inter alia*, the persons providing services used in acts for which it is suspected that they are infringing a right protected by the CRRA, such as information society service providers.[90] This request can be made in the form of a statement of claim or a preliminary injunction. The service provider can refuse to deliver the requested information, if it is permitted to refuse testimony under the rules of civil procedure. However, refusal to deliver the information without a justified reason makes the service provider liable for damages under the provisions of the OA.[91] Identical provisions exist also in the TA,[92] the IDA[93] and the PA,[94] regarding the protection of trademarks, industrial designs and patents, respectively.

As already discussed in this chapter, the TA, the IDA and the PA also contain provisions that allow the right-holder to file a lawsuit against the persons providing services used in acts infringing or threatening to infringe the protected right.[95] The plaintiff may ask the court to make a declaration of infringement,[96] to order the infringer to stop the infringement and prohibit such or similar infringement in the future,[97] and to stop the act which is a serious threat of infringement as well as to prohibit the infringement by such an act.[98]

It is important to emphasise that competent bodies of the Republic of Croatia other than courts can also act against service providers to help restrain wrongful conduct by others independent of the service providers being secondarily liable. A relatively wide authority in this respect is granted to inspectors under the State

---

However, the court may dispense with these conditions provided the person seeking the provisional measures provides a guarantee to indemnify any damage that might be suffered by the other party. See Article 344, paragraph 1, Article 346, paragraph 1 and Article 349 of the EA.

[87] See Article 79b of the TA.

[88] See Article 56d of the IDA.

[89] See Article 95j of the PA.

[90] See Article 187, paragraphs 1 and 2 of the CRRA.

[91] See Article 187, paragraph 5 of the CA.

[92] See Article 79a of the TA.

[93] See Article 56c of the IDA.

[94] See Article 95i of the PA.

[95] See Article 76 of the TA, Article 54 of the IDA and Article 95c of the PA.

[96] Article 76, paragraphs 1 and 4 of the TA, Article 54, paragraphs 1 and 4 of the IDA and Article 95c, paragraphs 1 and 4 of the PA.

[97] Article 76, paragraphs 2 and 4 of the TA, Article 54, paragraphs 2 and 4 of the IDA and Article 95c, paragraphs 2 and 4 of the PA.

[98] Article 76, paragraphs 3 and 4 of the TA, Article 54, paragraphs 3 and 4 of the IDA and Article 95c, paragraphs 3 and 4 of the PA.

Inspectorate Act (hereinafter, the STA).[99] The main task of the inspectors is to monitor the application of laws and other regulations in many areas, amongst others, in the areas of trade, provision of services, consumer protection and protection of intellectual property rights. In carrying out the inspection tasks, the inspectors are authorised to inspect the business premises, equipment, business books, registries, documents, contracts and other business documentation which provides insight into the business activities of the inspected person.[100] They are also authorised to temporarily seize the documentation and objects which can be used as evidence in misdemeanour proceedings or criminal proceedings.[101] Persons being inspected must allow the inspector to perform the inspection and provide accurate information under the threat of a substantial fine.[102] If the inspector considers that the inspection would not be as effective she or he might even choose not to inform the inspected person of the upcoming inspection.[103] Inspectors do not have to present any paperwork when performing the inspection, such as a warrant or court order—the authority is given to them by the STA. As a rule, the inspectors initiate the inspection proceedings *ex officio* and no complaint against the inspected subject is a necessary precondition for that purpose,[104] although very often a complaint is filed. The inspector can forward documentation, and reveal facts and data only to courts, state administration bodies conducting misdemeanour proceedings, and other state bodies at their elaborated written request, and exclusively for use in court and administrative proceedings within their respective jurisdiction.[105]

The authors of this Chapter have personal knowledge of cases in which a complaint by a third party was submitted to the State Inspectorate about the possible illegal commercial activities performed through a website.[106] Following such complaints, an inspector checked the business premises of the hosting provider, which hosted the website allegedly used for illegal commercial activities, and has

---

[99] Official Gazette of the Republic of Croatia 116/08, 123/08 and 49/11. The State Inspectorate as an independent body ceased to exist on 1 January 2014, when the Act Amending and Supplementing the Organization and Jurisdiction of Ministries and Other Central Bodies of State Administration came into effect. The latter Act was part of the reform aimed at reorganising the Croatian state administration. It did not eliminate or change the authorities of the inspectors under the State Inspectorate Act, but simply transferred the officials acting as inspectors to several government ministries, depending of the area which a particular inspector covered. For example, inspectors which were in charge of areas such as transactions of goods and services, intellectual property protection and prohibition and prevention of unregistered activities and illicit trade are now part of the Ministry of Finance, while the inspectors which were in charge of labour issues are now part of the Ministry of Labour and Pension System.

[100] See Article 27 of the STA.

[101] See Article 34 of the STA.

[102] See Article 30 and Article 61 of the STA.

[103] See Article 28 of the STA.

[104] See Article 33 of the STA.

[105] See Article 33 paragraph 3 of the STA.

[106] See Republic of Croatia, State Inspectorate, Regional Unit Zagreb, Minutes on the performed inspection of business activities of POSLuH d.o.o. Zagreb of 11 October 2010, Class: 336-01/10-02/15, Reg. no: 556-16-01-01/42-10-101.

requested, based on the STA, that the hosting provider immediately turns over the information about the website and its owner, i.e. the recipient of the hosting service. Even though it is obviously a practice of inspectors, it might be doubted whether the authority given to them by the STA should be interpreted so widely as to allow them to inspect the premises of hosting providers, regarding wrongful conduct by the users of the hosting services.

The Practice of Croatian Hosting Providers Concerning Takedown Requests

Croatian law does not contain specific provisions on "notice and takedown" proce-dures. It appears that the practice among Croatian hosting providers greatly varies when they receive a request to remove or disable access to the allegedly illegal data, which comes from the supposed right-holder and not an official state authority such as a court. Some hosting providers decide on a case-to-case basis whether it is likely that the court would find the reported activities or data illegal and decide whether or not to remove or disable access to the data based on their own opinion. Furthermore, there are hosting providers that refuse to remove any content absent a court deci-sion, while others are "over-compliant" and take down the allegedly illegal websites immediately upon receiving such requests. There is no explicitly prescribed penalty or disincentive for the conduct of either right-holders abusing Article 18 of the ECA or service providers being cautiously "over-compliant" with takedown requests according to the same Article. Nevertheless, both could be liable according to the relevant provisions of the OA to the party that sustained damage due to such conduct.

These differing reactions on the part of hosting providers are obviously the result of the current circumstances in which there is an obligation of hosting providers to remove or disable access to the data immediately upon obtaining the knowledge or becoming aware of the illegal activity or data, along with the lack of judicial prac-tice to clarify what would be the proper course of action for hosting providers when they receive a takedown request.

Conclusion

Lacking specific statutory provisions on secondary liability of Internet intermediar-ies, Croatian courts appear somewhat puzzled by the very idea of (and uncertain as to the legal basis for) such liability. Court practice is scarce, but based on the sole non-final judgment available on the issue it is possible to conclude that the basis for secondary liability of Internet intermediaries is in the OA and that it is based on fault. Further, liability is probably joint and several between the Internet intermedi-ary and the primary tortfeasor. There are four safe harbours from liability, depending on the type of service provided by the Internet intermediary. The lack of special rules on how to act in cases of the victim's notice for takedown results in differing treatment of such cases by different service providers.

References

BLAŽEVIĆ, Borislav, MATANOVAC, Romana and PARAĆ, Kamelija. 2004. Građanskopravna zaštita autorskog prava i srodnih prava. *Zbornik Hrvatskog društva za autorsko pravo* 5: 13–96.
BUKOVAC PUVAČA, Maja, SLAKOPER, Zvonimir and BELANIĆ, Loris. 2015. *Obvezno pravo. Posebni dio II., Izvanugovorni odnosi.* Zagreb: Novi informator.
Enciklopedija imovinskog prava i prava udruženog rada, vol. 1, Beograd: Službeni list SFRJ, 1978.
GORENC, Vilim, et al. 2014. *Komenatar zakona o obveznim odnosima.* Zagreb: Narodne novine.
HENNEBERG, Ivan. 2001. Autorsko pravo, 2nd ed., Zagreb: Informator.
KALEB, Zorislav. 2008. *Djelovanje kaznene presude na parnični postupak.* Zagreb: Vizura.
KLARIĆ, Petar and VEDRIŠ, Mladen. 2008. *Građansko pravo*, 11th ed., Zagreb: Narodne novine.
MATANOVAC, Romana. 2006. Građanskopravna zaštita prava intelektualnog vlasništva u odnosu prema Direktivi 2004/48/EZ o provedbi prava intelektualnog vlasništva – analiza stanja i nagovještaj promjena. In *Hrvatsko pravo intelektualnog vlasništva u svjetlu pristupanja Europskoj uniji*, ed. MATANOVAC, Romana 115–168. Zagreb: Narodne novine/Državni zavod za intelektualno vlasništvo.
SLAKOPER, Zvonimir, et al. 2005. *Susdska praksa 1980.-2005 - i bibliografija radova uz Zakon o obveznim odnosima.* Zagreb: RRIF.

Chapter 11
Information Society Between Orwell and Zapata: A Czech Perspective on Safe Harbours

Radim Polčák

The Czech Republic is an EU member state, so the safe harbour regime for information society service providers in the E-Commerce Directive applies as elsewhere in the EU. This Chapter explains and critically discusses the Czech harmonisation of the directive and related case-law. In addition, it discusses the impact of the adoption of the new Czech Civil Code. In particular, it focuses on ways in which the concept of legal liability of ISPs was transformed from past Orwellian safety to present Zapatist uncertainty.

Czech Private Law: A Revolution Going Nowhere

The Czech legal system belongs to the civil law tradition. The main areas of law, such as private law, labour law, criminal law etc. are primarily defined by codes (codexes) that are further supplemented by particular Acts and bylaws. Case-law has the same role as in any other civil law country: it does not serve as an officially recognised source of law, but it is effectively binding under principles of equality and legal certainty.

The sorts of legal liability that are of utmost importance to the information society, i.e. general liability for damage, for fault or for default, are grounded in private law. The Civil Code that serves as its central source was adopted in 2012 (and came into effect in 2014) and replaced the former regulatory regime that was based on two main sources, the Civil Code of 1964 and the Commercial Code of 1991.

This reform of Czech private law was inevitable. The Civil Code of 1964 originated under communism and was significantly amended in the 1990s to cater for

R. Polčák (✉)
Faculty of Law, Institute of Law and Technology, Masaryk University, Brno, Czech Republic
e-mail: radim.polcak@law.muni.cz; cyber.law.muni.cz

© Springer International Publishing AG 2017   255
G.B. Dinwoodie (ed.), *Secondary Liability of Internet Service Providers*,
Ius Comparatum – Global Studies in Comparative Law 25,
DOI 10.1007/978-3-319-55030-5_11

standard transactional mechanisms. As those amendments were made quickly and without any strategic vision, the result was extremely problematic. In a sense, the situation created by the Civil Code of 1964 and its numerous amendments was similar to what George Orwell describes in *1984*. In that book, Orwell pictures a society that is governed by totalitarian rules providing for almost no space for individual liberties. Unlike communist Czechoslovakia, Orwellian totality is hardly noticed by those who are being oppressed. People in that situation either live under a genuine, yet artificially created, impression of liberty or, even more likely, they simply do not care. That makes the Orwellian totality even more complicated than the communist regime, because people are neither actively hoping for a change nor able to understand the reasons for it. Similarly, the Civil Code of 1964, after being amended more than a hundred times during the 1990s, seemed to the public, prima facie at least, to be working. It provided for basic functional grounds for standard private transactions. However, it was in many respects too restrictive, it lacked clarity as well as logical systematics, and it contained a number of particular functional defects.

In this context, it was quite difficult to reshape Czech private law into its regular pre-communist liberal nature – partly due to the lack of actual public (and political) call for change and partly due to the fact that those who were meant to lead the work on the new codification were raised in the system created by the 1964 Civil Code. As such, substantial resources had to be spent first in order to create something that might be called a revolutionary spirit and to focus public and political attention on the need for a substantial reform of private law. While piercing the Orwellian veil of ignorance was quite successful (the Civil Code was adopted in 2012 relatively smoothly), the overall substantive and technical quality of the new code is not very impressive. Its corpus was taken primarily from a 1920s draft, while a number of fundamental provisions and institutes were copy-pasted from different civil codes, including German Bürgerliches Gesetzbuch, Austrian Allgemeines bürgerliches Gesetzbuch, French and Quebeqoise Code Civil, as well as from the civil codes of Argentina, Italy and Switzerland.

The patchwork character of the 2012 Civil Code is, however, not its main substantive difficulty. The true problem is that a number of provisions are technically newly codified (i.e. they are substantially different from the 1964 Civil Code), but they often originate in times before the invention of a telephone, a combustion engine or even before the first electricity power plant was built. While French or German private law had more than a century to adapt to technological and social developments, through extensive case-law and doctrinal publications, Czech private law is now grappling with codified rules that were meant for entirely different social and economic circumstances. As such, the 2012 Civil Code tackles in detail issues like the ownership of caves, use of pastures, or situations when bees accidently settle outside of their hives, while totally neglecting domain names, digital identities, communication devices, digital documents etc. Consequently, resolving standard legal issues of contemporary information society became instantly a matter of highly uncertain application of general principles of private law and (even more uncertain) analogy (see Polčák 2011a).

In that respect, it might seem as an appropriate solution for Czech private law to borrow foreign case-law and doctrine in the same manner as it borrowed the words of respective civil codes. That solution, however, is not available for the Czech legal practice. One reason is that the legislative copy-pasting was never complex and so it is difficult or even impossible to transplant complex foreign case-law or reputable teachings to an entirely different regulatory situation. Another reason that specifically applies on issues related to secondary liability is that some elements in the Czech private law remained, despite the announced revolutionary nature of the new codification, the same as in the 1964 Civil Code.

(No) Secondary Liability Under Czech Law

Unlike other legal systems, Czech private law does not work with secondary liability or with any similar conception that would make a distinction between different levels of liability. The only concept that somewhat works with different levels of liability is the one of aiding or abetting in criminal law (that is, however, an entirely different kind of concept, as criminal law is in Czech legal system entirely distinct from private law). In private law, the foundations of liability are based on some form of involvement of the liable person in the unlawful conduct, or inaction where action is required. Liability is construed as joint; all persons contributing to damage are liable simultaneously. Thus, any of offenders might be liable up to the full amount if demanded by the damaged person. Parties ordered to pay the full amount have a right of recourse against other offenders, to reclaim their fair contributions – these are called regressive claims (see Hurdík and Lavický 2010).

The general basis of liability of multiple offenders is laid down in § 2915 paragraph 1 of the Civil Code, which states: 'Should there be multiple offenders obliged to compensate for damage, they are liable jointly and severally; if any of the offenders is liable according to a special act only to a certain extent, he or she is jointly liable together with other offenders only to that extent. The same applies even in case when multiple offenders commit independent offences that might highly probably lead to the same damage and when it is impossible to determine who actually caused that damage.' The general rule of apportionment as between those who are jointly liable is defined in § 2916 of the Civil Code as follows: 'A person liable for damage together with others, shall settle the damages with them according to the level of involvement in causing the damage.' As there are situations when a person may be liable for damage that was caused entirely by the fault of others, there is a provision defining subsequent causes of action for the liable person against the person whose fault caused the damage to reclaim in full the amount previously awarded to the damaged party[1]. This might also be applicable in the case of strict liability of ISPs, as is discussed below.

---

[1] The provision is contained in § 2917 of the Civil Code, which states 'Who is liable for a damage caused by the fault of another person, shall have an action against that person.'

This regime of joint liability serves as a general rule for any case of private legal liability (Lavický 2004). However, in particular exceptional cases, the law allows for a distinction between liable persons based upon their contribution to the damage. This distinction makes it procedurally and factually a bit more difficult for the plaintiff to claim damages. The plaintiff has to divide her claims among different persons and some of them might be unable to be physically found or be amenable to damages claims (e.g. for being insolvent). Therefore, the apportionment has to be specifically approved by the court. The relevant provision is contained in § 2915 para 2 of the Civil Code, which states (informal translation): 'In exceptionally notable cases, the court might award damages according to the proportion of the participation of an offender in causing the damage; if a particular proportion cannot be stated, it is to be determined according to the respective level of probability. This does not apply in a case of intentional participation of an offender on a damage caused by another offender or in a case when the offender induced or supported the offence or in a case when the whole damage is accountable to any of participating offenders regardless of their independent acting or if an offender is obliged to recover a damage that was caused by his or her helper and the helper is also liable to recover the damage.' This provision is particularly important for cases when a person is liable for someone else's fault (typically in cases of strict liability of ISPs for copyright infringement, defamation etc., as discussed below). As it would be often a bit too harsh to order full remedies against such a person and to make her subsequently claim them against the real offender, courts tend to accept arguments as to the reasonability of distinct (not joint) liability, such that the strictly liable person will be liable only to relatively small extent.

This regime is similar to the 1964 Civil Code. It is a bit more particular and descriptive in details, mentioning, for instance, that joint liability applies also in cases where causes of action for the same damage are based on different laws or that distinct liability is not available for cases of malicious intention or inducement.

There is a general distinction between two sorts of liability with respect to their elements, i.e. liability for fault (referred to in the Czech doctrinal publications as 'subjective liability') and strict liability (referred to as 'objective liability'). The elements of subjective liability include:

1. Subject, i.e. legal capacity of a person sought to be held liable.
2. Object, i.e. legal capacity of something to be damaged (typically property, privacy, various individual rights etc. - see Ryška 2011).
3. Objective facts, i.e. Roman law-based objectively provable facts consisting of damage, illegal conduct (or inaction) and a causal nexus between the aforementioned two (i.e. the fact that the damage was caused by the illegal acting or non-acting). The causal nexus does not distinguish between different forms of causation. Thus, it is easily possible to state that there is a causal nexus between the activity of a user who submits a defamatory post on a discussion server and the damage of the defamed person as well as that there is the same kind of causal nexus in the case of the activity of the provider of that server (using the logic that if the server is not running, the defamatory statement would not be causing any harm).

4. Subjective facts, i.e. fault. The types of fault recognised in Czech law include direct or indirect intent (*dolus directus* or *dolus indirectus*) and conscious or unconscious negligence (*culpa lata* or *culpa levis*). Czech criminal law further works with one more special type of conscious negligence, termed 'gross negligence'. It applies in cases when a person negligently causes harm despite being not only properly informed but also repeatedly warned about the risk.

Subjective liability is the general form of liability in Czech private law. The Civil Code states, in § 2909 and 2910, that: '§ 2909 – An offender that causes damage by intentional breach of good manners is liable to recover it; if the offender acts in defense of his or her rights, he or she is liable only in cases where the damage is the purpose of such acting.' And, in '§ 2910 it provides – 'An offender who by his or her fault breaches a legal obligation and infringes an absolute right of a damaged person is liable to recover whatever was caused by such acting. Such liability emerges also if an offender infringes another legal right of a damaged person by faulty infringement of a legal obligation that is laid down in order to protect that respective right.'

Subjective liability is also a general form of liability in Czech administrative and penal law, in which cases there is a need to prove fault (i.e. negligence or malicious intent) to establish liability. However, there are a number of administrative offenses where fault is not an essential requirement (see Polčák 2011c). Also, the criminal liability of corporate units (legal persons) does not work with fault but rather with the concept of accountability.

Objective liability (i.e. strict liability) does not require fault. It is referred to in doctrinal publications as the 'liability for the result', meaning that an objectively liable person is liable for resulting damage regardless of the way in which it occurred. This kind of liability has no general foundation but is always defined specifically for each application, i.e. every sort of objective liability has its specific definition in the statutory law. That is, for example, true in the case of copyright, privacy or personal data protection (see Mates 2002, 2006), causes of action arising out of unfair competition etc. It also implies that the form and level of accountability required to establish objective liability might differ from one type to another; there might also be different reasons for relief. For example, some types of objective liability might be totally strict, some might be defied by *vis maior*, while others can be avoided by proving that the damage occurred despite taking all possible preventive measures.

It is possible that illegal action might result in multiple types of liability for different legal causes. This is relatively common, for example, with the mishandling of copyrighted works (see Myška 2009), which gives rise to liability under copyright law and the law against unfair competition (see Polčák 2005, 2008b). In such cases, all causes of actions might be used simultaneously (see Hajn 2011). However, we must distinguish the cause of action and the calculation of damages (or other remedies). The damages are in these cases calculated commonly for all causes of actions, meaning that the damage is recovered just once. As Czech law is based on a general principle of recovery of real damages (i.e. damages or other remedies are calculated upon proof of their factual existence and value), there are very few cases when different causes of actions might provide for different methods of calculation of remedies. One such example is statutory or punitive damages under copyright law. When copyright infringement serves as one of the causes of action, the court (or sometimes the damaged party) must choose the method of calculating damages.

In any case, joint liability is a universal concept that applies to the whole body of Czech private law. Various areas of law work with specific causes of actions using different elements of subjective or objective liability (see above), but the general approach to cases when damage is caused by multiple persons at the same time is universal.

The Concept of a Service Provider

Czech law has adopted the concept of "service provider" from Directive 2000/31/EC, which refers to Art. 1(2) of Directive 98/34/EC as amended by Directive 98/48/EC that defines information society service as 'any service normally provided for remuneration, at a distance, by electronic means and at the individual request of a recipient of services.' Further, Czech law uses the concept of a 'provider of services of electronic communications,' based on the definition of electronic communication service laid down in Art. 2(c) of Directive 2002/21/EC. As the concept of a 'provider of services of electronic communications' has almost no relevance as to the liability for illegal actions by users, we will refer to the definition of 'service provider' or ISP, as laid down in Art. 2(a) and 2(b) of Directive 2000/31/EC, in connection with the definition of information society service, as laid down in Art. 1(2) of Directive 98/34/EC as amended by Directive 98/48/EC (see Polčák 2008a).

It must be stressed that the definition of a service provider corresponds to a service rather than to a person (see Polčák 2007b). It means that one person (individual or corporate) can be classified differently with regards to different services that it provides. For example, a telecommunication company that runs a discussion board is to be regarded as a mere conduit provider with respect to telecommunications service and as a hosting provider with regard to the discussion board service.

The definition in Act No. 480/2004 Sb, on certain information society services, and on the amendments of certain acts (act on certain information society services), reads as follows (informal translation): '§ 2 For the purposes of this Act: (a) information society service means any service provided by electronic means on individual request of a user submitted by electronic means, provided regularly for a remuneration; a service is provided by electronic means if it is transmitted through network of electronic communications and received by a user with an electronic device for storage of data... (d) service provider means any natural or legal person that provides some of information society services.'

The definition appears not to be limited to online providers, but extends to any services that are provided electronically through some kind of network, including closed networks not connected to the Internet. In any case, the definition uses the term 'transmitted through network of electronic communications', with the term 'network of electronic communications' being defined in Directive 2002/21/EC and subsequently by the Czech Electronic Communications Act (Act No. 127/2005 Sb.) – it means that if a service provider is to be regarded as information society service provider, the respective service has to be provided using at least some of the services that fall under the definition of service of electronic communications (see Polčák 2009a, b).

Apart the general definition of a service provider, the Act No. 480/2004 Sb. follows the structure of the definitions of three types of service providers contained in the Directive 2000/31/EC, i.e. mere conduit, caching and hosting (see Polčák 2007a). It is to be stressed that all three definitions are based on the general definition of a service provider (*supra*), so it is necessary for a provider, in order to fall under the scope of any of the three classes, to fulfil at the same time the prerequisites laid down in the general definition. As with the Directive, the problem with this structure is that particular definitions are linked with immunity provisions, while there are no general immunity provisions for the general definition. Thus a service may fall within the scope of the general definition, but it is difficult to place it within any of the three particular categories to which immunity attaches. Consequently, it is theoretically possible to have an information society service provider that does not fit within any safe harbour. However, in practice, no cases of this type have come before the Czech courts.

A mere conduit is defined in § 3 of the Act No. 480/2004 Sb; in Czech this provision is designated as dealing with the 'liability of a service provider for the content of transmitted information'. It is defined as a 'service consisting of the transmission in a communication network of information provided by a recipient of the service, or the provision of access to a communication network provided to transmit information'. Examples of service providers falling under this category include electronic communications service providers, i.e. network operators (fixed or mobile networks) and Internet access providers (cable, ADSL, LTE, WiFi).

Caching is defined in § 4 of the Act No. 480/2004 Sb; in Czech this provision is designated as dealing with the 'liability of a service provider for content of temporarily automatically intermediately stored information. Caching activities are defined as a 'service consisting of the transmission of information provided by a recipient that are automatically temporarily intermediately stored.' Another part of this definition is contained in § 2 (h) of the Act. No. 480/2004 Sb., which provides for a more particular definition of automated temporary intermediate storage, and reads as follows: 'automated temporary intermediate storage [means for the purpose of this Act] storage of user information made to most efficiently transmit this information on a request by other users.' Typical examples of caching services include providers of mirror storage capacities that are used by multinational enterprises to optimise geographically the use of communication links. Despite the fact that these services are relatively widely used in the Czech Republic (because the Czech Republic is used as a Central European base for a number of multinational ICT corporations), the liability of their providers does not represent any practical legal issue – there have been no significant cases reported where caching provider liability was at stake (see Maisner et al. 2011).

The most important of the three sorts of service providers addressed is those that engage in hosting, which is designated and defined in § 5 of the Act No. 480/2004 Sb as a service consisting of storage of content provided by users. This type of information society service is by far the most important for Czech law, as it includes a broad variety of the most popular user-generated services including social networks, webhosting services, blogs, discussion boards, file exchange services, cloud storage services etc.

Some of most important information society services are those of search engines and online marketplaces. However, such providers do not constitute any specific category in the legislation. Despite the existence of legal disputes concerning these services, there is also no significant case-law that gives a clear answer as to the classification of these providers. The reason is that in all significant legal disputes, parties have chosen (unlike for example in the case of domain names – see below) not to litigate against service providers.

It is difficult to explain this situation in the case of search engines. We might speculate that potential plaintiffs expect that the court would in such cases deny the joint liability of the search engine provider and award remedies upon the exceptional principle of distinct liability (meaning that the level of liability of the search engine provider is probably almost none); and so plaintiffs decide that it is not efficient to include search engine providers among the defendants (see Polčák 2009c). The primary reason for the lack of court cases involving online marketplaces is the fact that these marketplaces regularly have their own dispute settlement procedures, and the dominant Czech service, *Aukro.cz*, has a relatively generous system of compensation for its users in cases of fraud. Therefore, typical disputes arising from online P2P sales are either settled or compensated by the provider, and are then presented to the police and courts not by the user but by the provider. Technically, the provider first compensates its clients and then seeks remedies against the offenders without its liability being at stake.

In any case, we have good reasons to believe that if a Czech court is to classify search engine providers or online marketplace providers, it would treat them as hosting providers in line with recent case-law of the Court of Justice of the EU in C-324/09 (*L'Oréal and Others v eBay*) and in joined cases C-236/08 (*Google France SARL and Google Inc. v Louis Vuitton Malletier SA*), C-237/08 (*Google France SARL v Viaticum SA and Luteciel SARL*) and C-238/08 (*Google France SARL v Centre national de recherche en relations humaines (CNRRH) SARL and others*).

Liable Only If...

There is no specific legal basis for the establishment of liability of service providers under Czech law. The only regulatory regime specifically targeting some ISPs and making them obliged to act in response to illegal activities of their users was adopted in the form of the Cybersecurity Act (Act No. 181/2014 Sb.). This Act applies to ISPs that deal with national cybersecurity (i.e. ISPs administering critical information infrastructure, public information systems of high importance or back-bone networks). In normal situations, these ISPs are obliged to apply standard security measures and procedures and to report security incidents. In the case of a cybersecurity incident of national significance, these ISPs have specific duties to react to orders issued by the Governmental or National Response Teams (CERT). The Act also defines an extraordinary state of cyber-emergency. It can be declared by the government in situations when the functioning of critical information infra-structure is seriously threatened. Under the state of cyber-emergency, the duties to act upon orders of the CERTs apply to all kinds of providers of services of electronic communications (i.e. not only to those listed in previous paragraph).

Apart this regulatory framework, the only part of Czech law that specifically tackles the liability of information society service providers is Act No. 480/2004 Sb.[2] This Act implemented the Directive 31/2000/EC with regards to liability limitations (see Smejkal and Švestka 2005). Safe harbours are laid down with respect to mere conduit, hosting and caching providers, as discussed above.

The above provisions are, save for the difference in wording pointed out *supra*, based on Art. 12–15 of the Directive 2000/31/EC. The fact the Directive shaped this part of the Act No. 480/2004 Sb. can be seen from the inclusion of § 6, which provides that 'Service providers... shall not be obliged to monitor the contents of the information which they transmit or store, actively seek facts or circumstances indicating illegal contents of information'. For a civil law country, where legal obligations must always arise from a statute, to have a provision of this kind is unusual. The fact that ISPs do not have a duty to monitor could simply be implied from mere non-existence of explicit statutory definition of such a duty.

A problem with the Act arises from the fact that, despite being translated literally from the Directive, it uses slightly different wording for the liability of particular sorts of service providers. For example, the limits on liability of hosting providers are laid down in the Directive in Art. 14(1) as follows: 'Where an information society service is provided that consists of the storage of information provided by a recipient of the service, Member States shall ensure that the service provider is not liable for the information stored at the request of a recipient of the service, on condition that . . .', while the same provision in § 5 para 1 of the Czech Act reads as follows (informal translation): 'A provider of a service consisting of storage of information provided by a user, is liable for the content of information stored on a request of a user, only if...' (emphases added).

Despite the purpose of both provisions being the same (i.e. to limit the liability of hosting providers), the Czech version might be interpreted as providing specific grounds for the liability of service providers for user conduct (the same structure of the provision is used also in the case of mere conduit and hosting providers - see Říha 2008). This unfortunate technical mistake has already misled Czech courts

---

[2] Some court cases that involved service providers have been decided before the Act came into force, but they did not establish any broader standard or general practice (see for example the decision referred to as obaly.cd available at http://www.itpravo.cz/index.shtml?x=102705).

in the case *Prolux Consulting* (Higher Court in Prague, 3 Cmo 197/2010–82, available in full text at www.husovec.blogspot.com), where the *ratio decidendi* included a reference to § 5 para 1 of the Act as the sole reason for the liability of a discussion board service provider who, despite receiving a notice from a defamed corporation, refused to remove the defamatory statement. The problem with this judgment was not, as was acknowledged by scholars, in the mere outcome but rather in the legal reasoning. Instead of founding liability on the provisions of Act No. 480/2004 Sb., the court should have relied on specific provisions of the Civil Code on the protection of corporate goodwill and defamation while stating that § 5 para 1 of the Act No. 480/2004 Sb. could prevent the provider from being held liable because the notice was served and no takedown followed.

Courts also decided cases of illegal domain name registration or use where the domain authority (CZ.NIC) was a defendant; in those cases, the courts regularly applied the concept of distinct liability (see Pelikánová and Čermák 2000). As a result, the courts ordered the cancellation of the registration and costs against the domain authority, while all other remedies (namely real damages and satisfactory damages) were ordered against the offender. It is difficult to derive a general standard from this, however, because it applied only in the relatively narrow field of law of domain name disputes (see for example the landmark decision referred to as ceskapojistovna.cz, available at http://www.itpravo.cz/index.shtml?x=218427). Contrarily, in a case concerning a file-sharing service that did not remove an illegally-posted movie upon receiving a notice by the rights-holder, the court did not apply distinct liability and ordered damages to be jointly paid by the service provider and the offender (see Husovec 2014a). It is difficult, however, to draw any general conclusions from this case either, because the ISP did not defend itself – see *Mestsky soud v Praze*, 31 C 72/2011–33 (referred to as 'share-rapid.cz', and it is available at http://blog.eisionline.org/2013/09/01/rozhodnutia-prazskych-sudov-vo-veci-share-rapid-cz/).

There is no black-letter limit as to the scope of the safe harbour created by Act No. 480/2004 Sb. Consequently, it is arguable that the limitation on liability contained therein is universal and thus that it immunises ISPs in private law as well as in other fields of Czech law (including administrative law and criminal law). However, the scope of the safe harbours has not yet been judicially tested. One might argue in favour of confining the safe harbour to private law delicts, by pointing to the fact that all references in Act No. 480/2004 Sb. are to the 1964 Civil Code or the 1991 Commercial Code (both of which were amended by the 2012 Civil Code), and to the fact that there is no reference to either administrative or criminal law. Moreover, Directive 2000/31/EC and Act No. 480/2004 Sb. were adopted before the adoption of the Lisbon Treaty, i.e. at the time when the reach of EU law was considerably more limited compared to the post-Lisbon state of affairs.

On the other hand, the directly applicable source of Czech law is Act No. 480/2004 Sb. Thus, contemporaneous limits of the EU (at that time EC) legislation do not matter if they do not form a part of the Act. As the applicable wording of the Act does not contain any limits to the safe harbours (whereas references made in footnotes only to the Civil and Commercial Code are not binding, because they are not considered an official part of Czech legislation), there is reason to think that the limits on liability contained therein are universal in nature and that they apply to all sorts of liability, including private law delicts (copyright infringement, defamation, contract breaches etc.), administrative law delicts (minor or administrative offenses) as well as to criminal law delicts (crimes). This reading also implies that the safe harbours, if applicable, protect ISPs from all forms of liability (i.e. real damages that are used to liquidate actual damage, duties to act, satisfactory damages that have compensative purpose etc.)

The availability of two sorts of remedies, however, remains unclear: desistance duties and disclosure of information in the case of copyright infringement, and preventive duties arising from the Civil Code (see Polčák 2011b). This is because the priority in application of competing statutory provisions is neither apparent from legislative provisions nor has yet been addressed by the courts. Desistance duty and disclosure of information in cases of copyright infringement is regulated by § 40 para 1 of the Copyright Act (Act No. 121/2000 Sb.), which reads as follows (informal translation taken from www.mkcr.cz):

> An author whose rights have been infringed or whose rights have been exposed to danger of infringement may claim, in particular: (a) Recognition of his authorship; (b) Prohibition of the exposure of his right, including impending repetition of exposure, or of the infringement of his right, including, but not limited to, the prohibition of the unauthorized production, unauthorized commercial sale, unauthorized import or export of the original or reproduction or imitation of his work, unauthorized communication of the work to the public, as well as its unauthorized promotion, including advertising and other forms of campaigns; (c) Disclosure of details concerning the way and extent of unauthorized use, the origin of the illicitly made reproduction or imitation of his work, the way and extent of the use thereof, the price thereof, the price of the service related to the unauthorized use of the work, and the identity of the persons involved in the unauthorized use, also including the persons for whom such reproductions or imitations were designated for the purpose of the provision thereof to a third party; the author may claim his right to information under this provision from the person who infringed or exposed his right and also, in particular, from the person who: (1) possesses or possessed an illicitly made reproduction or imitation of the author's work for the purpose of direct or indirect economic or commercial advantage; (2) makes or made use, for the purpose of direct or indirect economic or commercial advantage, of any service that infringes or infringed the author's right, or that exposes or exposed it to danger; (3) provides or provided, for the purpose of direct or indirect economic or commercial advantage, a service used within activities that infringe or infringed the author's right, or that expose or exposed it to danger; (4) has been identified by a person referred to under Items 1, 2 or 3 above as a person involved in the obtaining, production or distribution of a reproduction or imitation of the work or as a person involved in the provision of services that infringe the author's right, or that expose it to danger.

The applicability of these remedies against ISPs in the case of copyright infringement depends on the question of *lex specialis* versus *lex generalis*. As Act No. 480/2004 Sb. is special with regards to the scope of its application as compared to the Copyright Act, we have a reason to believe that safe harbour liability limitation should apply also to duties listed in Art. § 40 of the Copyright Act. Similar conclusions may apply to para 1, letter c) of the same provision which refers to all kinds of third persons that might be liable in case of copyright infringement next to the main perpetrator. In other words, if a third person listed under § 40 para 1, letter c) of the

Copyright Act is an ISP, it should be exempt from the liability for copyright infringement caused by users of its information society services(see Polčák 2010).

Similar points may be made as regards preventive duties that are laid down in the Civil Code. General preventive duties are laid down in § 2900, which reads as follows: 'If so demanded by the circumstances or by usages of private life, everyone shall be obliged to act to prevent unjustified damage to others' freedom, life, health or property.' Preventive duties impose liability for inaction in cases when a person has information and the ability to prevent damage but fails to act. Preventive duties are a traditional concept in continental European law, but have been neglected in Czech practice. As a result, case law has not yet determined whether the safe harbour provisions apply or whether such duties can be invoked against ISPs.

There is a reason to think that liability arising from preventive duties does not substantially differ from other sorts of liabilities that are covered by limitations in Act No. 480/2004 Sb. and so the safe harbours should apply to them as well. Section 6 of Act No. 480/2004 Sb. states that '[s]ervice providers referred to in Sections 3 to 5 shall not be obliged to monitor the contents of the information which they transmit or store, or to actively seek facts or circumstances indicating the illegal nature of such information.' This can be interpreted as directly excluding preventive duties. On the other hand, Czech law is in many respects influenced by German case-law, and courts in recent German decisions such as *Rolex v. Ricardo* (BGH 11.3.2004, I ZR 304/0d or *Rapidshare.de* (OLG Dusseldorf 22.03.2010, Az I-20 U 166/09), found an ISP liable for its failure to act despite the existence of safe harbours laid down upon the Directive 2000/31/EC. Thus, whether the safe harbours are universally applicable is unclear (see Polčák et al. 2012).

Act No. 480/2004 Sb. does not lay down any kind of immunity of ISPs from administrative decisions or judgments. If an order is issued by a court or in some cases even by an executive body (e.g. the police, customs, local administration bodies etc.), there is no special treatment of service providers as to their execution (see Polčák et al. 2007).

For example, when there is a reason to prevent mere conduit services from communicating certain content – especially if that content is outside of the Czech jurisdiction) – an ISP might be obliged by an order to act regardless of its involvement in the illegal conduct. In that case, an order might be issued to block access to data under general rules of civil, administrative or criminal procedure. However, cases such as this are not frequent and those that have been issued include blocking orders against Nazi propaganda or child pornography. The reason for the scarcity of blocking orders under Czech law may be that memories of Communist Czechoslovakia are still somewhat alive. Czech authorities are thus extremely cautious whenever the execution of their powers might lead to anything that looks like like censorship; the Supreme Administrative Court and the Constitutional Court especially tend to be extremely protective. In a situation when blocking orders have to be issued based upon generally defined competences (as there is no particular provision or procedure specifically for blocking orders), there is a need to argue every case extensively and this turns out to be far too complicated for everyday frequent application.

Apart from blocking orders, ISPs have regular procedural duties. These duties flow from rules that vest state authorities (namely, courts, state prosecution, police or customs) with the power to investigate and to request information, evidence or assistance. ISPs are not exempt from complying with procedural obligations; general rules regarding disclosure, handling of evidence or assistance apply to them. For example, the Code of Civil Procedure (Act No. 99/1963 Sb.) states in § 128 that '[e]veryone is obliged to communicate, without a right for remuneration, to the court facts that might be of some importance for the case and for the decision.' Some procedural rules have specific provisions for disclosure of information, or for the handling of specific types of evidence that is regularly available only through ISPs (for example, orders for wiretapping or for transfers of traffic data from electronic communications being processed upon data retention obligations).

Proportionality

Regardless of the nature of the damage, the liability of intermediaries always centres on the conflicting interests of the actual offender (i.e. the user acting in breach of the law), the damaged party and the ISP. The discussion above described ways in which Czech law deals with these issues on the statutory level. Due to the fact that these interests are grounded in fundamental rights, it is also possible to consider the matter at the constitutional level.

Although there are no cases addressing ISP liability as a constitutional matter in the Czech Republic, there is good reason to anticipate them in the near future. Thus, it is appropriate to explain a method that the Czech Constitutional Court uses whenever there is need to engage in proportionate balancing of constitutionally-protected rights (e.g. property, freedom of speech, privacy etc.). Moreover, the doctrine of proportionality is gradually being applied in different types of cases by regular courts and even by administrative authorities. Very recently, the Municipal Court in Prague issued a decision in a case that was similar to the *Delfi* judgement of the ECHR (Grand Chamber, application No. 64569, 16 June 2015), in which it applied this doctrine to the statutory assessment of the liability of a provider of a discussion board for xenophobic user comments (Mestsky soud v Praze, c.j. 66C 143/2013– 510, 12 January 2015). Consequently, there is a good prospect that the same doctrine will soon find its way to similar cases when ISP liability is related to freedom of speech or similar rights protected by statutory law, based on their fundamental constitutional character.

The methodological approach to proportionality of rights was established in a decision of the Constitutional Court No. P. ÚS 4/94. In this case, the court was assessing the constitutionality of anonymous witnesses and had to find a proportional balance between the witness's protection and the fair trial rights of the accused (see Holländer 2006). With respect to proportionate balancing of rights, the court ruled that[3]:

> When considering the possibilities of restricting a basic right or freedom for the benefit of another basic right or freedom the following conditions can be stipulated governing the priority of one basic right or freedom:
>
> The first condition is their mutual comparison, the other is the requirement to examine the substance and the sense of the fundamental right or freedom being restricted (Art. 4 para 4 of the Charter of Fundamental Rights and Freedoms).
>
> The mutual comparison of colliding fundamental rights and freedoms is based upon the following criteria:
>
> The first is the criterion of applicability, i.e. a reply to the question whether the institute restricting a certain basic right allows the achievement of the desirable aim (the protection of another basic right). In the given case the legislator can be affirmed in that the institute of anonymous witness allows to achieve the aim, i.e. to guarantee the inviolability of his person.
>
> The second criterion for measuring basic rights and freedoms is the criterion of necessity residing in the comparison of the legislative means restricting some basic right or freedom with other provisions allowing to achieve the same objective, however, without impinging upon fundamental rights and freedoms. The reply to the fulfilment of the criterion of necessity in the second case is not unambiguous: in addition to the legislative construction allowing the anonymity of the witness the government can use also other means for his protection (such as the utilization of anonymous testimony as a criminalistic means for further examination, offering protection to the witness etc.).
>
> The third criterion is the comparison of the importance of both conflicting basic rights. In the case under consideration one of them is the right of fair trial ensuring the right for personal freedom, the other is the right of personal inviolability. These basic rights are prima facie equal.
>
> The comparison of the importance of colliding basic rights (after having fulfilled the condition of appropriateness and necessity) resides in weighting empirical, systemic, contextual and value oriented arguments. As an empirical argument the factual seriousness of a phenomenon can be understood that is connected with the protection of certain fundamental right (in the case under consideration this is the increasing number of cases of threatening and terrorising of witnesses by organized crime - see Polčák and Gřivna 2008). A systemic argument means considering the sense and the classification of the respective fundamental right or freedom within the system of basic rights and freedoms (the right to fair trial in this connection is part of the general institutional protection of basic rights and freedoms). As contextual argument also further adverse impacts of the restriction of one fundamental right due to the favouring another right can be understood (in the given case the possibility of misusing the institute of anonymous witness in the criminal procedure). The value argument represents considering the positive aspects of the conflicting fundamental rights as regards the accepted hierarchy of values.
>
> Part of comparing the relative weight of the conflicting basic rights is also considering the utilization of legal institutes minimizing the intervention into one of them, supported by arguments.

---

[3] Decision No. Pl. ÚS 4/94, English translation available at http://www.usoud.cz/en/decisions/?tx_
ttnews%5Btt_news%5D=611&cHash=f69da5fcba1a2e433d74385371b3a196.

In effect, the Constitutional Court established a three-step test that consists of the following parts:

1. Suitability – whether the limitation of a fundamental right is able to serve the desired purpose.
2. Necessity – whether there might be alternative ways to achieve the desired effect without a need to limit the respective fundamental right(s).
3. Proportionality *stricto sensu* – whether there is a reason ad hoc to prefer one fundamental right over another.

If a limitation of a fundamental right is able to pass this test, the court must additionally assess whether that limitation is not more than necessary. This assessment, known also as the limited proportionality test, is in many cases crucial. In the case of anonymous witness discussed above, as well as in the case of a challenge to data retention rules, the instruments in question satisfied all three parts of the proportionality tests (i.e., the court stated that the measures were fit for purpose, there we no reasonable alternatives, and there was a reason to prefer certain fundamental rights over others). However, the court held that the instruments led to greater than necessary impact on fundamental rights. In other words, the Constitutional Court regularly holds that some instruments are proportionate as such, but that they need to be structured using less intrusive measures or implementing more safeguards or balances.

This Czech doctrine of proportionality is to a large extent inspired by German constitutional practice as well as by German scholarship. Consistent with those sources, no one fundamental right is per se superior to any other. Thus, it is impossible to state that, for example, privacy is in general more relevant than freedom of speech – on the contrary, every conflict of fundamental rights (or, in general, constitutional principles) has to be assessed on a case-by-case basis (see Kühn 2011).

An illustrative example of balancing of interests (although the full proportionality test was not mentioned in the *ratio decidendi*) can be seen in *Prolux Consulting* (Higher Court in Prague, 3 Cmo 197/2010–82, available in full text at www.husovec.blogspot.com). In this case, an anonymous user commented on the performance of a Czech real estate agency, saying that 'They lie like a swine.' When the provider of the discussion board service refused to take down this post upon a notice by Prolux, the agency in question, the court of first instance ordered the ISP to remove the post and to pay costs. On appeal, however, the court considered the statement to be partly true (Prolux really lied to its clients) and minimised the effect of the takedown order by ordering the ISP only to remove the indecent word 'swine' (see Husovec 2014b).

ISP Concerted Practices

As the liability of ISPs does not represent a major topic for Czech associations or guilds of service providers, no system of best practices has been established. Even highly influential associations like the ICT Unie or NIX.CZ have not developed policies designed to tackle liability limitations.

Rather, what have developed are concerted practices in specific fields of e-commerce and user-generated services dealing with the formation of service contracts or the reporting of illegal information. As for end-user service contracts or terms of services, Czech law does not allow for disclaimers (there is not even a word in Czech legal language for disclaimers) that would limit the liability of a provider to third persons or to its users. However, it is possible to stipulate between the user and the ISP for remuneration of any damages from the former that the latter had to pay to a third party for illegal activities of the user. (The ISP can recover the damage in full if the court decides a case on the basis of general joint liability and the plaintiff turns to the ISP for all the damages).

Another typical practice that we observed in the case of contracted (and often paid) services is the emergence of takedown clauses. These clauses enable the ISPs to take the respective user information down not just upon its illegality (this right of the ISP is implied by law), but upon receipt of a notice that prima facie appears to be justified. This partly relieves ISPs of the factual duty to determine the objective legality of user behaviour and gives them an opportunity to decide about takedown upon more formal criteria. It helps in unclear cases when it is difficult for the ISP to determine whether some information submitted by a user violates, for example, privacy protection, rules against unfair competition or some other rules that require extensive interpretation and whose application is never entirely certain (see Matejka 2013).

A different kind of concerted practice of ISPs can be seen in cases when the services are provided without service contracts (typically in the case of anonymous discussion boards, chat rooms, picture sharing services etc.) and there is a regular risk of anonymous illegal activities constituting defamation, privacy violations, hate speech, racist propaganda etc. In these cases, ISPs regularly use automated reporting systems allowing users to point with ease to allegedly illegal information by simply clicking on a 'report violation' button appearing next to each post. Removal of reported information is then normally swift and in some cases even automatic – the reason is that there is greater risk for the service provider of possible joint liability arising from illegally posted information than the risk of annoying a client whose information was removed upon such an announcement.

Webhosting or similar services that are directly paid by contracted users obviously cannot use such simplified techniques for removal of user information, because of high risk of valuable client disputes. In these cases, another highly problematic concerted practice consists of hiding the contact details of the ISP in order to make serving a notice as difficult as possible. Whilst the Directive 2000/31/E has strict identification requirements, there are no corresponding provisions in Czech law. Consequently, it is often extremely difficult to find the provider of a hosting service on whom to serve a notice (and this state of affairs seems to be, at least in a part, deliberately created by ISPs).

Concluding Remarks

The Czech ICT industry is considered one of most advanced in Europe and statistical figures of Czech e-commerce are far beyond the EU average. This should motivate the Czech government to take proper care of the legal regulatory background of the development and use of ICT. However, there are, apart from current initiatives in the field of cybersecurity, almost no efforts to introduce laws that specifically cater to the needs of information society. The Civil Code of 2012, which should serve as the main basis for everyday liability claims, deals in great detail with pets and hives, while almost entirely avoiding issues of utmost interest to information society. In result, there is a need to apply general principles of private law by analogy in order to resolve everyday claims arising out of the use of information society services.

This situation is unfortunate because of the accompanying high level of uncertainty for ISPs. It is not only difficult for them to predict possible liability risks, but also causes the very limited availability of efficient standard compliance mechanisms or risk-insurance products. In that sense, it is even difficult to argue that the revolutionary departure from the Orwellian certainty of the Civil Code of 1964 was a good move (see Polčák 2012).

While there is great level of uncertainty as to the foundations of liability of ISPs, the situation as regards its limitations, or of safe harbours, is considerably better. This is caused by the implementation of the e-commerce directive and by subsequent case-law of the Court of Justice[4].

Due to the relative lack of Czech case-law, there is also remaining uncertainty as to the application of the safe harbour regime to claims based upon preventive duties. There are pending Czech cases that might clarify whether or to what extent safe harbours cover general preventive duties, but decisions in those cases will probably take months or years to become final.

References

Hajn, P.K. 2011. souběhu práva autorského a práva proti nekalé soutěži (A concourse of unfair competition and copyright). In *Karlovarské právnické dny*, ed. V. Zoufalý. Praha: Leges.

Holländer, P. 2006. *Filosofie práva* (Philosophy of Law). Plzeň: Aleš Čeněk.

Hurdík, J., and P. Lavický 2010. *Systém zásad soukromého práva* (System of principles of civil law). Brno: Masarykova uneiverzita.

Husovec, M. 2014a. *Zodpovědnost na internete podla ceskeho a slovenskeho prava* (The internet liability under the Czech and Slovak law). Praha: CZ.NIC.

———. 2014b. Zodpovednosť poskytovateľa za obsah diskusných príspevkov (The provider liability for discussion posts). *Revue pro právo a technologie* 2(3): 40.

---

[4] In that respect, there are just concerns regarding the concept of the place of establishment of a service provider—in C-170/12 *Peter Pinckney v KDG Mediatech AG*, the Court of Justice indicated that the original principle of one-stop-shop for ISPs laid down in the e-commerce directive might be shifted in favour of copyright holders.

Kühn, Z. 2011. Ochrana soukromí v Internetové době (Privacy in the internet age). In *Právo na soukromí*, ed. V. Šimíček. Brno: Mezinárodní politologický ústav.

Lavický, Petr 2004. *Solidární závazky* (Solidary Obligations). Praha: C. H. Beck.

Maisner, M., J. Černý, J. Donát, Z. Loebl, T. Nielsen, R. Polčák, and J. Zeman. 2011. *Základy softwarového práva* (Foundations of the software law). Praha: Wolters Kluwer ČR.

Matejka, Ján 2013. *Internet jako objekt práva: hledání rovnováhy autonomie a soukromí* (Internet as an object of law: Finding a balance of autonomy and privacy). Praha: CZ.NIC.

Mates, P. 2002. *Ochrana osobních údajů* (Protection of personal data). Praha: Karolinum.

———. 2006. *Ochrana soukromí ve správním právu* (Privacy protection in administrative law). Praha: Linde Praha

Myška, M. 2009. The true story of DRM. *Masaryk University Journal of Law and Technology* 3 (2): 267–278.

Pelikánová, R., and K. Čermák 2000. *Právní aspekty doménových jmen* (Legal issues in domain names). Praha: Linde.

Polčák, R. 2005. Nekalosoutěžní agrese na Internetu (Unfair competitive aggression on the Internet). *Právní rozhledy: časopis pro všechna právní odvětví*. Praha: C.H.Beck, vol. 13, No 13, p. 473–477. ISSN 1210–6410.

———. 2007a. *Právo na Internetu – spam a odpovědnost ISP* (Law on the Internet – Spam and the Responsibility of ISPs). Brno: Computer Press.

———. 2007b. Správci on-line světa: Pojem a koncepce odpovědnosti poskytovatelů služeb informační společnosti (The administrators of the on-line world: The concept of responsibility of information services providers). *Právny obzor: časopis Ústavu štátu a práva Slovenskej akadémie vied, Bratislava: Slovenská akadémia vied* 90 (5): 447–460.

———. 2008a. Právní reflexe pojmu informační společnosti (Legal reflection of the concept of Information Society). *Časopis pro právní vědu a praxi, Brno: Masarykova univerzita, Právnická fakulta* 2008 (4): 350–358.

———. 2008b. Stone roots, digital leaves: Czech Law against unfair competition in the Internet Era. *Review of Central and East European Law, Leiden: Martinus Nijhoff Publishers* 33 (2): 155–180.

———. 2009a. Liability of information services providers as a new phenomenon (Liability of Information Services Providers as a New Phenomenon). In *Five years of EU membership – Case of Czech Republic and Slovenian law*, ed. R. Knez, N. Rozehnalová, and V. Týč. Maribor: University of Maribor.

———. 2009b. *Odpovědnost poskytovatelů služeb informační společnosti* (ISP) (Liability of Information Society Service Providers (ISP)). Právní rozhledy, Praha: C.H.Beck, vol. 2009, No 23, p. 837–841.

———. 2009c. *Právo a evropská informační společnost* (Law and European Information Society). Brno: Masarykova Univerzita.

———. 2010. The legal classification of ISPs (The legal classification of ISPs). *Journal of Intellectual Property, Information Technology and E-Commerce Law, Hannover: Leibniz Universität Hannover* 1(3): 172–177.

———. 2011a. European Policies for the Information Society. In *One or many? The law and the structure of the European Union and the United states*, 171–174. Rock Island: East Hall Press.

———. 2011b. Působení práva EU ve svéře českého práva a informační společnosti (The applicability of EU law on Czech Information Society). *Časopis pro právní vědu a praxi, Brno: Právnická fakulta Masarykovy univerzity* 19 (4): 392–396.

———. 2011c. Wiggum in cyberspace: Legal issues in Czech and EU cybersecurity (Wiggum in cyberspace: Legal issues in Czech and EU cybersecurity). In *Information security summit*, ed. R. Haňka, Z. Kaplan, V. Matyáš, J. Mikulecký, and Z. Říha, 149–157. Praha: Data Security Management.

———. 2012. *Internet a proměny práva* (Internet and metamorphoses of law). Praha: Auditorium.

Polčák, R., and T. Gřivna 2008. *Kyberkriminalita a právo* (Cybercrime and the Law). Praha: Auditorium,

Polčák, R., Z.G. Balogh, M. Bogdan, G.M. Riccio, D.J.B. Svantesson, and A. Wiebe. 2007. *Introduction to ICT law (Selected issues)*. Brno: Masarykova Univerzita.

Polčák, R., J. Čermák, Z. Loebl, T. Gřivna, J. Matejka, and M. Petr. 2012. *Cyber Law in the Czech Republic (Cyber Law in the Czech Republic)*. *1. vyd*. Alphen aan den Rijn: Kluwer Law International.

Říha, J. 2008. *Odpovednost provideru se zamerenim na odpovednost host-providera a access-providera* (Responsibility of providers with special focus on the responsibility of host-provider and access-provider). Acta Universitatis Carolinae Iuridica, vol. 4.

Ryška, M. 2011. Ochrana vnitřního kruhu i jeho okolí v praxi práva na ochranu osobnosti. In *Právo na soukromí* (Right to privacy), ed. V. Šimíček. Brno: Mezinárodní politologický ústav.

Smejkal, V., and J. Švestka 2005. Odpovědnost za škodu při provozu informačního systému, aneb nemalujme čerty na zeď (Liability for damage caused by operation of an information system). *Právní rozhledy* 13(19).

Chapter 12
Website Blocking Injunctions under United Kingdom and European Law

Jaani Riordan

Introduction

A person harmed by tortious material on the internet can seek to enforce his or her rights by three alternative means. The first is well known: action may be taken against the primary wrongdoer who was responsible for creating or disseminating the material. However, direct enforcement may be impracticable where the primary wrongdoer is anonymous, judgment-proof, or domiciled abroad—or where the wrongdoing is simply too widespread for action against individuals to be worthwhile. The second method therefore seeks to remove tortious material at its source by involving network and application-layer intermediaries—for example, sending a notice to the platform or host from which material is made available to internet users or, less commonly, alleging that the intermediary should be held liable for the material under national doctrines of secondary or accessory liability. In Europe, effective notice will preclude the recipient from relying upon the storage safe harbour unless it removes the material expeditiously.[1] However, an intermediary may simply ignore such a notice, or the primary wrongdoer may shift to a less scrupulous or uncontactable host.

The third method is to block or restrict end users from accessing the material. This remedy seeks to curtail the dissemination of harmful content in a particular jurisdiction by making it more difficult to locate and download. This may involve

---

[1] See Directive 2000/31/EC of the European Parliament and of the Council of 8 June 2000 on certain legal aspects of information society services, in particular electronic commerce, in the Internal Market [2000] OJ L 178/1 art 14(1) ('E-Commerce Directive'); *Electronic Commerce (EC Directive) Regulations 2002* (UK) regs 19(a)(i), 22 ('*E-Commerce Regulations*').

J. Riordan (✉)
Barrister, 8 New Square, Lincoln's Inn, London, UK
e-mail: Jaani.Riordan@8newsquare.co.uk

© Springer International Publishing AG 2017

G.B. Dinwoodie (ed.), *Secondary Liability of Internet Service Providers*,
Ius Comparatum – Global Studies in Comparative Law 25,
DOI 10.1007/978-3-319-55030-5_12

co-opting search intermediaries to remove hyperlinks to the material (thereby making it harder to find), or it may use technical means to prevent individual computers from retrieving the material using the services of a network-layer intermediary, such as an internet service provider ('ISP'). In such a case, the object of a blocking remedy is neither the primary author nor host of the material complained of, but rather a local intermediary which carries the tortious material to its users within the jurisdiction.

Until relatively recently, blocking remedies were largely unknown in Europe. Blocking technology was expensive, ineffective and largely confined to corporate networks, university campuses and authoritarian states. However, blocking is now emerging throughout Europe as a complementary remedy to notice-and-takedown: if tortious material cannot be removed at its source, it might yet be targeted and curtailed at its destinations. Claimants argue that blocking remedies offer a cheap and reasonably effective means of enforcing judicial determinations of primary liability against offshore infringers. However, intermediaries and internet users question whether the economic and social costs of blocking are outweighed by modest reductions in the overall level of infringement of claimants' rights.

This Chapter examines the current framework for website blocking remedies within Europe, with a particular focus on the transposing provisions enacted in the United Kingdom. It is divided into four parts. First, this Chapter summarises the relevant parts of the European framework governing injunctive remedies against internet intermediaries, and identifies the main elements and limitations of website blocking injunctions. Second, this Chapter examines how these rules have been transposed in the United Kingdom, and considers how blocking remedies have been approached in judicial decisions involving copyright and other intellectual property rights. Third, this Chapter analyses factors relevant to the proportionality of website blocking remedies, including the substantiality of primary wrongdoing, their technical feasibility, effectiveness, dissuasiveness, alternatives, effect upon lawful content and trade, technical and procedural safeguards, cost and complexity.

Finally, this Chapter considers several emerging issues, including the availability of blocking remedies against search engines and other intermediaries, their application to new species of primary wrongdoing, their invocation against new categories of target websites, interim relief, and network neutrality regulation.

The rise of website blocking remedies in Europe is consistent with a worldwide shift towards non-monetary relief against internet intermediaries. Most website blocking remedies have been granted in cases involving commercial scale copyright infringement, but their application is widening to encompass other forms of civil wrongdoing, such as websites dealing in counterfeit goods. However, perhaps due to the nature of the claims brought before national courts to date, courts have struggled to articulate meaningful limits on blocking remedies and to consider their relationship with other enforcement options.

The European Legal Framework

Injunctions against Intermediaries

Copyright and Related Rights

Article 8(3) of the Information Society Directive provides:

> Member States shall ensure that rightholders are in a position to apply for an injunction against intermediaries whose services are used by a third party to infringe a copyright or related right.[2]

The word 'intermediaries' is not defined by the Directive, but recital (59) confirms that the provision was intended to grant relief against those 'best placed' to bring infringing activities to an end:

> In the digital environment, in particular, the services of intermediaries may increasingly be used by third parties for infringing activities. *In many cases such intermediaries are best placed to bring such infringing activities to an end.* Therefore, without prejudice to any other sanctions and remedies available, rightholders should have the possibility of applying for an injunction against an intermediary who carries a third party's infringement of a protected work or other subject-matter in a network. This possibility should be available even where the acts carried out by the intermediary are exempted under Article 5. The conditions and modalities relating to such injunctions should be left to the national law of the Member States.[3]

This recital reflects two cardinal features of article 8(3). First, it is directed at any party 'who carries a third party's infringement ... in a network', even if it has the benefit of an exemption from primary liability.[4] It is unnecessary that the intermediary be a primary wrongdoer: in *L'Oréal SA v eBay International AG* the Court made clear that injunctions are available against an intermediary (in that case a marketplace operator) 'regardless of any liability of its own in relation to the facts at issue'.[5] Thus, the intermediary does not need to exercise actual or legal control over the use of its service; it is sufficient merely to supply the connection or network over which rights are infringed.[6]

Second, recital (59) confirms the policy underlying article 8(3), which is to recognise an effective remedy against facilitators who are least-cost avoiders. In other words, the remedy exists to reduce overall enforcement costs by requiring those in the best position to take efficient precautions as gatekeepers to do so. However, implicit in the phrase '[i]n many cases such intermediaries are best placed' is a

---

[2] Directive 2001/29/EC on the harmonisation of certain aspects of copyright and related rights in the information society [2001] OJ L 167/10 art 8(3) ('Information Society Directive').

[3] Ibid recital (59) (emphasis added). See also Directive 2004/48 on the enforcement of intellectual property rights [2004] OJ L 195/16 recital (23) ('Enforcement Directive').

[4] See, e.g., Information Society Directive art 5(1)(a).

[5] Case C-324/09, EU:C:2011:474, [127] ('*L'Oréal*').

[6] Case C-557/07, *LSG-Gesellschaft zur Wahrnehmung von Leistungsschutzrechten GmbH v Tele2 Telecommunication GmbH*, EU:C:2009:107 [2009] ECR I-1227, [43]–[46].

recognition that a particular intermediary may not *always* be the least-cost avoider of harm to the claimant. Blocking is not an absolute obligation, but something to be assessed in light of the role played by an intermediary, and the costs and benefits of blocking.

In *UPC Telekabel Wien GmbH v Constantin Film Verleih GmbH*, the Court of Justice began with the proposition that 'the term "intermediary" used in Article 8(3) covers any person who carries a third party's infringement of a protected work or other subject-matter in a network'.[7] However, it added that this will include placing protected subject-matter (such as copyright works) on the internet and making it available in circumstances where the ISP is an 'inevitable actor in any transmission of an infringement'.[8] This appears to reflect a causation-based justification for the grant of blocking remedies: if a service provider *would* be a necessary actor in a transmission of infringing material that has been made available online, then it will be an intermediary with respect to that material.[9]

Other Intellectual Property Rights

Article 8(3) of the Information Society Directive applies only to 'infringements of the rights and obligations set out in [that] Directive', meaning copyright and related rights. Article 11 of the Enforcement Directive is not so limited. The third sentence of that paragraph states:

> Member States shall also ensure that rightholders are in a position to apply for an injunction against intermediaries whose services are used by a third party to infringe an intellectual property right, without prejudice to Article 8(3) of Directive 2001/29/EC.

The Enforcement Directive applies to all forms of intellectual property. That phrase is not given an exhaustive definition in EU law, but includes 'industrial property rights'.[10] According to the European Commission, the definition should be understood to include copyright and related rights, *sui generis* database right, semiconductor topographies, trade marks, design right, patents and supplementary protection certificates, geographical indications, utility models, plant varieties, and trade names insofar as 'protected as exclusive property rights'.[11] Recital (13) confirms that the definition should be construed 'as widely as possible in order to encompass all the intellectual property rights covered by Community provisions in this field

---

[7] Case C-314/12, EU:C:2014:192, [30] (*'UPC Telekabel'*).

[8] *UPC Telekabel*, [32].

[9] *UPC Telekabel*, [36]. In the case of copyright, services may be used whether or not any person has actually accessed the protected subject-matter via that service provider. This appears to be on the basis that an infringement occurs the moment material is made available to the public, irrespective of whether it is actually accessed and downloaded by anyone in particular.

[10] Enforcement Directive art 1, second sentence.

[11] European Commission, 'Statement Concerning art 2 of Directive 2004/48/EC' (2005/295/EC) [2005] OJ L 94/37.

and/or by the national law of the Member State concerned'. This suggests that the
scope of application is in part determined by national law.

Minimum Standards

The Enforcement Directive obliges member states to provide effective, proportionate and dissuasive remedies to enforce intellectual property rights.[12] As Article 3(1)
states:

> Member States shall provide for the measures, procedures and remedies necessary to ensure
> the enforcement of the intellectual property rights covered by this Directive. Those mea-
> sures, procedures and remedies shall be fair and equitable and shall not be unnecessarily
> complicated or costly, or entail unreasonable time-limits or unwarranted delays.

These obligations reflect the minimum standards provided for by the *TRIPS
Agreement*, which requires members of the World Trade Organization to confer
enforcement procedures 'so as to permit effective action against any act of infringement' and which 'constitute a deterrent to further infringements'.[13] Such procedures
must also be 'fair and equitable', 'not be unnecessarily complicated or costly, or
entail unreasonable time-limits or unwarranted delays'.[14] These minimum standards
clearly apply to website blocking injunctions, both under article 8(3) of the
Information Society Directive and article 11 of the Enforcement Directive.

At the European level, injunctions against intermediaries are recognised as
important remedies. In its Final Report on the Enforcement Directive, the European
Commission concluded that 'the currently available legislative and non-legislative
instruments are not powerful enough to combat online infringements of intellectual
property rights effectively.'[15] Although the report did not directly mention website
blocking, the Commission made two recommendations which support more powerful forms of injunctive relief. First, although the details of how and when an injunction is available are left to member states, national laws should clarify that injunctive
relief against an intermediary does not depend on substantive liability for an
infringement. Second, it said that intermediaries should be more closely involved in
preventing and terminating online infringements.[16]

---

[12] Enforcement Directive art 3(2).

[13] *Marrakesh Agreement Establishing the World Trade Organization*, opened for signature 15 April
1994, 1867 UNTS 3 (entered into force 1 January 1995), annex 1C (*Agreement on Trade-Related
Aspects of Intellectual Property Rights*) art 41(1) ('*TRIPS Agreement*').

[14] *TRIPS Agreement* art 41(2).

[15] European Commission, *Application of Directive 2004/48/EC of the European Parliament and
the Council of 29 April 2004 on the Enforcement of Intellectual Property Rights*, COM(2010) 779
final, 7.

[16] Ibid.

Maximum Standards

EU law also specifies certain limitations applicable to intellectual property remedies, including injunctions against intermediaries. These are set out in articles 3(1) and 3(2) of the Enforcement Directive, article 15 of the E-Commerce Directive, the *Charter of Fundamental Rights of the European Union*,[17] and general principles of EU law. They function as upper limits on the scope and consequences of an injunction, and fall to be considered by national courts as part of the exercise of their discretion as to whether (and on what terms) to grant injunctive relief.

'Effective, Proportionate and Dissuasive'

Article 3(2) of the Enforcement Directive states:

> Those measures, procedures and remedies shall also be effective, proportionate and dissuasive and shall be applied in such a manner as to avoid the creation of barriers to legitimate trade and to provide for safeguards against their abuse.

Again, most of these limitations derive from the *TRIPS Agreement*, which recognises requirements of fairness, proportionality,[18] legitimate trade, and safeguards in almost identical terms. In *Productores de Música de España (Promusicae) v Telefónica de España SAU*, the Court of Justice explained that these limitations, as expressed in the Enforcement Directive and Information Society Directive, must be applied by national courts to shape the application of domestic remedies:

> the authorities and courts of the Member States must not only interpret their national law in a manner consistent with those directives but also make sure that they do not rely on an interpretation of them which would be in conflict with [fundamental rights protected by the EU legal order] or with the other general principles of [EU] law, such as the principle of proportionality …[19]

General Monitoring Duties

A further limitation is contained in article 15(1) of the E-Commerce Directive, which prevents any general monitoring duties from being imposed upon an information society service provider. That provision states:

> Member States shall not impose a general obligation on providers, when providing the services covered by Articles 12, 13 and 14 [mere conduit, caching and hosting], to monitor the information which they transmit or store, nor a general obligation actively to seek facts or circumstances indicating illegal activity.

---

[17] [2010] OJ C 83/389 ('*Charter*').

[18] See, e.g., *TRIPS Agreement* arts 46 (referring to 'the need for proportionality between the seriousness of the infringement and the remedies ordered') and 47.

[19] Case C-275/06 [2008] ECR I-271, [68] (citations omitted).

This limitation is not directly transposed into national law in the United Kingdom, but functions as a further limit on the scope of injunctions that may be granted against intermediaries. For example, an intermediary may not be ordered to carry out 'an active monitoring of all the data of each of its customers in order to prevent any future infringement'.[20] Conversely, monitoring is permissible 'in a specific case', as confirmed by recital (47) of the E-Commerce Directive. Examples of permitted monitoring may be obligations to identify or terminate access to a particular wrongdoer or infringement, though others can be envisaged.

Finally, it is worth noting recital (45) of the E-Commerce Directive, which confirms that the safe harbours established under that Directive do not affect the availability or scope of injunctions available against intermediaries:

> The limitations of the liability of intermediary service providers established in this Directive do not affect the possibility of injunctions of different kinds; such injunctions can in particular consist of orders by courts or administrative authorities requiring the termination or prevention of any infringement, including the removal of illegal information or the disabling of access to it.

Recital (45) expressly acknowledges the possibility of ordering intermediaries to disable access to 'illegal information' and to prevent 'any infringement'. Accordingly, it is beyond doubt that blocking remedies are compatible with the E-Commerce Directive, provided that they do not impose general monitoring obligations. These limitations are discussed further below.

Blocking Remedies in the United Kingdom

Website blocking can be sought in the United Kingdom in at least four ways. First, blocking injunctions are available against service providers as statutory remedies for third parties' infringement of copyright, performers' right, and database right. Second, such injunctions have been held to lie within the courts' general jurisdiction to grant injunctive relief ancillary to other kinds of wrongdoing. Third, blocking injunctions may be sought to give effect to data subjects' rights of blocking and erasure under the *Data Protection Act 1998* (UK), including the so-called 'right to be forgotten'. Finally, ISPs undertake blocking of certain forms of unlawful content on a voluntary basis pursuant to self-regulatory codes of conduct. These mechanisms and their application are discussed below.

---

[20] *L'Oréal*, [139].

Copyright

First, ss 97A and 191JA of the *Copyright, Designs and Patents Act 1988* (UK) (*'1988 Act'*) create statutory blocking remedies consequent upon a finding that a third party has infringed the applicant's copyright or performers' right, respectively, using the services of the respondent. These provisions transpose article 8(3) of the Information Society Directive and they are, in substance if not in form, 'an order under Article 8(3)'.[21] The legislative purpose of s 97A is essentially the same as article 8(3), and does not require proof of substantive liability.[22] The scope of s 97A is therefore much broader than doctrines of authorisation and joint liability under English law.[23]

Legislative History

Sections 97A and 191JA were introduced by reg 27 of the *Copyright and Related Rights Regulations 2003* (SI 2003/2498) (UK). Their enactment appears to have reflected a degree of uncertainty about whether or not injunctions could be sought directly against intermediaries who were not wrongdoers. As the Patent Office stated in the consultation paper which preceded the regulations:

> Regarding Article 8.3, it is already possible under UK law to seek injunctions against inter-mediaries. It is also possible to notify an intermediary of an injunction served on an infringer so that the intermediary is liable for contempt of court proceedings if he aids and abets an infringer. It is considered that this meets the requirements of Article 8.3.[24]

In other words, the Government appears to have assumed that existing copyright law would permit injunctions against primary infringers, while the *Spycatcher* prin-ciple could be used to bind intermediaries by giving notice of an injunction made against a primary infringer.[25] However, after the close of the consultation, the Government altered its position, concluding that, 'in order to avoid uncertainty' about whether or not blocking remedies were available, it would enact s 97A.[26]

---

[21] *Newzbin2 Order*, [35] (Arnold J).

[22] *Newzbin2*, [146] (Arnold J).

[23] For more detailed discussion of the differences between the scope of substantive secondary lia-bility for copyright infringement and injunctive relief, see Jaani Riordan, *The Liability of Internet Intermediaries* (2016) chs 6, 14.

[24] Patent Office, *Consultation Paper on Implementation of the Directive in the United Kingdom* (7 August 2002) 16.

[25] See *Attorney–General v Guardian Newspapers Ltd. [No 2]* [1990] 1 AC 109, 260 (Lord Keith), 281 (Lord Goff) (explaining that an equitable duty of confidence may arise upon the recipient of information being given notice of confidentiality). See also [1988] Ch 333, 375 (CA).

[26] Patent Office, *Consultation on UK Implementation of Directive 2001/29/EC on Copyright and Related Rights in the Information Society: Analysis of Responses and Government Conclusions* (2002) [8.4].

Elements of Relief

To succeed the applicant must satisfy four elements. First, the respondent against whom the order is sought must be a 'service provider'. This term is given the same broad definition as under regulation 2 of the *E-Commerce Regulations*,[27] and would therefore include most if not all retail ISPs, as well as many other categories of network and application-layer intermediaries. Second, there must have been a primary infringement of the relevant right by a third party.[28] Third, that infringement must have taken place using the respondent's services. Fourth, the respondent must have actual knowledge of that fact.

While the fourth element appears to set a high threshold requiring subjective knowledge of a specific infringer, the statutory language is broad and it has been so interpreted. In assessing whether the required standard of knowledge is met, the Court is obliged to consider all relevant circumstances, including whether the respondent received a notice of infringement containing the relevant particulars.[29] Even if the Court has jurisdiction under s 97A, it retains discretion whether or not to make an order (or to do so subject to conditions). The most salient factor going to the question of discretion is whether the order would be proportionate.

A number of decisions have considered the application of s 97A to ISPs and website operators in the United Kingdom. Several dozen blocking orders have also been made against the five main ISPs, who together account for around 92% of the market for fixed line broadband services in the United Kingdom.[30] These orders are summarised in Table 12.1, and discussed below.

Newzbin1

The first case in which a blocking injunction was sought was *Twentieth Century Fox Film Corp v Newzbin Ltd*, where Kitchin J granted an injunction against a website operator to restrain the future publication of NZB files which aggregated links to infringing copies of the claimants' copyright works hosted on third parties' Usenet servers.[31] This was an injunction against a website operator (who also happened to be a primary infringer),[32] rather than an ISP. The Court rejected the defendant's

---

[27] *1988 Act* s 97A(3).

[28] Note that the relevant infringements extend to database right by operation of regulation 23 of the *Copyright and Rights in Databases Regulations 1997* (UK) (which extends certain remedies under the *1998 Act* to database right).

[29] *1988 Act* s 97A(2).

[30] Ofcom, *The Communications Market Report* (6 August 2015) 292.

[31] [2010] FSR 21 ('*Newzbin1*').

[32] The defendant was the operator of a website on which large quantities of hyperlinks to copyright-infringing files were aggregated from Usenet news group servers and rendered searchable to users. The defendant was held to be liable for authorising copyright infringement and jointly liable with users who downloaded infringing copies using the facilities provided: *Newzbin1*, [102]–[112] (Kitchin J).

Table 12.1 Summary of blocking orders obtained by applicants under section 97A

Case	Applicant	Target websites	Primary infringement	Outcome
Newzbin1	BPI	Newzbin	Index of Usenet binaries	Blocked
Newzbin2	BPI	Newzbin2	Index of Usenet binaries (successor website)	Blocked
Dramatico [No 1] *Dramatico [No 2]*	BPI	The Pirate Bay	BitTorrent tracker and index	Blocked
EMI	BPI	KAT, H33T, Fenopy	BitTorrent trackers and indices	Blocked
Paramount [No 1]	MPA	EZTV	BitTorrent tracker and index	Blocked
Paramount [No 2]	MPA	Free-TV-Video-Online	Video streaming	Blocked
Paramount [No 3]	MPA	Primewire and 2 others	Video streaming and downloads	Blocked
Paramount [No 4]	MPA	YIFI Torrents	BitTorrent tracker and index	Blocked
FirstRow	FAPL	FirstRow	Sports streaming	Blocked
Paramount [No 5]	MPA	SolarMovie, TubePlus	Video streaming	Blocked
Paramount [No 6]	MPA	Movie2K, Download4All	Video streaming	Blocked
Paramount [No 7]	MPA	Megashare and 3 others	Video streaming	Blocked
Dramatico [No 3]	BPI	Filestube and 11 others	Meta-search engine, file upload	Blocked
Dramatico [No 4]	BPI	1337x and 8 others	BitTorrent trackers and indices	Blocked
1967 [No 1]	BPI	BitTorrent.am and 20 others	BitTorrent trackers and indices	Blocked
Paramount [No 8]	MPA	Watchseries.to and 6 others	Video streaming	Blocked
Rojadirecta	FAPL	Rojadirecta and 2 others	Sports streaming	Blocked
Bloomsbury [No 1]	Publishers Association	Ebookee.org and 6 others	Commercial book downloads	Blocked
1967 [No 2]	BPI	MP3 Monkey and 16 others	MP3 downloads	Blocked
Popcorn Time	MPA	Popcorn Time and 12 others	Video streaming using downloaded software	Blocked

argument that it lacked notice of infringement, not having received a valid notice under s 97A(2); receipt of a notice in that form is not a precondition of actual knowledge, but merely one factor to be considered.[33]

Newzbin2

In *Twentieth Century Fox Film Corporation v British Telecommunications plc*, the High Court issued an injunction against the respondent ISP ('BT') to prevent its services from being used to infringe the applicants' copyrights via the Newzbin website and its phoenix-like successor, Newzbin2.[34] Building on the findings made by Kitchin J in *Newzbin1*, Arnold J held that the Court had jurisdiction to grant the order, because (1) the operators of the website (in combination with BT's subscribers) used BT's services to infringe the applicants' copyrights;[35] and (2) BT had the necessary knowledge.

As to use of BT's services, Arnold J reasoned that the relevant question was whether the users or operators of Newzbin2 used BT's network equipment to transmit the infringing acts:

> In my view, it is important to consider the nature of the infringing act and its relationship with the service in question. In the present case, the infringing acts by the users consist of making digital copies of the Studios' films and television programmes on their computers. Each of those digital copies is made by assembling thousands of packets received via BT's service. No complete copy is necessarily made on BT network equipment at any single point in time, but transient copies of all the packets that in aggregate make up a complete copy will be made on BT network equipment. Does that amount to use of BT's service to infringe copyright?[36]

Relying on the CJEU's decision in *LSG-Gesellschaft zur Wahrnehmung von Leistungsschutzrechten GmbH v Tele2 Telecommunication GmbH*,[37] the Court concluded that BT's services were being used to transmit infringing material between subscribers and third parties. Its subscribers 'actively download' the infringing material by accessing the Newzbin2 website and thereby used BT's services to infringe. The Court further held that the operators of Newzbin2 used BT's services: although Newzbin2 was itself a service provider, it did not follow that it could not also use the services of BT. They did so in two ways: first, as accessories to users' infringements; second, by making available the claimants' copyright works 'in such a way that users can access them over BT's network (among others).'[38]

---

[33] *Newzbin1*, [135] (Kitchin J).

[34] [2011] EWHC 1981, [11], [204] (Arnold J) ('*Newzbin2*'). The precise form of order was determined in [2011] EWHC 2714, [56] (Arnold J) ('*Newzbin2 Order*').

[35] *Newzbin2*, [113] (Arnold J).

[36] *Newzbin2*, [103] (Arnold J).

[37] EU:C:2009:107 [2009] ECR I-1227 ('*Tele2*').

[38] *Newzbin2*, [113] (Arnold J).

As to knowledge, it was sufficient that BT had been notified that at least one person was using its service to infringe copyright by means of the evidence served in support of the claim.[39] Unlike traditional forms of monetary secondary liability, s 97A does not require knowledge of 'a specific infringement of a specific copyright work by a specific individual.'[40] This reflects the broader statutory language of s 97A, the necessity of a broad interpretation to make the remedy achieve its stated purpose, and the more flexible context of equitable injunctions rather than the imposition of substantive liability.

This approach is nevertheless a highly permissive interpretation of the threshold conditions applicable to s 97A in four respects. First, it is relatively easy for a service provider to discover (or be alerted to) a single infringement on a website which occurred using one its allocated IP addresses; the test pays no regard to the volume of infringement or the proportion of non-infringing use (these matters instead forming part of the proportionality analysis). Second, s 97A does not require the service provider to know the identity of the person who is engaging in the infringing activity. That person may be located abroad and may not even be a subscriber of the service. Third, the applicant need not establish which copyright work is being infringed. Fourth, Arnold J observed in *obiter* that actual knowledge is capable of being conferred by notice of facts arising from the receipt of a 'sufficiently detailed notice and a reasonable opportunity to investigate the position'.[41] This appears to be so whether or not the ISP did actually investigate the position. In other words, knowledge extends to being put on notice of facts that would suggest to a reasonable person that an infringement of copyright was taking place.[42] This comes very close to holding that the phrase 'actual knowledge' encompasses constructive knowledge. Arguably, the phrase should go no further than wilful blindness, which is more likely to be consistent with the limitations imposed by article 15(1) of the E-Commerce Directive.

On the facts, Arnold J had no hesitation in concluding that BT had actual knowledge that third parties were using its service to infringe copyright. In some respects, *Newzbin2* was a special case, since BT accepted that it had actual knowledge of all the facts as found by Kitchin J in *Newzbin1*, which had already ruled on the liability of Newzbin and its members in considerable detail.

The scope of the injunction granted in *Newzbin2* was the subject of considerable contention between the parties. BT argued that blocking should be limited to the particular infringements of which the service provider had actual knowledge; the applicants argued that it should not be so limited. Relying on the broad statutory language, context and purpose of s 97A, Arnold J held that the injunctive relief should not be limited to the identified infringements. It may include within its ambit

---

[39] *Newzbin2*, [157] (Arnold J).

[40] *Newzbin2*, [148] (Arnold J).

[41] *Newzbin2*, [149] (Arnold J).

[42] This reflects the approach taken in a line of authorities on secondary infringement: see *Albert v Hoffnung & Co Ltd.* (1921) 22 SR (NSW) 75, 81 (Harvey J); *RCA Corporation v Custom Cleared Sales Pty Ltd.* [1978] FSR 576.

the prevention of future infringements of the same kind, even relating to different copyright works.[43]

By the time the final blocking injunction had been implemented, Newzbin2's operators had already released proxy circumvention software designed to allow BT's subscribers to continue accessing Newzbin2. In an attempt to defeat the operation of this software and guard against the future possibility that the website would simply relocate to another IP address, the injunction provided for a mechanism by which additional IP addresses and URLs could be added to the blocking measures. Such addresses or URLs must have the 'sole or predominant purpose' of enabling or facilitating access to Newzbin2. The applicants bore responsibility for checking the notified data met this condition.[44] Subsequently, similar provisos were included in the orders made against the other major British ISPs.

The Pirate Bay

In *Dramatico Entertainment Ltd v British Sky Broadcasting Ltd* the High Court granted a website blocking injunction in relation to the popular BitTorrent tracker and index, The Pirate Bay ('TPB').[45] In this case the claimants, comprising nine English record companies who were members of British Phonographic Industry Ltd, sought a blocking injunction against six English ISPs[46] to prevent their subscribers from accessing TPB. In a preliminary decision, Arnold J held that the users of that website infringed copyright by copying and communicating the claimants' works to the public, while the operators of TPB were liable as joint tortfeasors and for authorising users' infringements. This satisfied the first prerequisite of the Court's jurisdiction to issue a blocking order.

In a second decision, the Court held that the ISPs had the required knowledge of infringement from their involvement in the initial proceedings, and that blocking was proportionate for the same reasons as in *Newzbin2*. The scale and proportion of infringing activity made for an even stronger case for blocking than *Newzbin2*. Because the defendant did not share an IP address with other websites, it was appropriate to require IP address blocking.[47] Similar provision was made for the order to be extended and adapted in response to circumvention attempts by the website operator.

---

[43] Cf *L'Oréal*, [141].

[44] *Newzbin2 Order*, [12] (Arnold J).

[45] [2012] EWHC 268 ('*Dramatico*').

[46] Two of the respondents (Telefónica UK Ltd. and British Sky Broadcasting Ltd) later merged, so all subsequent orders have been made against five respondents.

[47] *Dramatico Entertainment Ltd. v British Sky Broadcasting Ltd. [No 2]* [2012] EWHC 1152 (Ch), [13] (Arnold J) ('*Dramatico [No 2]*').

KAT, H33T and Fenopy

In *EMI Records Ltd v British Sky Broadcasting Ltd* the Court issued injunctions under s 97A against the six main British ISPs requiring them to block (or attempt to block) access to three BitTorrent trackers.[48] This is the first reported case in which the applicants sought injunctions against more than one target website simultaneously, and in which issues of *prima facie* liability and the terms of relief were decided together. This highlights the evolution of s 97A blocking remedies into a *de facto* system of internet content regulation by courts, in which applicants seek progressively wider and more flexible forms of relief.

In large part, this procedural economy was possible because the trackers functioned similarly to TPB and their operators were liable in much the same way as primary and secondary infringers. Importantly, the Court concluded that at least the infringements of copyright of website users who uploaded content to the target websites took place in the United Kingdom, since the communications originated there. However, it was unclear whether the target websites' communications to the public were targeted there, as much of their content was not related to the United Kingdom.

The uploading users and target website operators used the ISPs' services to infringe in the same way as Newzbin2 and TPB, with some 4.3m instances of infringement attributable to the ISPs' subscribers. In the circumstances, the use of the respondents' services was 'substantial'. Accordingly, Arnold J had no hesitation in concluding that the court had jurisdiction to make the orders sought.[49]

As to proportionality, it was relevant that the applicants had previously attempted to remove material from the target websites by means of notices to hosts alleging infringement. However, the majority of notified material was not removed from KAT and H33T, while a minority remained on Fenopy. There was no evidence that access to non-infringing materials would be impaired. As such, the blocking order was proportionate in the circumstances.

FirstRow

Section 97A was used to block access to the FirstRow website, which is a portal that indexes and aggregates live-streamed broadcasts of sporting events.[50] This claim represented a new class of target website, which was not an index of binary files or torrent files, but rather of live video streams. By visiting the website, it was possible for members of the public to stream live coverage of sporting events including football.

The claimant was a sporting association responsible for organising Premier League football matches in the United Kingdom. In granting the blocking orders

---

[48] *EMI Records Ltd. v British Sky Broadcasting Ltd.* [2013] EWHC 379 (Ch) (*'EMI'*).

[49] *EMI*, [87], [102]–[107] (Arnold J).

[50] *The Football Association Premier League Ltd v British Sky Broadcasting Ltd* [2013] EWHC 2058 (Ch) (*'FirstRow'*).

sought, the Court commented that FirstRow was responsible for the communications of unauthorised video footage, even though the video streams originated from third party servers and were merely aggregated by the target website.[51] It was also a joint tortfeasor with third parties who uploaded the live streams and notified them to FirstRow. Blocking was said to be proportionate in light of the large scale of infringement, the impracticability of identifying and bringing proceedings against the operators of FirstRow, and the need to uphold the public policy objectives served by restricting the live broadcast of football matches (such as encouraging stadium attendance).[52] In this respect, *FirstRow* provides one of the few examples of English courts engaging with the policy objectives underlying the protected interests and using those objectives to shape the proportionality analysis in a website blocking claim. Further, the fact that the ISPs did not oppose the application and had agreed the terms of the draft order supported a finding of proportionality.

The Court made orders requiring IP address blocking of the entire FirstRow website (which was hosted on a dedicated server), and re-routing and URL blocking for any websites with shared IP addresses. However, a new safeguard was introduced (first adopted by Mann J in the *EZTV* decision): the Court granted permission for any affected website operator to apply to set aside the blocking order.[53]

SolarMovies

Blocking applications succeeded against ISPs in relation to two movie streaming websites called SolarMovie and TubePlus.[54] In that case, Arnold J referred to the previous cases and described the applicable principles as 'now settled at this level'.[55] There the website operators communicated the films to the public by linking to them, which (at the very least) intervened to make those films available to a new audience. They were also jointly liable with the third party hosts of the films and liable as authorisers of the resulting infringements. The blocking orders were necessary and proportionate for essentially the same reasons as in the previous cases, and the defendant ISPs were the same.

---

[51] *FirstRow*, [42] (Arnold J).

[52] *FirstRow*, [55(ii)], [55(iii)] (Arnold J). See also *Rugby Football Union v Viagogo Ltd* [2012] 1 WLR 3333, [45] (Lord Kerr JSC) ('*Viagogo*').

[53] *FirstRow*, [57]–[59] (Arnold J).

[54] *Paramount Home Entertainment International Ltd v British Sky Broadcasting Ltd* [2013] EWHC 3479 (Ch) ('*Paramount [No 5]*').

[55] *Paramount [No 5]*, [2] (Arnold J).

1967

As courts became more familiar with the s 97A jurisdiction, claimants grew bolder, targeting a larger number of websites simultaneously and identifying new classes of primary infringer. In parallel, ISPs have stopped appearing in court to make submissions and now generally take a neutral position on claims for blocking injunctions— in large part because the claims relate to obviously infringing websites and the terms of blocking orders follow a template that has been agreed between claimants and ISPs in a number of previous cases, and are considered uncontroversial. As a result, claims for website blocking orders now tend to be considered on paper. This emerging practice illustrates the increasingly administrative nature of the s 97A blocking remedy, which is sought and granted as a matter of course and with limited if any opposition by affected third parties.

For example, in *1967 Ltd v British Sky Broadcasting Ltd*, Arnold J made blocking orders against the main English ISPs requiring them to block access to 21 target websites.[56] They were all BitTorrent trackers which were 'broadly the same in their operation' as TPB, KAT, H33T, and Fenopy. The ISPs did not appear or make submissions to the Court.[57] The Court had little hesitation in concluding that the websites and their users infringed copyright and used the ISPs' services to do so, and made orders in similar terms requiring the use of IP address and URL blocking.

Popcorn Time

In *Twentieth Century Fox Film Corp v Sky UK Ltd*, Birss J made blocking orders in respect of 9 websites which provided access to the claimants' films and television programmes.[58] The target websites divided into three sub-categories: streaming sites (analogous to those previously considered in *Paramount [No 1]*); BitTorrent indices (analogous to those considered in *Dramatico*); and a new category: 'Popcorn Time type sites'. In broad terms, these target websites involved a downloadable application which enabled access to copyright protected content (television shows and films) which could be streamed or downloaded using BitTorrent. The application was obtained from a website which did not itself make available any content. The application worked by aggregating torrent files indexed by third party hosts, with a list of BitTorrent trackers being obtained from a further 'source' website. Once content had been selected by the user to be played or downloaded, it was transmitted using the BitTorrent protocol by the usual process of assembling its constituent parts from third party peers in the 'swarm'. It was clear that Popcorn Time was not used to watch lawful material but rather infringing content.

The analysis of primary infringement for the PopcornTime-type sites was complicated by the fact that the website providing the downloadable application did not

---

[56] [2014] EWHC 3444 (Ch), [3] (Arnold J) ('*1967*').

[57] *1967*, [7] (Arnold J).

[58] [2015] EWHC 1082 (Ch) ('*Popcorn Time*').

itself commit any act of communication to the public (the Popcorn Time application was not an infringing work). Similarly, the operators of 'source' websites could not be said to communicate works to the public; all they provided was a data file with a list of feeds (which also were not copyright works). From the perspective of a user of the Popcorn Time application, the 'source' websites were invisible. Consequently, the primary infringements were divided between multiple parties, none of whom individually dealt in all the pieces of a copyright work but who collectively supplied a seamless service to the end user.

Although it might have been argued that the application suppliers had authorised infringements by users, the claimants did not make such an argument and instead relied upon authorisation by the hosts of the websites. The difficulty with that submission is that there was no evidence about their relationship with the application suppliers, so the allegation failed. However, a similar allegation based on joint tortfeasorship succeeded: the website operators intentionally procured and induced the infringements, and formed a common design with the operators of the host websites to communicate the claimants' works to the public.[59]

In these circumstances, it was clear that the ISPs 'had an essential role in the infringements' because the ISPs' networks were used to transmit the Popcorn Time application, the source lists, and the transmitted copyright works.[60] As such, the jurisdictional preconditions for a blocking order were satisfied. On proportionality, this was a 'clear case'; the claimants were seeking both to prevent access to the Popcorn Time application and to inhibit the usage of copies that had already been downloaded. Blocking orders directed at both the application suppliers and the source websites were likely to achieve those objectives and it was proportionate to make the order.[61]

Other Intellectual Property Rights

Non-transposition of the Enforcement Directive

Unlike copyright and related rights, the United Kingdom has not taken any specific steps to transpose article 11 of the Enforcement Directive, meaning that there is no statutory provision providing for blocking injunctions for other kinds of intellectual property rights. The reasons for this are unclear. When it was considering the implementation of the Directive, the Government stated in relation to article 11: 'No action is required.' It added that injunctions could already be sought 'against the infringer' where goods have been found to be infringing, but this appears to relate to the first two sentences of article 11 (not the third, which concerns intermediaries). Later in the document, the Government cited s 97A and explained that an injunction

---

[59] *Popcorn Time*, [55]–[56] (Birss J).

[60] *Popcorn Time*, [59] (Birss J).

[61] *Popcorn Time*, [61]–[62] (Birss J).

is already available against an intermediary whose services are used to infringe 'an intellectual property right'.

It appears that the Government may have assumed that s 97A covers the field, as there is no reference to other kinds of infringement. In light of the failure to implement article 11 for all intellectual property rights, Arnold J observed in *L'Oréal* that it is 'not entirely clear ... that English law is fully compliant with that provision'.[62]

Inherent Powers to Grant Injunctions

In *Cartier International AG v British Sky Broadcasting Ltd*, the High Court held that it had an inherent jurisdiction to grant blocking injunctions against ISPs whose services could be used to access websites selling counterfeit jewellery, pens, and watches.[63] In that case, the claimants were members of the Richemont group of companies who sought injunctions to require the five main British ISPs to block access to six e-commerce websites. The evidence was that the websites sold no (or *de minimis*) lawful goods. There was no evidence that any particular person had ever accessed the websites using the services of the respondent ISPs, but the claimants argued that this could be inferred from the market-share of the ISPs, taken in combination. The claimants sought blocking injunctions either pursuant to the Court's inherent jurisdiction, or by relying on article 11 of the Enforcement Directive directly.

At trial, Arnold J granted the injunctions. Their doctrinal basis was said to be s 37(1) of the *Senior Courts Act 1981*, which states in relevant part:

> The High Court may by order (whether interlocutory or final) grant an injunction ... in all cases in which it appears to the court to be just and convenient to do so.

The Court accepted that it had an 'unlimited' power[64] to grant injunctions and that there would be a principled basis for exercising that power where four elements were satisfied:

(i)   First, the respondent must be an 'intermediary' within the meaning of article 11 of the Enforcement Directive.[65]

(ii) Second, the users or the operators (or both) of the target website in question must be infringing the applicant's rights (in that case, trade marks).

(iii) Third, such infringement must be carried out using the services of the respondent.

(iv) Fourth, the respondent must have actual knowledge of that fact.[66]

---

[62] *L'Oréal SA v eBay International AG* [2009] EWHC 1094 (Ch); [2009] RPC 21, [447] (Arnold J).

[63] [2014] EWHC 3354 (Ch) ('*Cartier*').

[64] See *Cartier*, [101], [110]–[111] (Arnold J).

[65] Interestingly, this suggests that the power is not limited to internet infringement and may also apply in the case of 'offline' intermediaries, such as auctioneers, warehousemen, and carriers.

[66] See *Cartier*, [141] (Arnold J).

These conditions flowed from the language of article 11 of the Enforcement Directive and from the need to avoid imposing a general monitoring duty upon intermediaries (who could not, consistently with article 15(1) of the E-Commerce Directive be required to block websites without actual knowledge of infringement).

The Court relied upon an analogy with the duties of parties who come into possession of infringing goods under the equitable protective jurisdiction. Such parties, although not wrongdoers, 'must not aid the infringement by letting the goods get into the hands of those who may use them or deal with them in a way which will invade the proprietor's rights.'[67] Arnold J expressed the analogy as follows:

> Although this principle is inapplicable to the circumstances of the present case, it is not a long step from this to conclude that, once an ISP becomes aware that its services are being used by third parties to infringe an intellectual property right, then it becomes subject to a duty to take proportionate measures to prevent or reduce such infringements even though it is not itself liable for infringement.[68]

The dividing line between the scope of the equitable protective principle and possible analogies with, and extensions to, that principle is unclear. However, the reasons why the principle was said to be 'inapplicable to' the facts of *Cartier* may have included that the target websites were not goods, and that the ISPs could not be in possession of data relating to the target websites, since data was incapable of possession.[69]

Significance of the Decision in *Cartier*

The decision in *Cartier* reflects a broad and pragmatic view of s 37—in marked contrast to comparatively recent House of Lords authorities which held that an injunction is not an independent cause of action, and so could not be sought and obtained against a wholly innocent party.[70] It is arguable that the approach in *Cartier* creates powers to grant injunctive relief which are even wider than under the doctrine of *Störerhaftung* in German civil law, since they may be exercised even if the intermediary has acted without fault and independently of pursuing any claim against the primary wrongdoer.[71]

---

[67] *Norwich Pharmacal Co v Customs & Excise Commissioners* [1974] AC 133, 145–6 (Buckley LJ).

[68] *Cartier*, [106] (Arnold J).

[69] See, by analogy, *Your Response Ltd v Datateam Business Media Ltd* [2014] EWCA Civ 281, [23] (Moore-Bick LJ) (Davis and Floyd LJJ agreeing).

[70] See *The Siskina* [1979] AC 210; *South Carolina Insurance Co Ltd v Assurantie Maatschappij De Zeven Provincien NV* [1987] AC 24; *Channel Tunnel Group Ltd v Balfour Beatty Construction Ltd* [1993] AC 334; *Mercedes-Benz AG v Leiduck* [1996] AC 284; *Fourie v Le Roux* [2007] 1 WLR 320.

[71] Cf 158 BGHZ 236 — *Internet-Versteigerung*; 2004 NJW 3102; 2004 GRUR 693, 695 — *Schöner Wetten* (suggesting that intermediaries' duties are limited what it is reasonable to inspect and control). See Gerald Spindler and Matthias Leistner, 'Secondary Copyright Infringement — New Perspectives in Germany and Europe' [2006] *International Review of Intellectual Property and Competition Law* 788, 797, 801.

The result in *Cartier* is particularly significant because the target websites in question were far removed from the profligate emporia of infringement that were blocked in earlier cases under s 97A. Although the target websites undoubtedly held the claimants' rights in contumelious disregard, they were small-time operators in a much larger ocean of infringement: for example, the claimants had identified 239,000 other similar websites which were candidates for blocking.

Second, the target websites were very unpopular: the most frequented was 'carti-erloveonline.com' which had a global Alexa ranking of 5,575,490.[72] Two other target websites had global rankings of 6,837,762 and 15,003,668, but the others had such low levels of visitors that a ranking could not be calculated; similarly, none of the target websites had enough United Kingdom visitors to calculate a local ranking.[73]

Third, there was no evidence that subscribers of the respondents had ever purchased any counterfeit goods from the target websites (and in view of the unpopularity of the websites it is difficult to justify the result on a *quia timet* basis). Fourth, the level of wrongdoing occurring in the United Kingdom by means of the target website was to all appearances very insubstantial. Fifth, very limited attempts had been made to disable the target websites at their source (for example, by sending take-down notices to their hosts and payment providers) or to de-index them from search engines. The Court considered that the claimants were 'open to criticism' for failing to do this, but it would not disentitle them from injunctive relief.[74]

Nevertheless, the inherent jurisdiction recognised in *Cartier* is qualified in a number of ways. First, the content of the duty is 'to take proportionate measures to prevent or reduce … infringements' committed using the intermediary's services.[75] This is somewhat duplicative of the requirement of proportionality, which delimits the scope of the duty while also forming a precondition of relief. The effect is similar to the formulation considered by the Court of Justice in *UPC Telekabel*, where the ISP could avoid liability under the terms of the national court's injunction 'by proving that he has taken all reasonable measures.'[76] Such a duty is not absolute, and an intermediary could not be ordered 'to make unbearable sacrifices'[77]—in other words, 'the addressee of such an injunction has the possibility of avoiding liability, and thus of not adopting some measures that may be achievable, if those measures are not capable of being considered reasonable'.[78]

---

[72] An 'Alexa' ranking was used by the Court as a proxy measure of each website's popularity. Rankings are calculated by sampling the browsing activity of users who have installed the Alexa toolbar. The methodology is necessarily imprecise.

[73] *Cartier*, [247] (Arnold J).

[74] *Cartier*, [201] (Arnold J).

[75] *Cartier*, [106] (Arnold J).

[76] *UPC Telekabel*, [53].

[77] *UPC Telekabel*, [53].

[78] *UPC Telekabel*, [59].

Second, as an equitable remedy, an injunction against an intermediary remains discretionary and it will not always be appropriate. It appears that the most important, and perhaps only, question is whether or not the relief sought is proportionate in the circumstances. This is necessarily a multifactorial inquiry whose breadth and flexibility confer significant discretion upon the trial judge hearing the claim.

International Treatment

The approach in *Cartier* has been cited with approval by courts in other jurisdictions in connection with website blocking orders against intermediaries. In *Equustek Solutions Inc v Jack*, the Supreme Court of British Columbia held that it was appropriate to grant an interim injunction against Google Inc to require it to exclude the defendants' websites from search results generated by Google search engines worldwide.[79] The defendants had failed to comply with court orders requiring them to cease operating various websites on which they sold products which were alleged to misuse the claimants' confidential information.

At trial, Fenlon J held that the Court had subject-matter jurisdiction to grant such an injunction pursuant to the Court's inherent powers, which were expressed in similar terms to s 37(1). On appeal, the Court of Appeal for British Columbia upheld the decision, citing *Cartier* in support of the trial judge's approach to jurisdiction and noting that 'Arnold J's conclusions with respect to the jurisdiction of English courts to grant injunctions are equally applicable to the Supreme Court of British Columbia' due to the common origin of those jurisdictions' statutory provisions governing injunctions.[80]

However, the injunctions in *Equustek* were subject to a number of procedural and substantive limitations. First, the jurisdiction to grant an injunction was parasitic 'on the existence of a justiciable issue between the parties to the litigation' (that is to say, the claimant and the defendant), without which no order could be made against an innocent third party.[81] Second, the order must be 'necessary', for example if it was 'the only practical way' to impede wrongdoing by the defendants.[82] Third, there must be a real and substantial connection to the forum—in that case, the ongoing proceedings and the claimant's business—and appropriate judicial self-restraint before granting an injunction with extraterritorial effects.[83]

In Ireland, *Cartier* was distinguished in a case involving an injunction requiring an ISP to notify its subscribers of allegations of copyright infringement.[84] In Australia, the Explanatory Memorandum to the Copyright Amendment (Online

---

[79] [2014] BCSC 1063.

[80] *Google Inc v Equustek Solutions Inc* [2015] BCCA 265, [75] (Groberman J) ('*Equustek*').

[81] *Equustek*, [80] (Groberman J).

[82] *Equustek*, [105] (Groberman J).

[83] *Equustek*, [56] (Groberman J).

[84] *Sony Music Entertainment (Irl) Ltd v UPC Communications Irl Ltd [No 2]* [2015] IEHC 386, [22] (Cregan J).

Infringement) Bill 2015 expressly referred to *Cartier* as an example of the kinds of powers and safeguards that should be available under a newly-enacted statutory blocking remedy for copyright infringement.[85]

Future Developments

In light of the expansive approach to s 37(1) taken in *Cartier*, it appears to be arguable that an equivalent power exists to grant a blocking injunction for other kinds of civil or criminal wrongdoing, even where the respondent is not a wrongdoer. This would be on the basis that the intermediary had actual knowledge that its services were being used to facilitate primary wrongdoing, such as defamation, misuse of private information, dissemination of unlawful material, 'phishing' of users' credit card details, or potentially any other kind of wrong. Although the decision of the Court of Appeal broadly confirms the approach taken at first instance in *Cartier*,[86] it remains to be seen how the jurisdiction will develop in future cases.

Data Protection

Website blocking orders are also available against intermediaries to enforce data subjects' rights under the Data Protection Directive and the transposing provisions of the *Data Protection Act 1998* ('*1998 Act*').[87] Three rights are of particular relevance to blocking by intermediaries.

First, data subjects are guaranteed the right to obtain from a data controller 'the rectification, erasure or blocking of data the processing of which does not comply with the provisions of [the] Directive, in particular because of the incomplete or inaccurate nature of the data'.[88] This right is transposed by s 14 of the *1998 Act*, which creates a statutory cause of action for an injunction where (1) personal data are 'inaccurate' and contain an expression of opinion based on the data, or (2) there has been a contravention of the Act by a data controller, the applicant has suffered damage thereby, and there is a substantial risk of further contraventions.

In light of the decision in *Google Spain*,[89] it appears that many intermediaries will be data controllers with respect to personal data that they store, process or transmit. The definition of 'processing' is potentially broad enough to include transmissions of personal data by an ISP, which may give rise to the possibility that

---

[85] Parliament of the Commonwealth of Australia, Copyright Amendment (Online Infringement) Bill 2015, Explanatory Memorandum, [41].

[86] See *British Sky Broadcasting Ltd v Cartier International AG* [2016] EWCA Civ 658. The decision is currently on appeal to the Supreme Court in relation to the issue of compliance costs.

[87] Directive 95/46/EC on the protection of individuals with regard to the processing of personal data and on the free movement of such data [1995] OJ L 281/31 ('Data Protection Directive').

[88] Data Protection Directive art 12(b).

[89] *Google Spain SL v Agencia Española de Protección de Datos* [2014] 3 WLR 659.

a data subject could seek an injunction under s 14 to block access to personal data whose transmission from a third party website would involve unlawful processing. The contrary argument is that an ISP does not determine the purposes for which transmitted data may be processed, other than by allowing the transmission. However, the availability of a s 14 blocking remedy against an ISP has not yet been considered by any English court.[90]

In *Mosley v Google Inc*, the High Court refused an application by the defendant for summary dismissal of claims for an order under s 14 which would have required the de-indexing of certain websites from Google's search engine in the United Kingdom.[91] It was common ground that the defendant was a data controller that was processing personal data contained in images and material depicting the claimant's private sexual activity. Google argued that it was exempted from compliance with any injunction by the operation of the safe harbours under the E-Commerce Directive. However, this defence was rejected, both because the data protection regime was unaffected by safe harbours,[92] and because injunctions were still available consistently with the safe harbours.[93] The case settled.

The second relevant right of data subjects is the right to object to the processing of personal data about them if there are 'compelling legitimate grounds',[94] including because of the data subject's fundamental rights and freedoms under articles 7 and 8 of the *Charter*. This involves conducting a balancing exercise to strike a fair balance between the rights and freedoms of the data subject and those of the data controller and third parties.[95]

The right to object is transposed in a more limited way in the United Kingdom: s 10 of the *1998 Act* creates a right to object to processing of personal data if its purpose or manner 'is causing or is likely to cause substantial damage or substantial distress', and such damage or distress is 'unwarranted'. However, the right to object does not apply where any of four lawfulness conditions is met: consent; performance or formation of a contract with the data subject; compliance with a legal obligation; or protection of the vital interests of the data subject.[96] Where a data subject notifies the relevant data controller with a well-founded objection, and the controller fails to comply with the request, a court may by injunction order the data controller to take such steps to comply with the notice as it thinks fit.[97] This right potentially supplies a further basis for seeking an injunction requiring an intermediary to block access to a website on which personal data are processed in a harmful or distressing way, where it is proportionate to do so.

---

[90] See *Hegglin v Persons Unknown & Google Inc* [2014] EWHC 3793 (QB) in relation to search engines.

[91] [2015] EWHC 59 (QB) (*'Mosley'*).

[92] *Mosley*, [45]–[48] (Mitting J).

[93] *Mosley*, [55] (Mitting J). See E-Commerce Directive recital (45).

[94] Data Protection Directive art 14(1)(a).

[95] *Google Spain*, [74].

[96] *1998 Act* s 10(2), sch 2, paras 1–4.

[97] *1998 Act* s 10(4).

Voluntary Website Blocking

It is unclear to what extent ISPs may voluntarily intervene in their subscribers' communications to block requests for unlawful material. Section 3 of the *Investigatory Powers Act 2016* (UK) prohibits any person from intentionally intercepting any communication in the course of its transmission by means of a public or private telecommunication system (such as an ISP's network) in the United Kingdom where that person does not have lawful authority to do so.[98] 'Interception' is defined to include monitoring transmissions and interfering with the operation of the telecommunication system, if the effect is to make some or all of the content of a communication available to a person (other than the sender or intended recipient) while the communication is being transmitted or stored.

On its face, this prohibition could be argued to preclude an ISP from voluntarily examining HTTP requests to determine whether they relate to infringing or other tortious materials, since to do so would involve interfering in subscribers' communications and analysing the content of the communication (for example, the specific URL being accessed). However, this issue has not yet been considered by an English court. A similar prohibition on interfering in network transmissions exists in EU law, in the form of article 3(3) of the Network Neutrality Regulation, which is discussed below.

Notwithstanding s 3, all major British ISPs implement the blacklist of the Internet Watch Foundation ('IWF') to block websites containing indecent images of children. The contents of this blacklist are, for obvious reasons, unpublished, but it is believed to contain around 2000 URLs. The list is administered centrally by the IWF, and updates are notified to ISPs automatically and blocked using a combination of URL blocking and deep packet inspection. The IWF sets out a series of principles which it describes as 'Blocking Good Practice', which include the use of URL-level blocking and the display of an appropriate warning message when blocking occurs. However, administration of the IWF blacklist has been criticised for lacking transparency and oversight, and has been responsible for several widely publicised instances of over-blocking.[99]

Although IWF blocking is voluntary in the sense that it is not undertaken pursuant to a court order, the ISP industry faces significant commercial and regulatory pressure to block such material. All signatories of the ISPA Code of Practice are 'encouraged' to become members of the IWF. Additionally, clause 5.4 of the Code requires signatories to remove specific web pages or Usenet articles they host when notified by the IWF that they contain material which the IWF's editorial board

---

[98] The general prohibition on interception is subject to numerous exceptions, including acts by the providers of telecommunications services undertaken for purposes relating to the 'provision or operation of the service' or to the enforcement of any enactment relating to the content of transmitted communications: see *Investigatory Powers Act 2016* (UK) s 45.

[99] See, e.g., Tom Espiner, 'IWF Chief: Why Wikipedia Block Went Wrong' (20 February 2009) http://www.zdnet.com/iwf-chief-why-wikipedia-block-went-wrong-3039616171/.

considers to be illegal. If removal is not possible, signatories are required by the Code to notify the IWF of the reasons as soon as reasonably practicable. Almost all UK ISPs are signatories to the ISPA Code. Additionally, the United Kingdom Government has threatened to legislate if ISPs did not implement IWF blocking.[100]

The Code of Practice for the Self-Regulation of Content on Mobiles applies to mobile carriers.[101] Paragraph 3 of this Code obliges mobile service providers to work with the IWF and law enforcement agencies to report unlawful content of third parties, and to remove any hosted unlawful content in accordance with notice and take-down procedures. Additionally, signatories are obliged to block customers from accessing URLs on the IWF blacklist. The four main mobile carriers are signatories.

Conversely, the Broadband Stakeholder Group, an ISP industry body, has endorsed the principle in the Open Internet Code of Practice that the 'open internet' requires that legal content, applications and services should not be blocked.[102] Many of the UK's largest ISPs are also signatories to this Code.

Proportionality

The main balancing mechanism applied by English courts in website blocking cases is the concept of 'proportionality'. This concept has been shaped by EU law in cases involving the E-Commerce Directive, the Enforcement Directive and the *Charter*, which require any injunctive relief against an intermediary to be proportionate.[103] Proportionality functions as an analytic tool used to weigh competing interests on a policy fulcrum. However, it has a tendency to be deployed in ways that avoid real engagement with the underlying policy issues, and which lead to inconsistent and unpredictable results.

Proportionality is described in different ways depending on the context in which it is being assessed. In the context of injunctive relief between private litigants, the basic approach has been outlined in a number of cases,[104] and is as follows:

---

[100] See, e.g., Press release, 'PM Announces New global Action to Deal with Online Child Abuse' (Prime Minister's Office, 11 December 2014).

[101] Mobile Broadband Group, 'UK Code of Practice for the Self-Regulation of Content on Mobiles: Version 3' (1 July 2013) http://mobilebroadbandgroup.com/documents/UKCodeofpractice_mobile_010713.pdf.

[102] Broadband Stakeholder Group, Open Internet Code of Practice: Voluntary Code of Practice Supporting Access to Legal Services and Safeguarding against Negative Discrimination on the Open Internet (May 2013) http://www.broadbanduk.org/wp-content/uploads/2013/06/BSG-Open-Internet-Code-of-Practice-amended-May-2013.pdf.

[103] Enforcement Directive arts 3(2), 11(2); Information Society Directive art 8(1); *Charter* art 52. See also *L'Oréal*, [139]; *Scarlet*, [36].

[104] See *Re S (a child)* [2005] 1 AC 593; *Campbell v MGN Ltd* [2004] 2 AC 457.

First, the claimants' copyrights are property rights protected by Article 1 of the First
Protocol to the ECHR and intellectual property rights within Article 17(2) of the *Charter*.
Secondly, the right to privacy under Article 8(1) ECHR/Article 7 of the *Charter* and the
right to the protection of personal data under Article 8 of the *Charter* are engaged by the
present claim. Thirdly, the claimants' copyrights are 'rights of others' within Article 8(2)
ECHR/Article 52(1) of the *Charter*. Fourthly, the approach laid down by Lord Steyn where
both Article 8 and Article 10 ECHR rights are involved in *Re S* at [17] is also applicable
where a balance falls to be struck between Article 1 of the First Protocol/Article 17(2) of the
Charter on the one hand and Article 8 ECHR/Article 7 of the *Charter* and Article 8 of the
Charter on the other hand. That approach is as follows:

  (i)   neither Article as such has precedence over the other;
  (ii) where the values under the two Articles are in conflict, an intense focus on the compara-
        tive importance of the specific rights being claimed in the individual case is necessary;
  (iii) the justifications for interfering with or restricting each right must be taken into
        account;
  (iv) finally, the proportionality test — or 'ultimate balancing test' — must be applied to
        each.[105]

Although formally applied in claims for website blocking injunctions against
intermediaries, this exercise has resulted in few practical constraints being imposed
upon relief. Usually, this is because the primary wrongdoing is obvious and widespread, as in the case of most s 97A claims involving file-sharing websites. However,
it remains to be seen whether the proportionality inquiry will provide meaningful
limits in more borderline cases.

A number of factors are relevant to the proportionality inquiry. These factors
were helpfully summarised by Arnold J in *Cartier*:

in considering the proportionality of the orders sought by Richemont, the following considerations are particularly important:

  (i)   The comparative importance of the rights that are engaged and the justifications for
        interfering with those rights.
  (ii) The availability of alternative measures which are less onerous.
  (iii) The efficacy of the measures which the orders require to be adopted by the ISPs,
        and in particular whether they will seriously discourage the ISPs' subscribers from
        accessing the Target Websites.
  (iv) The costs associated with those measures, and in particular the costs of implementing
        the measures.
  (v)   The dissuasiveness of those measures.
  (vi) The impact of those measures on lawful users of the internet.

In addition, it is relevant to consider the substitutability of other websites for the Target
Websites.

In addition … Article 3(2) of the Enforcement Directive requires remedies to be applied in
such a manner as to 'provide for safeguards against their abuse'.[106]

Despite the large number of factors that are relevant to proportionality, relatively
few English cases have considered how these considerations apply to website
blocking claims. It is clear that proportionality is necessarily 'a context-sensitive

---

[105] *Golden Eye (International) Ltd v Telefónica UK Ltd* [2012] EWHC 723 (Ch); [2012] RPC 28,
[117] (Arnold J) (citations omitted); approved in *Viagogo*, [45] (Lord Kerr JSC).
[106] *Cartier*, [189]–[191] (Arnold J).

question'.[107] However, in all the cases considered above, English courts had very little hesitation in concluding that the orders sought were proportionate. Conversely, courts in other member states have reached almost diametrically opposed conclusions concerning the proportionality of blocking access to some of the same websites.[108] This is regrettable, since it indicates that either (1) courts' understanding of the costs and benefits of blocking remedies differ between member states despite use of essentially similar technology by ISPs, or (2) the factors considered in the proportionality analysis are not being consistently applied. These factors are summarised below.

Gravity of Wrongdoing

Initially, English claimants sought blocking orders rarely and against 'the more egregious infringers'.[109] This partly reflected the finite enforcement resources available to claimants; it may also embody an understandable desire on the part of copyright industries to ensure that the principles of this emerging jurisdiction developed in obviously meritorious cases. For ISPs, this meant that the number of blocking applications remained relatively steady from 2009–2016 and focussed on the most serious causes of infringement, such as TPB and other notorious platforms.

This focus on the most serious cases of primary infringement meant that it was largely unnecessary for courts to grapple with the question of how much wrongdoing is needed before a blocking injunction can be proportionate. More recent claims—such as *Cartier*—have targeted websites of very marginal popularity in the United Kingdom, and have thrown this question into stark relief.[110]

One possible approach to this issue may be to draw an analogy with defamation actions, where there must be proof of substantial publication within the jurisdiction and a 'real and substantial tort'.[111] This requirement was designed to prevent libel tourism and limit access to English courts for the adjudication of disputes involving foreign publication on internet platforms; if the harm caused by publication was less

---

[107] *EMI*, [100] (Arnold J).

[108] See, e.g., *Ziggo BV v Stichting Bescherming Rechten Entertainment Industrie Nederland (BREIN)* (Court of Appeal of The Hague, decision of 28 January 2014) (presently under appeal and the subject of a pending reference to the CJEU); Case No I ZR 3/14, *Keine Störerhaftung für Access-Provider—3dl.am* (decision of 26 November 2015, German Federal Supreme Court); Case No I ZR 174/14, *Goldesel.to* (decision of 26 November 2015, German Federal Supreme Court).

[109] *Newzbin2*, [199] (Arnold J).

[110] In *Cartier* itself, the most popular target website had a local ranking of 5,575,490 (compared to TPB, which was ranked 43rd in the UK at the time of blocking: *Dramatico [No 1]*, [26]). See further nn 73–74.

[111] *Jameel v Dow Jones & Co Inc* [2005] QB 946, [48] ('*Jameel*'). See also *Defamation Act 2013* (UK) s 9.

than the costs of a proceeding or granting relief, then it would be disproportionate to allow the claim to continue.[112] This principle also applies to copyright claims.[113]

The Court of Justice has accepted that the seriousness of wrongdoing is a relevant factor when considering the proportionality of an injunction.[114] This will involve considering the popularity of the target websites in the jurisdiction, and the nature, scale and degree of wrongdoing. By analogy with cases such as *Jameel* and *Promusicae*, it might be suggested that if blocking would entail costs that are out of all proportion to what would be achieved, then blocking will 'not have been worth the candle' and the claim should be refused as an abuse of process.[115] However, English courts have not yet considered this question in blocking claims.

Technical Feasibility

All of the injunctions made to date in the United Kingdom have been premised on the assumption that technical means to comply with them were possessed by the respondent ISPs—typically because they had already developed blocking capabilities in order to implement the IWF blocklist. Indeed, since *Newzbin2* the terms of blocking orders have been largely agreed between claimants and defendants. This is a factor said to militate in favour of the proportionality of blocking, but is insufficient on its own since third parties' rights and interests are also affected by blocking.[116] It is an open question whether it could ever be proportionate to make a blocking order that required its addressee to develop a blocking system that did not already exist. This may be a particularly important issue if website blocking injunctions are sought in future against smaller ISPs or classes of intermediaries that do not already have blocking mechanisms at their disposal.

Effectiveness and Dissuasiveness

Effectiveness is an important consideration when evaluating the proportionality of website blocking; in general, the more a measure fails to achieve its intended aim, the less likely it is to be proportionate in light of its associated cost and adverse consequences.[117] Any blocking system carries a risk of 'under-blocking', or false

---

[112] See, eg, *Sheffield Wednesday Football Club Ltd v Hargreaves* [2007] EWHC 2375 (QB), [17] (HHJ Parkes QC).

[113] *Sullivan v Bristol Film Studios Ltd* [2012] EWCA Civ 570, [32]–[37] (Lewison LJ) (Etherton and Ward LJJ agreeing).

[114] *Promusicae*, [118].

[115] *Jameel*, [54], [69]–[70] (Lord Phillips MR).

[116] *FirstRow*, [55(i)] (Arnold J).

[117] Cf Broder Kleinschmidt, 'An International Comparison of ISP's [sic] Liabilities for Unlawful Third Party Content' (2010) 18 *International Journal of Law and Information Technology* 332, 353–4.

negatives, which occur where the system fails to prevent access to the targeted material by users. To determine whether a blocking remedy is likely to be effective to prevent wrongdoing by an intermediary's users, a number of considerations are relevant. The starting point is, as Consumer Focus has argued,[118] the technical question of to what degree access to the targeted website can actually be interdicted. Secondarily, and perhaps more importantly, the wider effects of blocking on the growth of legal markets and the availability of lawful alternatives must be considered, including the signalling function of blocking orders as expressions of the courts' findings as to the unlawfulness of blocked websites.

As to technical effectiveness, English courts have tended to focus on the localised question of whether access to the targeted URL or domain name can be materially reduced, rather than the wider question of whether infringements of the protected subject matter will fall overall. Whichever approach is taken, the difficulty of quantifying effectiveness is compounded by the limited data available concerning the popularity of blocked websites. According to Alexa traffic statistics, overall access to Newzbin and Newzbin2 fell by 49% in the 12 months after the major UK ISPs began blocking, and the website's global traffic rank fell by 900% before it ceased operation for want of revenue.[119] Newzbin was formerly a predominantly Anglophone service, but 1 year after blocking UK visitors accounted for just 8.5% of the website's overall traffic. Importantly, traffic to Newzbin mirror websites did increase over the same period, but not enough to cancel out the overall reduction entirely.[120]

Similar data for TPB led the Court in *FirstRow* to conclude that:

> While they are unlikely to be completely efficacious, since some users will be able to circumvent the technical measures which the orders require the [ISPs] to adopt, it is likely that they will be reasonably effective.[121]

This reflects the fact that blocking increases the cost of infringement by adding to consumer search costs (finding a mirror or alternative website), configuration costs (for example, having to download and install proxy or anonymisation software) and in some cases service costs (paying for a VPN provider or new ISP). While different consumers will evaluate these added costs with different price sensitivities, it can be expected that overall levels of infringement will fall. Eventually, the cost of infringement will rise to a point where it becomes more efficient to pay for content from a legitimate source. Thus, as Arnold J reasoned in *Newzbin2*:

---

[118] Consumer Focus, *Response to 'Proposal for Code of Practice Addressing Websites that are Substantially Focused on Infringement'* (19 September 2011) 1.

[119] Alexa Internet Inc, 'Statistics Summary for Newzbin.com' (26 March 2012) http://www.alexa.com/siteinfo/newzbin.com#.

[120] See, eg, Alexa Internet Inc, 'Statistics Summary for Demonoid.me' (26 March 2012) http://alexa.com/siteinfo/demonoid.me; 'Statistics Summary for Torrentreactor.net' (26 March 2012) http://alexa.com/siteinfo/torrentreactor.net (showing growth in line with previous trends).

[121] *FirstRow*, [55(v)] (Arnold J).

> If, in addition to paying for (a) a Usenet service and (b) Newzbin2, the users have to pay for (c) an additional service for circumvention purposes, then the cost differential between using Newzbin2 and using a lawful service … will narrow still further. This is particularly true for less active users. The smaller the cost differential, the more likely it is that at least some users will be prepared to pay a little extra to obtain material from a legitimate service.[122]

Such an argument relies on the availability of blocking technology which is sufficiently robust that most users require a circumvention service that entails some material additional cost for them. Even if most users are willing to invest some additional time and effort to evade a website blocking injunction, for a sufficiently robust form of blocking many users will simply not possess the technical skill required to restore access. In any case, it is difficult to predict how many users would actually take steps to circumvent a block, as Kenneth Parker J noted in *R (British Telecommunications plc) v Secretary of State*:

> It is not disputed that technical means of avoiding detection are available, for those knowledgeable and skilful enough to employ them. However, the central difficulty of this argument is that it rests upon assumptions about human behaviour. … In theory, some [users] may cease or substantially curtail their unlawful activities, substituting or not, for example, lawful downloading of music; others may simply seek other means to continue their unlawful activities, using whatever technical means are open. The final outcome is uncertain because it is notoriously difficult accurately to predict human behaviour …[123]

For this reason, English courts have largely ignored Ofcom's conclusion that circumvention of a blocking injunction is 'technically a relatively trivial matter' which is 'not technically challenging and does not require a particularly high level of skill or expertise.'[124]

In order to ensure that blocking orders stand the best chance of being effective to prevent infringement of intellectual property rights, the English courts have supported three important practices. First, injunctions tend to be drafted in terms which accommodate the addition of URLs and IP addresses for mirrored services. Second, blocking is accompanied by educative messages which clearly explain why access to material has been restricted. This message is intended to support the dissuasive effect of blocking upon internet users; however, in *Cartier*, Arnold J recognised that, despite the display of such warnings, 'consumers are very likely to turn to other websites if the Target Websites are blocked'.[125]

Third, blocking is not normally treated with a set and forget mentality, but rather made the subject of regular 'update notifications' from claimants. Such notifications list new mirror sites, proxies and domain names. As Clayton points out:

> the effectiveness of any blocking system, and the true cost of ensuring it continues to provide accurate results, cannot be properly assessed until it comes under serious assault.[126]

---

[122] *Newzbin2*, [196] (Arnold J).

[123] [2011] EWHC 1021 (Admin), [232] (Kenneth Parker J) ('*BT*').

[124] Ofcom, '"Site Blocking" to reduce Online Copyright Infringement' (27 May 2010) 51.

[125] *Cartier*, [258].

[126] Richard Clayton, 'Failures in a Hybrid Content Blocking System' in George Denezis and David Martin (eds), *Privacy Enhancing Technologies* (2005) 78, 90.

Blocking is thus an ongoing process of maintenance, countermeasures and circumvention whose effectiveness will change over time. Left unmaintained, circumvention will progressively rise by both website operators and users, diminishing the effectiveness of blocking unless viable countermeasures are applied.

Alternatives

In *FirstRow* and *EMI*, the Court considered that the difficulties faced by the claimants in identifying and pursuing the operators of infringing websites was a factor supporting the proportionality of a blocking injunction against those websites.[127] The evidence was that a variety of other enforcement methods had been pursued by the claimants—including notice-and-takedown, letters to the website operators, and letters to their hosting companies—but that these had failed to result in the content complained of being wholly removed.

In contrast to courts of other member states, English courts have not considered website blocking to be a remedy of last resort,[128] or subject to a principle of subsidiarity. In *Cartier*, for example, the Court rejected the submission that, before blocking could be considered proportionate, the claimants should be required to exhaust other enforcement mechanisms and pursue a claim against the primary wrongdoer.[129] However, it may be relevant to consider whether there are alternative measures which are less onerous than website blocking but at least as effective. In that case, the Court discounted notice-and-takedown, de-indexing, domain name seizure, and notices to payment intermediaries on the basis that they could not be considered to achieve more than short-term disruption of the target websites, or were at least as onerous for the claimants as a website blocking injunction.

Effect on Lawful Content

In *Scarlet*, Advocate General Villalón observed that systems of content filtering

> inevitably affect lawful exchanges of content [and] therefore have repercussions for the content of rights protected under Article 11 of the *Charter*, if only because the unlawful or lawful nature of a given communication, which depends on the scope of the relevant copyright, varies from country to country and is therefore impossible to grasp through technical means.[130]

---

[127] See *FirstRow*, [55(ii)] (Arnold J).
[128] Cf European Commission, *Public Hearing on Directive 2004/48/EC and the Challenges Posed by the Digital Environment* (Brussels, 7 June 2011) 1.
[129] *Cartier*, [198]–[217] (Arnold J).
[130] Case C-70/10, *Scarlet*, Advocate General's Opinion, [86].

'Over-blocking', or false positives, involves the inadvertent denial of access to lawful material by a blocking system. This is one of the most important considerations relevant to the proportionality of a blocking order. However, no case has yet arisen in England in which over-blocking concerns have meant that a blocking injunction was disproportionate.

The risk of over-blocking is usually affected by two considerations: first, the type of blocking being sought (for example, IP address, URL, DNS, or DPI blocking); and second, the safeguards contained in the blocking order. When implemented alone, IP blocking tends to be the least granular method, with the greatest potential for false positives if lawful material is hosted on the same server as blocked material. IP blocking is rarely implemented in this way; when it is, English courts have required applicants to certify (in a formal manner supported by a statement of truth) that all content hosted on the targeted IP address is 'unlawful'. However, the difficulty with this approach is that it delegates to claimants the authority to determine whether or not information is unlawful, and requires intermediaries to follow that determination, in most cases without it ever being reviewed by a court.

More commonly, intermediaries are ordered to redirect matching traffic to a second-stage filter which uses DNS, URL or DPI-based blocking methods. DNS blocking also carries a risk of over-blocking, since it blocks an entire domain name. For this reason, it tends only to be used where an entire website is held to consist of tortious material (or contain *de minimis* lawful material). URL filtering offers the highest granularity—and the lowest risk of false positives—because it allows specific files or webpages to be blocked without affecting other resources on the same website or server. Similar considerations apply to DPI-based blocking methods, which tend to be the most accurate, but subject traffic to systematic analysis. The impact of such analysis on internet users' rights under article 8 of the *Convention* and article 7 of the *Charter* has been largely ignored. Most English ISPs use a two-stage blocking system which minimises the risk of over-blocking.

False positives in a blocking system are most problematic when internet users are unaware that a website's inaccessibility is deliberate. This is because users' natural assumption is that a website is experiencing a technical glitch or has ceased operating; they may try again later, search for information about the outage, or simply find an alternative source of information. Among the most important safeguards provided for by English court orders is notification in the form of 'splash pages' which must inform users of the reason for blocking, the court order pursuant to which blocking is carried out, and the party who applied for that order. The 'error 451' initiative is another valuable mechanism for ensuring transparency in blocking remedies,[131] thereby minimising the consequences of over-blocking.

---

[131] See Open Rights Group, '451 Unavailable: Site Blocked for Legal Reasons' (2015) http://www.451unavailable.org/.

Safeguards

English courts have progressively recognised a number of safeguards in orders providing for website blocking. These safeguards are now standard features of blocking injunctions against ISPs.

First, as noted above, intermediaries are required to publish a notice to affected users when they attempt to access blocked material. This important provision aims to improve the transparency of blocking and its educative and deterrent functions.

Second, intermediaries, their subscribers, and affected third parties (such as the operators of target websites or other lawful websites) have a general permission to apply to the Court in the event that there is any material change in circumstances, or if it appears that their fundamental rights and freedoms are being unduly impacted. Such a review mechanism is important because it provides a *de facto* appeal mechanism whereby interested parties can ask the Court to reconsider whether blocked material properly falls within the scope of a blocking order, or whether blocking remains proportionate. However, the obscurity of this procedure means it is little used; for those who are aware of it, the risk of being ordered to bear the other parties' costs is likely to constitute a serious deterrent.

Third, a sunset clause is now included in blocking orders which provides that blocking will automatically cease 2 years after the date of the order, subject to any further order of the Court. This sensible limitation reflects the relatively short half-life of many websites associated with infringing activity, and the desirability that blocking 'not endure longer than necessary'.[132]

In other cases, it may also be appropriate to require the applicant to notify the operator of the websites targeted by an application, and provide an opportunity to remove the tortious material, provide an undertaking as to damages or make submissions to the Court as to the lawfulness of the material. In many cases it will prove impossible or pointless to contact target website operators. However, notification ensures transparency and affords an opportunity to respond to allegations which would not otherwise be given in the context of an uncontested—and in substance *ex parte*—hearing between the applicant and an ISP.

Fair Balance

In the context of intellectual property, the Court of Justice has made clear that both monetary and non-monetary remedies must 'strike a fair balance' between the fundamental rights and freedoms of claimants, intermediaries, and internet users.[133]

---

[132] *Cartier*, [262]–[265] (Arnold J).

[133] *Promusicae*, [68]; *L'Oréal*, [143]; *UPC Telekabel*, [63]. See also *Charter* art 52(1).

This principle serves as an important limit on remedies in cases involving *Charter* rights within the fields coordinated by EU law. In *Promusicae*, the Court of Justice approached this balancing exercise by enumerating and weighing the relevant primary rights and any fundamental rights of individuals that would be engaged by the remedy being sought. Similarly, in *Scarlet Extended SA v Société des Auteurs, Compositeurs et Éditeurs SCRL*, the Court of Justice held that a Belgian blocking injunction which required an ISP to monitor all electronic communications in perpetuity failed to strike a fair balance between the claimant's right to property, the service provider's freedom to conduct its business, and users' rights to private life and freedom of expression.[134] That is, at least formally, the approach taken by English courts in claims for blocking injunctions.

In practice, the outcome of this analysis is either obvious or not explored in any detail. In part this is because claimants have tended to focus on the most egregious cases of infringement in selecting targets for blocking claims. This has created a tendency for courts to regard interferences with other rights as justified by reference to the scale and illegitimacy of infringements of the claimants' rights, rather than by reference to the benefits of blocking in particular cases or compared to the available alternatives. For example, in *Newzbin2*, the parties accepted that a website blocking would engage and interfere with users' freedom of expression, and that such interference, to be lawful, must be proportionate to the need to protect the claimants' property rights. For perfectly understandable reasons in light of the scale of wrongdoing, Arnold J held that the claimants' rights 'clearly outweigh' those of users to access unlawful material.[135]

Underlying this assessment of the relative importance of the parties' rights is the obviously correct proposition that any freedom to express tortious information is comparatively weak, since such material by its nature infringes the rights of others and accordingly falls within an exception to freedom of expression. Consequently, English cases reflect the assumption that a fair balance between claimants and users is essentially coterminous with the risk of over-blocking. However, this focus risks drawing attention away from the other aspects of the proportionality inquiry and other rights of intermediaries, internet users and third parties.

Intermediaries' freedom to conduct a business under article 16 of the *Charter* is infrequently mentioned in English website blocking cases. Its main relevance appears to be to require courts to consider the cost and technical feasibility of implementing a measure. English courts have concluded that compliance with blocking injunctions would not be especially costly in light of the blocking systems already available to ISPs, and so respect the essence of their freedom of business:

> The [blocking] orders would not interfere with the provision by the ISPs of their services to their customers. The orders would not require the ISPs to acquire new technology: they have the requisite technology already. ... *The main effect of the orders would be to impose additional operating costs on the ISPs*. It is true that there is a small risk of the ISPs being attacked either by hackers or by operators of the Target Websites, but in my judgment this

---

[134] *Scarlet*, [46]–[49]. See *Charter* arts 11(1), 16, 17(2).

[135] *Newzbin2*, [200] (Arnold J).

risk is not a significant one. It is also true that there is a risk of reputational damage to the ISPs, particularly in the event of overblocking, but again I do not consider this risk a significant one.[136]

This conclusion followed the decision of the Court of Justice in *UPC Telekabel* that a blocking injunction requiring 'reasonable measures' did not infringe 'the very substance' of the freedom.[137] However, English blocking injunctions are founded upon the assumption that the respondent is able to comply with the order using its existing blocking systems. It is therefore an open question whether a blocking order would be compatible with EU law if it required an intermediary to develop and adopt new capabilities which did not already exist.

Compliance Costs

Blocking injunctions can impose costs upon intermediaries which, although modest in individual cases, may be substantial in aggregate. In *Newzbin2*, the evidence was that the respondent would incur costs of around £5000 to comply with the order and around £100 per subsequent notification received when a target website moved to a new URL or IP address. Arnold J concluded that costs on this scale were 'modest and proportionate'.[138] However, if each target website moves, on average, 20 times in a year, and 50 blocking orders are obtained each year against 200 websites, it is easy to see how the aggregate costs can grow. Further and much more substantial costs may be required to upgrade a blocking system that reaches capacity or is placed under strain by a large volume of blocked traffic under orders.

Although English courts have agreed that the aggregate costs are relevant to proportionality,[139] they have so far imposed the costs of defending and implementing blocking orders upon intermediaries rather than claimants. In *Newzbin2*, the Court rejected an analogy with *Norwich Pharmacal* orders and held that BT should bear the costs of implementing the order. In effect, because BT's services created the problem, it should be liable to pay for the solution:

> BT is a commercial enterprise which makes a profit from the provision of the services which the operators and users of Newzbin2 use to infringe the [applicants'] copyright. *As such, the costs of implementing the order can be regarded as a cost of carrying on that business.* It seems to me to be implicit in recital (59) of the *Information Society Directive* that the European legislature has chosen to impose that cost on the intermediary. ... The cost of implementing the order is a factor that can be taken into account when assessing the proportionality of the injunction ...[140]

---

[136] *Cartier*, [195] (Arnold J) (emphasis added).
[137] *UPC Telekabel*, [48]–[56].
[138] *Newzbin2 Order*, [32] (Arnold J). See also *EMI*, [102] (Arnold J); *FirstRow*, [55(i)] (Arnold J).
[139] *Cartier*, [242] (Arnold J).
[140] *Newzbin2 Order*, [32] (Arnold J) (emphasis added).

Arnold J accepted that the intermediary had committed no legal wrong but held that it was proportionate for it to bear all implementation costs.[141] However, the judge left open the possibility that, in other cases, it would be appropriate for applicants to do so. To date, no English case has taken a different approach. Whether this approach is correct is among the issues to be decided by the Supreme Court in the pending appeal in *Cartier*.

Emerging Issues

Target Websites

The general trend in English claims under s 97A has been an expansion in the kinds of websites targeted by claimants and a progressive streamlining of the procedure required to obtain and maintain such orders. This expansion can be expected to continue now that the basic principles and procedures have been tested. As most well-known BitTorrent trackers have been blocked, the next generation of blocking orders seems likely to focus on new targets. Some of them may raise new issues, especially in relation to proportionality.

First, 'cyberlockers' and other cloud storage services will highlight the unsettled relationship between blocking and other enforcement methods, and will test the courts' approach to over-blocking. Many such services host large volumes of infringing material, but also host significant quantities of lawful material. It is doubtful whether domain name level blocking or IP address blocking would be proportionate if it would prevent access to that lawful material, or prevent the cloud service from providing lawful paid services. Moreover, such services may enjoy *prima facie* safe harbour protection from primary liability.

Second, unlicensed foreign music and film services will present difficult targets for blocking injunctions. Such services may be licensed abroad but accessible in the United Kingdom and unlicensed by collecting societies such as PRS and MCPS; however, such websites may well claim to be lawful, or not be specifically targeted at United Kingdom users. These more marginal cases will require new balancing tools, fair procedures and careful scrutiny to ensure that primary wrongdoing is properly substantiated and that affected third parties are not unfairly prejudiced from defending claims.

---

[141] Cf *Totalise plc v Motley Fool Ltd* [2002] 1 WLR 1233 (awarding the innocent party its costs of providing disclosure of a wrongdoer's identity and its legal costs to have the question determined).

Search Engines

It is an open question whether blocking orders under s 97A (or pursuant to the courts' inherent jurisdiction) are available against search engines such as Google and Bing. Such parties are undoubtedly 'service providers'; the crucial issue is likely to be whether their services are being used by a third party to infringe the relevant right. One potential argument may be that such service providers, by hyper-linking to sources of infringing material, make its transmission possible to users who would otherwise be unable or unwilling to locate it. This may be a question of evidence. For example, if it can be shown that a large proportion of an infringing website's visitors are referred from search results, then it may be arguable that the search engine's services are being used by the website operator or its users in order to transmit the infringing material. Alternatively, it might be argued that by storing and caching website material, search engines are themselves carrying the infringe-ments on a network. However, the issue remains unresolved and it seems likely to press at the boundaries of the concept of services being 'used'.

Primary Wrongdoing

As noted above, English courts have begun to recognise blocking remedies against intermediaries for types of wrongdoing other than copyright infringement. One important issue that has not yet arisen for decision is whether the inherent jurisdic-tion could be used to support a claim for an injunction in cases that do not fall within the scope of article 11 of the Enforcement Directive—for example, a claim for defa-mation or misuse of private information, or to enforce a statutory duty.

It is submitted that, if the courts have the inherent power to grant orders in respect of intellectual property wrongs, then there is no good reason why that power could not also be exercised to prevent other kinds of wrongdoing. It would be necessary to establish that the wrongdoing was being facilitated by the intermediary, who would otherwise be a mere bystander and not amenable to an injunction. However, it would not necessarily follow that a blocking injunction would always be proportionate: the effectiveness of the order, any reasonably practicable alternatives, and any wider consequences, would need to be considered by reference to the nature of primary wrongdoing and the particular material in question.

Interim Remedies

An effective blocking remedy may sometimes require relief to be available rapidly and on an interim basis—for example, to prevent the unauthorised streaming of sporting events and other live content, and access to pre-release films, music and books. If an unauthorised stream is operative, much of the economic damage may be done to the copyright owner within a matter of hours. Given that it can be impossible to remove such materials at their source within this timeframe, blocking may have a future role to play in reducing unlicensed streaming.[142]

At least one interim s 97A injunction has been granted against a blogging platform.[143] However, two difficulties presently stand in the way of blocking orders as interim remedies against ISPs. First, provisions such as s 97A require proof of at least one instance of infringement, which makes it difficult to use blocking as a preventative measure.[144] Second, the accepted practice is for blocking injunctions (and updates) to be implemented by ISPs within 10 days, which reflects the extensive manual processes and testing needed to ensure reliable implementation. However, as ISPs' blocking systems continue to develop and automation techniques improve, the implementation time can be expected to fall.

Network Neutrality

It is a widely accepted principle of internet governance that ISPs and other intermediaries should not be arbiters of the legality of content. Website blocking is sometimes criticised as compromising this neutrality, since it requires ISPs to intervene in the content of transmissions based on their source or destination. This is said to encourage a move towards private censorship and restrict public internet spaces without appropriate scrutiny or review.[145]

Although these criticisms are intrinsically less forceful in the context of website blocking injunctions which are adjudicated by courts, there remains an intractable tension between the neutrality expected of intermediaries (for example, to enjoy safe harbour protection as mere conduits or hosts) and their expanding duties to block or disable access to material.

This tension is embodied in the Network Neutrality Regulation,[146] which requires internet users to 'have the right to access and distribute information and content ... via their internet access service', without prejudice to national and EU law related

---

[142] Substantial improvements to blocking technology may be required before interim blocking will become viable; currently, ISPs have up to 10 business days to implement court orders, and network engineers may be unable to implement a blocking order at short notice.

[143] *Jirehouse Capital v Google Inc* [2014] (Unreported, High Court of Justice, Birss J, 6 February 2014, Claim No HP14 E00462).

[144] See also Ofcom, n 124, 48.

[145] See, eg, European Commission, above n 128, 1.

[146] Regulation 2015/2120 laying down measures concerning open internet access and amending Directive 2002/22/EC on universal service and users' rights relating to electronic communications networks and services and Regulation No 531/2012 on roaming on public mobile communications networks within the Union [2015] OJ L 310/1 ('Network Neutrality Regulation').

to the lawfulness of such content.[147] Equally, ISPs are required to 'treat all traffic equally', irrespective of the content being accessed or distributed.[148] Although ISPs may implement 'reasonable traffic management measures', they are prohibited from going further except in limited circumstances. Article 3(3), third paragraph, provides:

> Providers of internet access services ... shall not block, slow down, alter, restrict, interfere with, degrade or discriminate between specific content, applications or services, or specific categories thereof, except as necessary, and only for as long as necessary, in order to:
>
> (a) comply with Union legislative acts, or national legislation that complies with Union law, to which the provider of the internet access services is subject, or with measures that comply with Union law giving effect to such Union legislative acts or national legislation, including with orders by courts or public authorities vested with relevant powers; ...

The effect of article 3(3)(a) is unclear: on its face, it preserves the ability of ISPs to comply with court orders; however, such orders appear to be limited to measures which give effect to EU law or national legislation, such as s 97A, and in circumstances where the measures do not affect lawful content, applications or services. Additionally, article 3(3) appears to prevent ISPs from blocking 'specific content' voluntarily. Even if a court order requires blocking of specific content, doing so will only be permissible 'as necessary, and only for as long as necessary'. For example, sunset clauses of the kind incorporated into English blocking orders may now be mandatory. It remains to be seen whether the Regulation will affect the scope of blocking obligations in other ways.

Harmonisation

Despite a common framework and aims, national approaches to website blocking orders vary significantly within Europe. Websites such as TPB are inaccessible in some European jurisdictions through some service providers, but not others. There are many reasons for this divergence, including incomplete transposition of the Enforcement Directive and differing constitutional norms concerning freedom of expression and remedial innovation by courts. However, this divergence distorts competition, undermines the objectives of copyright harmonisation and risks fracturing the European internet into a set of inconsistent national networks within which content is subjected to a patchwork of incompatible and overlapping regimes for injunctive remedies. Ultimately, some degree of harmonisation of blocking remedies and procedures may be desirable to promote common remedial policies throughout the EU.

---

[147] Network Neutrality Regulation art 3(1).
[148] Network Neutrality Regulation art 3(3)(i).

Conclusion

The growing use of statutory and equitable blocking powers in the United Kingdom reflects the incremental improvement of ISPs' network technologies and gradual recognition that injunctive relief against intermediaries can be at least as effective as remedies against primary wrongdoers. It is, in some respects, surprising that claimants have waited so long to seek blocking remedies against ISPs. The focus on monetary liability, which occupied much of the preceding decade, may have distracted claimants and regulators from asking when and how ISPs and other intermediaries can be ordered to block access to tortious materials even where they face no primary or secondary liability.

Because blocking injunctions do not impose monetary liability upon ISPs, the contributory conduct required for a remedy to be granted is lower. This reflects the distinct status of such injunctive remedies as forms of secondary liability which are aimed at preventing third parties' wrongdoing, as compared to monetary remedies which may be aimed at compensating primary wrongdoing by that defendant or at disgorging his or her unjust gains. An intermediary may thus be ordered to block even though it would never itself be liable for copyright infringement.

The more lenient approach to blocking injunctions makes these remedies an attractive option for claimants who wish to prevent tortious materials from being accessed by consumers in a valuable market but are unable to identify a primary wrongdoer. If recent trends continue, claimants will continue invoking blocking powers to reduce the impact of offshore havens for piracy and other material which cannot be removed using notice-and-takedown; the classes of targeted material, and the kinds of intermediaries required to block access to such material, are likely to continue expanding.

However, blocking entails significant ongoing costs for ISPs and the internet-consuming public, which makes it imperative to assess each application carefully by reference to clear and consistent principles. Proportionality remains the main criterion against which claims are assessed, and it offers tools and safeguards to prevent the technical capabilities of blocking from being abused by claimants or used in circumstances which carry unacceptable consequences for the privacy or freedom of expression of internet users. It will be necessary for European courts to reach a degree of consensus about when blocking will be appropriate, which websites should be blocked and how this should be achieved.

Website blocking is not a panacea for all forms of wrongdoing on the internet. Although filtering technologies are steadily improving, they cannot be relied upon to prevent access to tortious materials without error. Other remedies and enforcement tools—including notice-and-takedown, disclosure, de-indexing and claims against primary wrongdoers—may be more proportionate alternatives in particular cases. It is regrettable that English courts have not required claimants to exhaust the available alternatives before seeking blocking injunctions.

Blocking must be treated with appropriate caution. Its natural tendency is to compromise the end-to-end nature of the internet by requiring intermediaries at one network layer to interfere with activities taking place at other layers. Although the English cases demonstrate that, in many cases, blocking can be implemented in an effective and proportionate manner, these remedies merit close assessment of their effects upon intermediary neutrality, chilling effects upon speech and internet communications, procedural fairness, and wider effects on competition and innovation. Only by careful and ongoing assessment of the proportionality of website blocking can courts fashion a remedy which is both effective and supportive of the future development of internet technologies.

Chapter 13
The Liability of Internet Intermediaries and Disclosure Obligations in Greece

Georgios N. Yannopoulos

Establishing Liability for Internet Intermediaries

Since the expansion of Web 2.0, large quantities of illegal and harmful material have come to occupy parts of the Internet. Why is it important to establish the liability of Intermediaries for that material (Yannopoulos 2013a, 9)? First, the technical contribution of certain actors is indisputable, since without access and host providers there would be no Internet. They are the only entities that can be easily traced in order to seek compensation and sometimes act as scapegoats,[1] even though they are only the messengers. Second, as things stand, it seems that Internet Intermediaries are the only entities able to enforce the methods (e.g. blocking/filtering/take down) by which to control Internet content, either by: (a) intervening during an allegedly illegal or harmful action; or (b) by taking preventive/dissuasive measures. It has been observed that such methods may be extremely effective in restraining certain types of illegal behaviour. Therefore, any analysis of the liability of Internet Intermediaries has not only to consider the *stricto sensu* responsibility, but also their ability to fully control information flows over the Internet: to prohibit or allow access, to define unilaterally the terms of that access, to block or facilitate users, to impose commercial, political or cultural rules of manipulation, and to guarantee the security and integrity of data.

The principle of secondary liability, widely accepted in common law jurisdictions, might not be applied straightforwardly under the civilian structure of Greek law. In theory, to establish such indirect liability, and hence the power to control, one has to examine the existence of a 'causal link' between the services offered by

---

[1] Since antiquity cf. verse 277 of Sophocles' Antigone : "*That no man loves the messenger of ill*" (transl. R. Jebb).

G.N. Yannopoulos (✉)
Law School, University of Athens, Athens, Greece
e-mail: gyannop@law.uoa.gr

© Springer International Publishing AG 2017   317
G.B. Dinwoodie (ed.), *Secondary Liability of Internet Service Providers*,
Ius Comparatum – Global Studies in Comparative Law 25,
DOI 10.1007/978-3-319-55030-5_13

the Internet Intermediary to third parties and the commitment of infringements/ unlawful acts. Legal attitude has mainly been influenced by common law theories underlying secondary liability for intellectual property infringement and defamation. However, under Greek legal doctrine the messenger may not be held liable for the message. In theory, it seems easy to target Internet Intermediaries in order to implement measures of control, but practice shows that Intermediaries tend not to commit illegal acts themselves, so it would be necessary to prove that they provide the aforesaid "causal link".

One possible basis for establishing secondary liability would be to follow the theory of 'adequate cause' and prove that the intervention of the Intermediary constitutes the indispensable link in the chain of events that has led to the illegal result. In terms of civil law liability the mental state (*intention/purpose* or *negligence*) and *adequate cause* must first be examined. The same methodology is followed, in the Greek system, for what under common law is defined as 'primary' and 'secondary' liability. The particular theory could provide adequate grounds for establishing liability in cases of copyright infringement, but seems rather weak in cases of defamation or similar harmful content. It is precisely the difficulty (Yannopoulos 2013a, 55) in connecting the behaviour of Intermediaries to the damage caused that led to the adoption of systems of immunity ('safe harbours' under US terminology) for Intermediaries, like the system of the Directive for Electronic Commerce (hereinafter ECD).[2]

If one tried to apply the principles of the above theory according to the rules of Greek contract law, *the mental state* must be examined, and either *intention (purpose)* or *negligence* (arts. 330, 334 of the Greek Civil Code) must be established. The unlawful behaviour must consist of an act or omission and a causal link must be shown between the harmful act or omission and the damage itself. Such intentional act or omission creates an obligation to compensate for damage when someone was bound to act by law, or by a legal obligation, or by the principle of good faith, as interpreted under Greek civil law doctrine. Under that doctrine, this liability to compensate will be established when someone who has provoked a harmful situation was under an obligation (*duty of care*) to take all the necessary steps in order to avoid such damage to a third party, either before or after the events that have caused the damage. In many of the contexts covered in this Chapter, it would have been difficult to prove that the Internet Intermediary has predicted or could have predicted the harmful situation or the illegal result.

Regrettably, article 926 of the Greek Civil Code establishing joint liability for more than one culprit in the case of tort, does not offer more help in the matter. In this case, it must be proven that the wrongdoer is using the Intermediary as the 'means' to accomplish his/her actions. Liability in this scenario requires, firstly, a

---

[2] Directive 2000/31/EC of the European Parliament and of the Council of 8 June 2000 on certain legal aspects of information society services, in particular electronic commerce, in the Internal Market ('Directive on electronic commerce'), Official Journal L 178, 17/07/2000, p. 0001 – 0016. Greece has implemented this Directive via Presidential Decree 131/2003, Official Gazette vol. A-116/12-5-2003.

relationship between the two 'culprits', or at least that the Intermediary is part of the activities of the wrong-doer and, secondly, the Intermediary receives a measurable profit. Certain types of tort may fall under the above scheme, but no general rule may be established. In the case of unjust enrichment generally, a causal link must be identified between the enrichment and the behaviour of the Intermediary. But in the case of civil lawsuits for insults to personality (art. 57 Greek Civil Code) and/or unjust enrichment (art. 904 Greek Civil Code) things may be easier because it is not necessary to examine culpability. Likewise in cases of unjust enrichment for copyright infringements, article 65 paragraph 3 of Greek law 2121/1993 (the Greek Copyright Law, implementing article 13 of Enforcement Directive 2004/48[3]), provides that establishing a causal link is not required.

From a technical point, it is not possible to monitor content efficiently. Similarly Greek law, implementing the system of immunity in the E-Commerce Directive (ECD), has repeated that there can be no general obligation to monitor content (art. 15 ECD). Under the ECD system, Greek legal theory has concluded (Yannopoulos 2013a, 57, fn 241) that the notion of liability does not only concern civil-type compensation but, in a broader sense, comprises all instances that Intermediaries may be held liable because of flows of information that fall within the meaning of 'illegal' or 'harmful' content.

Greek Law has Adopted the Immunity System of ECD

Based on the immunity regime provided for by ECD, the Greek system has introduced a horizontal approach governing all types of liability. Commentators agree that Internet liability, for the purpose of that provision, concerns all types of responsibility under civil, administrative and criminal law, which leads to general legal liability (Yannopoulos 2014, 790). In that vein, EU scholars,[4] have proposed that a 'special type of liability', corresponding to the liability defined in the 4th Section of ECD referred to as: *Liability* (*responsabilité, Verantwortlichkeit*), be introduced. In view of the consolidation of the internal market such horizontal regulation, no matter if sanctions are characterised as civil, administrative or criminal, simply means that the safe harbour is offered to Internet Intermediaries without any discrimination regarding the cause of liability.[5] The European (and the Greek) system is different to the US system, where regulation addresses separately each area e.g. copyright, defamation, trademarks etc.

---

[3] Directive 2004/48/EC of the European Parliament and of the Council of 29 April 2004 on the enforcement of intellectual property rights. OJ L 157, 30.4.2004, p. 45.

[4] See for example Ufer (2007, 38) and Schmoll (2001, 38).

[5] See in this vein the Introductory Report on German Law for Electronic Commerce (BT-Drs 14/6098 of 17-5-2001, Elektronischer Geschäftsverkehrgesetz – EGG), for the harmonisation of the ECD stating that immunity concerns also criminal law («...*die Beschränkungen der Vernatwortlichkeit gelten auch für den Bereich des Strafrechts...*»).

Still, the critical element in order to establish liability is 'knowledge': Internet Intermediaries are responsible for their own content, but for third party content they must have knowledge of the infringement/harmful/unlawful material in order to be held liable and, hence, to proceed with further action such as blocking /filtering/ removal of content. The Greek Presidential Decree implementing the ECD has similar wording and, while access providers are immune, caching and hosting intermediaries must act "expeditiously" upon obtaining such knowledge in order to remove or to disable access to the information.

The concept of a service provider is defined by law and is similar to that of ECD i.e. any natural or legal person providing an information society service (art. 1(b) PD 131/2003). In turn, 'Information Society Services' is defined as "services within the meaning of Article 1(2) of Directive 98/34/EC as amended by Directive 98/48/ EC" (art. 1(a) PD 131/2003). Under the wording of the Greek law, hyperlinks and Search Engines are not regulated (cf. art. 21 par. 2 ECD), but courts have tried to assess liability by extending the term 'Information Society Services'. In the case of copyright, article 28B of the Greek Copyright Law 2121/1993, implementing art. 5(1) of the EU Copyright Directive 2001/29, has adopted a similar term for 'intermediary' as the one stated in the Directive.[6]

Constitutional Limitations against the ECD System

The right of participation in the Information Society and the right of access to information is protected by the Greek Constitution (art. 5A). The provision is interpreted as the right of everyone to participate in the Information Society and, accordingly, the state has a responsibility to assist in the advancement of it, or, in other words, not to hinder the enjoyment of such right. Following this, under Greek law there is no specific *in toto* law addressing the control of illegal internet content. Nonetheless, the Greek Constitution refers *verbatim* to two limitations of the above right: the protection of privacy and personal data (art. 9 and art. 9A) and the protection of secrecy of communications (art.19). A third restriction, that of enforcing intellectual property rights, is grounded upon a combination of constitutional provisions (art. 5 par. 1 and 3, art. 14 par. 1, art. 16 par.1 and art. 17).

Furthermore, censorship and all other preventive measures are prohibited (art. 14 par. 2), as is the seizure of newspapers and other publications before or after circulation (art. 14 par. 3).[7]

---

[6] This term was interpreted by the CJEU in case C-557/07 *LSG – Gesellschaft zur Wahrnehmung von Leistungsschutzrechten GmbH v Tele2 Telecommunication GmbH,* Reports of Cases, 2009, I–01227.

[7] Exceptionally, seizure by order of the public prosecutor is allowed after circulation in case of: (a) an offence against the Christian or any other known religion; (b) an insult against the person of the President of the Republic; (c) a publication which discloses information on the composition, equipment and set-up of the armed forces or the fortifications of the country, or which aims at the violent overthrow of the regime or is directed against the territorial integrity of the State; or (d) an

Other standards, under which the liability of Intermediaries is assessed, can be found in the chapters of liability – contract and / or tort – of the Greek Civil Code. It should be noted that in 2003 the Greek Parliament implemented the ECD without adopting any specific regulations affecting directly the actions of Internet Intermediaries e.g. for blocking, filtering and taking down of illegal content. Following that analysis, the traditional framework of civil liability in Greece has added more arguments in favour of the system of immunity of ECD, which supports the prohibition of a general obligation to monitor content (art. 15 ECD, art 14 Greek PD). Under this system, Greek legal theory has developed the view[8] that the notion of liability does not only concern civil-type compensation but, in a broader sense, comprises all cases for which Intermediaries must take action. Furthermore, ECD immunity is not unlimited and courts and authorities may require preventive or restrictive measures on behalf of access and host providers within the scope of the above restrictions.

Adherence to International Instruments

In 2002, art. 348A was introduced into the Greek Penal Code in order to penalise child pornography. The particular article was recently amended,[9] so as to harmonise with the Child Pornography Directive.[10] Greece has also ratified and implemented the Optional Protocol to the Convention on the Rights of the Child on the sale of children, child prostitution and child pornography,[11] adopting the references to the Internet and emerging technologies. Equally, Greece has ratified and implemented the Convention on the Protection of Children against Sexual Exploitation and Sexual Abuse[12] (known as the Lanzarote Convention).

Recently, Law 4285/2014 harmonised Greek legislation with Council Framework Decision[13] on combating certain forms and expressions of racism and xenophobia by means of criminal law. The new law criminalises racist content, xenophobia, hate

---

obscene publication which is obviously offensive to public decency, in the cases stipulated by law (art. 14 par. 3 Greek Constitution).

[8] See Yannopoulos 2013a, 57 fn 241.

[9] By means of art. 8 of Law 4267/2014.

[10] Directive 2011/92/EU of 13.12.2011 on combating the sexual abuse and sexual exploitation of children and child pornography, and replacing Council Framework Decision 2004/68/JHA, OJ L 335, 17.12.2011, p. 1–14.

[11] Adopted and opened for signature, ratification and accession by UN General Assembly Resolution A/RES/54/263 of 25 May 2000, entered into force on 18 January 2002 (available at www.un-documents.net/a54r263.htm, ratified by Greek Law 3625/2007).

[12] Council of Europe Convention on the Protection of Children against Sexual Exploitation and Sexual Abuse, Council of Europe Treaty Series (CETS) No. 201, Lanzarote 25.10.2007 (available at www.coe.int/en/web/conventions/full-list/-/conventions/treaty/201, ratified by Greek Law 3727/2008).

[13] Council Framework Decision 2008/913/JHA of 28 November 2008 on combating certain forms and expressions of racism and xenophobia by means of criminal law, OJ L 328, 6.12.2008, p. 55.

speech, as well as the denial, gross minimisation, approval or justification of geno-cide or crimes against humanity. The latter two must be recognised as such, either by an international court or by the Greek Parliament. Article 3 of Law 4285/2014 pro-vides that when above actions are being committed through the Internet or via other means of communication, then the place of committing the crime (*locus delicti*) is considered to be the entire Greek territory, as long as access to the particular medium is achieved in Greece and irrespective of the place of establishment of the medium. Therefore, if the perpetrator is located in Greece he/she would be subject to the sanctions, no matter if the illegal content is hosted in hardware or software (e.g. website or cloud storage) situated outside Greece.

Greece has signed the Convention on Cybercrime,[14] and has implemented it in domestic law including the Additional Protocol to the Convention, concerning the criminalisation of acts of a racist and xenophobic nature committed through com-puter systems. The ratifying Greek Law 4411/2016 has also harmonised Directive 2013/40/EU on attacks against information systems. Greece has signed, but has not yet ratified the Convention on the Prevention of Terrorism.[15]

The Convention for the Protection of individuals with regard to Automatic Processing of Personal Data was ratified by Greece in 1992.[16] Greece has imple-mented the Data protection Directive 95/46/EC,[17] while a constitutional amendment in 2001 has awarded constitutional status to the protection of personal data (art. 9A of the Greek Constitution). It should be noted that existing legislation will be replaced by the new General Data Protection Regulation (EU) 2016/679 (GDPR), which comes into full force on 25th May 2018.

Legislation Imposing Obligations on Intermediaries

The system of immunity under the ECD did not operate in its pure form for long in Greece. A series of laws created a parallel framework of rules, which has either formed new bases of liability for Internet Intermediaries or has increased existing obligations. Of course, the attribution of full liability to intermediaries allegedly

---

[14] Council of Europe Convention on Cybercrime, Council of Europe Treaty Series (CETS) No. 185, Budapest, 23.11.2001 (available at http://www.coe.int/en/web/conventions/full-list/-/conventions/treaty/185).

[15] Council of Europe Convention on the Prevention of Terrorism, Council of Europe Treaty Series (CETS) No. 196, Warsaw, 16.5.2005, available at http://www.coe.int/en/web/conventions/full-list/-/conventions/treaty/196

[16] Council of Europe Convention for the Protection of individuals with regard to Automatic Processing of Personal Data, Council of Europe Treaty Series (CETS) No.108, Opened for signa-ture in Strasbourg on 28 January 1981 (available at http://www.coe.int/en/web/conventions/full-list/-/conventions/treaty/108, Greek Law 2068/1992).

[17] Directive 95/46/EC of the European Parliament and of the Council of 24 October 1995 on the protection of individuals with regard to the processing of personal data and on the free movement of such data, OJ L 281, 23.11.1995, p. 31–50 (harmonised by Greek Law 2472/1997).

implicates censorship or oppressive governments and legislators in Greece have carefully avoided explicit mention of such compromising terms. Instead, they have chosen to increase the obligations of providers via a grid of new pieces of legislation and to strengthen the grounds for intervention by several supervising Authorities. This attitude has led to a patchwork of rules and to fights of jurisdictional competence among various Authorities. In turn, users face difficulties in enjoying the constitutionally protected principle of access to networks and information, while providers must cope with disproportionate compliance costs. The following are the most important pieces of legislation in Greece that have imposed obligations on Internet intermediaries.

Legislation for Telecommunications, Electronic Networks and ECHR Principles

In 2006, the Greek Parliament fully harmonised the bundle of Telecommunications Directives (Maniotis et al. 2015, 38),[18] which was later revised by the Telecommunications Directive 2009/140.[19] Following the European Parliament compromise of 5 November 2009, the Greek Law 4070/2012, implementing Directive 2009/140, repeated[20] in article 3 the ambitious original wording, referring to article 10[21] of ECHR (freedom of expression), stating that:

---

[18] Directive 2002/19/EC of the European Parliament and of the Council of 7 March 2002 on access to, and interconnection of, electronic communications networks and associated facilities (Access Directive), OJ L 108, 24.4.2002, p. 7–20. Directive 2002/20/EC of the European Parliament and of the Council of 7 March 2002 on the authorisation of electronic communications networks and services (Authorisation Directive) OJ L 108, 24.4.2002, p. 21–32. Directive 2002/21/EC of the European Parliament and of the Council of 7 March 2002 on a common regulatory framework for electronic communications networks and services (Framework Directive) OJ L 108, 24.4.2002, p. 33–50. Directive 2002/22/EC of the European Parliament and of the Council of 7 March 2002 on universal service and users' rights relating to electronic communications networks and services (Universal Service Directive) OJ L 108, 24.4.2002, p. 51–77. Directive 2002/77/EC of 16 September 2002 on competition in the markets for electronic communications networks and services, OJ L 249, 17.9.2002, p. 21–26.

[19] Directive 2009/140/EC, of 25.11.2009 amending Directives 2002/21/EC on a common regulatory framework for electronic communications networks and services, 2002/19/EC on access to, and interconnection of, electronic communications networks and associated facilities, and 2002/20/EC on the authorisation of electronic communications networks and services, OJ L 337, 18.12.2009, p. 37–69.

[20] It did so with several ambiguities in relation to the original text of the Directive, see Yannopoulos 2013a, 189–190.

[21] See a similar reference to article 10 ECHR in Internet Recommendation CM/Rec (2008) 6, 26.3.2008 on Measures to Promote the Respect for Freedom of Expression and Information with Regard to Internet Filters (available at https://wcd.coe.int/ViewDoc.jsp?id=1266285), according to which users may object the use of filters. See also CoE document: Human Rights Guidelines for Internet Service Providers, H/Inf (2008) 9, available at http://www.coe.int/t/dghl/standardsetting/media/Doc/H-Inf(2008)009_en.pdf

> Measures taken by Member States... shall respect the fundamental rights and freedoms of natural persons, as guaranteed by the European Convention for the Protection of Human Rights and Fundamental Freedoms and general principles of Community law. Any of these measures ... liable to restrict those fundamental rights or freedoms may only be imposed if they are appropriate, proportionate and necessary within a democratic society, and their implementation shall be subject to adequate procedural safeguards in conformity with the European Convention for the Protection of Human Rights and Fundamental Freedoms and with general principles of Community law, including effective judicial protection and due process

The obscure wording of the law, however, does not answer directly the critical question: Are *private entities*, such as Internet Intermediaries allowed to restrict fundamental rights such as the right to access a network? Although the ECHR case-law is clear on several of the above matters, the particular legislation has not been fully tested by the Greek courts. A decision granting injunction for IP infringement (FIC Athens, 4658/2012—see *infra*), accepts only *in passim* and before harmonisation, that article 3 of Directive 2009/140 is not directly applicable,[22] while adjudicating a dispute between private entities.

Greek legislation also contains a certain obligation for access and service providers to register with the National Telecommunications and Post Commission (NTPC), to obtain a General License[23] and to provide technical information to the NTPC, as regards the quality and cost of services. The NTPC may request providers to disseminate information of 'public interest' to the users, such as warnings for data protection breaches and IP infringements, malicious software etc.

Finally, art. 37 of Greek Law 4070/2002 has implemented art. 1 par. 15 of Directive 2009/140, which introduces a Chapter regarding security and integrity of networks and services. The wording of the Greek law is trying to enhance the protection of networks' security and integrity. However, the execution of Greek law has been delegated among three different independent Authorities: (a) The Hellenic Authority for Communication Security and Privacy (HACSP); (b) The Hellenic Data Protection Authority (HDPA); and (c) the NTPC. Practice until now (see *infra*) has revealed insurmountable difficulties and bureaucratic stubborness between the three Authorities.

---

[22] The Greek Decision supports this argument by referring to CJEU decisions: Joint Cases C-397/01 to C-403/01, *Pfeiffer et al. v. Deutsches Rotes Kreuz, Kreisverband Waldshut eV*, Reports of Cases 2004 I-08835 and Case C- 91/92, *Paola Faccini Dori v Recreb Srl*. Reports of Cases 1994 I–03325.
[23] Article 18 of Law 4070/2012.

Data Protection Legislation

Apart from the Data Protection Directive, Greece has also implemented the E-Privacy Directive 2002/58[24] and the Data Retention Directive 2006/24.[25] However, since Directive 2006/24 has been cancelled,[26] the discussion is focused only on the fate of existing data gathered by providers: Whether they will be allowed to erase them or whether they should retain them will be answered by future legislation. Regarding standard data protection issues, it is worth mentioning that the HDPA[27] has followed the ruling of CJEU[28] and has accepted that the characterisation of the processing of personal data as legitimate, under the provisions of the Data Protection Directive, should be performed *ad hoc*, no matter if it concerns an original collection of data or any subsequent transmission of such data via electronic means. An individual may assert against host intermediaries the right of objection under Greek Data Protection Law.[29] If the Intermediary does not respond, the data subject may then refer the matter to the Data Protection Authority, who may impose a provisional suspension of the processing (of data) until the final decision. Greece has also harmonised the 'Cookies' Directive 2009/136,[30] and HDPA has published instructions for the provision of consent via electronic means.[31]

---

[24] Directive 2002/58/EC of 12.7.2002 concerning the processing of personal data and the protection of privacy in the electronic communications sector (Directive on privacy and electronic communications) OJ L 201, 31.7.2002, p. 37–47 (Greek Law 3471/2006).

[25] Directive 2006/24/EC of 15.3.2006 on the retention of data generated or processed in connection with the provision of publicly available electronic communications services or of public communications networks and amending Directive 2002/58/EC, OJ L 105, 13.4.2006, p. 54–63 (Greek Law 3917/2011).

[26] By CJEU Decision in joint cases C-293/12 *Digital Rights Ireland Ltd   v   Minister for Communications,   Marine   and   Natural   Resources   and   Others* and C-594/12 *Kärntner Landesregierung and Others*. Note that recently a Greek court has submitted to the CJEU a reference for a preliminary ruling (pending Case C-475/16) regarding the fate of the Greek Law 3917/11 that had harmonised the cancelled Data Retention Directive.

[27] See HDPA Decision 38/2008 regarding posts of Microsoft Hellas concerning legal action against software infringers, available (in Greek) at http://www.dpa.gr/APDPXPortlets/htdocs/document-Display.jsp?docid=217,199,0,137,106,230,234,134   (the   decision   was   later   upheld   for procedural reasons).

[28] See CJEU Decision C-73/2007 *Tietosuojavaltuutettu v Satakunnan Markinapörssi Oy and Satamedia Oy*, Reports of Cases 2008 I–09831.

[29] Article 13 of Law 2472/1997(cf. art. 14 Data Protection directive 95/46/EC).

[30] Directive 2009/136/EC of the European Parliament and of the Council of 25 November 2009 amending Directive 2002/22/EC on universal service and users' rights relating to electronic communications networks and services, Directive 2002/58/EC concerning the processing of personal data and the protection of privacy in the electronic communications sector and Regulation (EC) No 2006/2004 on cooperation between national authorities responsible for the enforcement of consumer protection laws OJ L 337, 18.12.2009, p. 11–36

[31] See www.dpa.gr. Existing HDPA Guideline 2/2011 for electronic consent to data processing does not apply to 'cookies'.

In the event of data protection breaches over networks, providers are obliged[32] to inform, without undue delay, the HACSP, the HDPA and the affected user. Providers may be relieved of this obligation if they prove that to avoid breaches they have taken the necessary technical and organisational measures "to the satisfaction of the competent Authority". This means that the HACSP and the HDAP, who are delegated to issue a common regulation regarding notifications,[33] are best suited to set the limits of responsibility.

Greek Law has exceeded the parameters set by the original text of article 15 paragraph 2 of the e-Privacy Directive 2002/58 and has introduced a regime of direct liability for providers in case of data protection breaches. Such liability to compensate may be established without any relevant decision of the HACSP or the HDPA. A similar system of direct liability applies to providers for trafficking of spam mail (art. 11 paras. 5 and 6, Law 3471/2006). The introduction of direct liability is questionable in view of the original text of the e-Privacy Directive 2002/58 (as amended by the Cookies Directive 2009/136). However, Greek Law has followed the view that by increasing compensation and, similarly expanding the range of those obliged to notify breaches, providers would become more responsible. This approach by the Greek Legislators coincides with the EU philosophy that "those who profit from the information revolution must respond to the public policy responsibilities that come with it",[34] in order to achieve a higher degree of confidentiality for users.

Intellectual Property Legislation

Greece has implemented both the Copyright (or Information Society) Directive 2001/29/EC[35] and the Enforcement Directive 2004/48/EC.[36] However, as explained above, under Greek law there is no distinction between primary and secondary infringement, which may establish a certain degree of liability under common law

---

[32] Greek Law 4070/2012, amending Law 3471/2006, has followed the wording of art. 3 par. 4 Directive 2009/136. However, the wording for 'identity theft' of Recital No. 61 of the Directive does not appear in the Greek translation.

[33] They have already issued Common Act 1/2013 regarding data retention. See also Commission Regulation (EU) No 611/2013 of 24.11.2013 on the measures applicable to the notification of personal data breaches under Directive 2002/58/EC of the European Parliament and of the Council on privacy and electronic communications.

[34] See p. 2–3 in the speech of the then Commissioner Viviane Reding, entitled "Securing personal data and fighting data breaches", delivered at the ENISA Seminar, Brussels, 23.10.2009, available at www.edps.europa.eu.

[35] Directive 2001/29/EC of the European Parliament and of the Council of 22 May 2001 on the harmonisation of certain aspects of copyright and related rights in the information society OJ L 167, 22.6.2001, p. 10–19 (harmonised by art. 28B of Greek Law 2121/1993).

[36] Directive 2004/48/EC of the European Parliament and of the council of 29 April 2004 on the enforcement of intellectual property rights (Text with EEA relevance) OJ L 157, 30.4.2004, p. 45–86 (harmonised by arts 64 and 64A of Greek Law 2121/1993).

systems. Before harmonisation, it was common in Greece to impose the criminal sanctions of copyright law for those importing, using, possessing or circulating illegal copies of protected works. Following harmonisation, Greek courts have directly applied the Information Society and Enforcement Directives on numerous occasions (see *infra*). A critical difficulty arises when applying article 8 of Enforcement Directive 2004/48 (cf. art. 63A of Greek Law 2121/1993) for the disclosure of 'information' by Internet Intermediaries. This privilege of the right-holder is subject to constitutional barriers safeguarding data protection and secrecy of communications (see *infra*).

Child Pornography Legislation

As explained above, Greece has harmonised the existing article 348A of the Greek Penal Code with the Child Pornography Directive. Nonetheless, article 25 of the Directive (entitled 'measures against websites') has been harmonised separately by article 18 of Law 4267/2014, stating that the competent Public Prosecutor (deciding either on the first degree or on appeal) is able to order the "elimination" (meaning: the "taking down") of a website hosted in Greece, containing or transmitting child pornography. Furthermore, when the website cannot be traced in Greece or elsewhere, the Prosecutor may order the temporary (for two months) deactivation of any Domain Name assigned in Greece, hosting or leading to such website. Finally, if the website is neither hosted in Greece, nor belonging to a domain name assigned in Greece, the Prosecutor may order the blocking of access to such websites. The Order must be individually and fully justified and must be addressed to the owner of the website and the NTPC. The NTPC must, in turn, notify all access providers registered in Greece as per Greek Telecommunications Law (Law 4070/2012). Apart from seeking compliance, the NTPC may demand that the providers take measures in order to notify the public about the illegal actions.

The law provides for an "individually and fully justified Order" that must be notified to the owner of the website to be removed (see art. 18 par. 1 of Law 4267/2014 harmonising art. 25 of child Pornography Directive). In this sense, Greek law is trying to comply with article 10 of ECHR, which has established that the restriction to the freedom of expression must be pursuing a legitimate goal and must be necessary in a democratic society. In art. 18 par. 2 of the same law, the owner of a deactivated domain name is entitled to a petition (a *quasi* appeal) to the Prosecutor for the domain to be re-activated. Again, both the initial Order and the 'appeal' decision must be "individually and fully justified" and notified to the domain owner. The same rule applies to the blocking of a website with child pornography content (art. 18 par. 3). It looks like the Greek Legislator is trying, inter alia, to comply with certain aspects of article 6 of ECHR (fair trial).

The option of issuing Prosecutor's Orders for blocking of sites is very recent and has not yet been tested. Similarly, Greek Law 4285/2014, which concerns xenophobia, racism and hate speech, has not yet been interpreted by the courts.

Thus, it is not clear how the courts will proceed with decisions regarding blocking/
filtering of websites, following the conviction of the owner for a criminal offense
based upon these laws.

Gambling Legislation

Gambling Law has introduced, for the first time, a regulatory regime[37] that imposes
an obligation on ISPs to block certain websites. The Greek Gaming Commission
(GGC) publicises from time to time a 'blacklist' of prohibited gambling sites.[38]
According to the wording of the law, all "ISPs" – apparently meaning all "access
providers" – operating in Greece and registered with the Telecommunications
Commission (see *supra*), according to the telecommunications legislation, must disable access to those 'blacklisted' sites. While the wording of the law states that
blocking must target "domain names" and "IP addresses" which are included in the
"blacklist", it is not clear which particular blocking method should be followed, or
if any other hardware or software identification of the 'blacklisted' site is required.
It is affirmed, however, in The Internet Gambling Regulation (art. 3.4 of GGC
Decision No. 51/26.4.2013), that blocking must take place when access is attempted
from "an IP address residing within the Greek territory", without any other indication concerning the identification of such "residence" of an IP address. Additionally,
ISPs must not allow "any action of commercial communication"[39] of illegal gambling providers. Again, it is not evident whether ISPs must block all types of advertisement, including for example frames and nested hyperlinks. It is also interesting
that the simple posting of the 'blacklist' on the GGC's website is considered to
constitute "adequate knowledge" and proof of evidence against ISPs. Furthermore,
according to the law, the GGC is also entitled to send to ISPs a list of 'key-words'
that indicate a connection to Internet gambling. If ISPs are asked to register a
Domain Name that includes any such prohibited 'key-word', then they should
within 15 days notify the GGC accordingly. In view of the heavy criminal and
administrative sanctions, most ISPs operating in Greece have abided by the law.

The law also includes a prohibition against banks transferring funds to similarly
'blacklisted' accounts (Fox 2003). Under the circumstances, before the imposition
of capital controls in Greece (June 2015), it would have been difficult to implement

---

[37] Articles 45–48 of Law 4002/2011 as amended.

[38] As prescribed by art. 48 par. 8 of Law 4002/2011, and backed by GGC Regulation for the
Conduct and Control of Internet Gambling (Decision No. 23/3/23.10.2012, Official Gazette
B-2952/5.11.2012, as amended by GGC Decision No. 51/3/26.4.2013, Official Gazette
B-1147/13.5.2013).

[39] See Yannopoulos 2013d, 46. Decision of FIC Athens 10586/2011, examining illegal advertisement of illegal gambling, refers to Recital 19 of the ECD in order to define the place of establishment, but, finally, has not applied the ECD claiming that according to the exception of art. 1 par.
5(d) "gambling activities" are not regulated by the ECD.

such prohibitions in respect of accounts in foreign banks, or to credit card transactions. It seems, generally, that the Greek Legislator has not taken into account the experience of other countries, and especially the United States (Morse 2007; MacCarty 2010). Fostering users' protection by means of national rules enforcement has a narrow margin of success.[40]

A question could be raised as to whether the blocking of 'illegal' gambling sites, according to the Greek Law, contradicts the principles of necessity and proportionality in connection with the enjoyment of property (ECHR, Protocol 1) or the economic freedom and transfer of services within the EU (cf. art. 16 of the EU Charter of Fundamental Rights), especially for gambling sites that have already obtained a license in other EU countries. However, since the restriction has been imposed by a law enacted by Greek Parliament, such considerations may only be answered when a case reaches the European Court of Human Rights or the CJEU.

Domain Abuse

The NTPC, in its capacity as the authority supervising the '.gr' top level domain, is entitled to deregister domain names that infringe existing intellectual property rights or trademarks, names that have been registered in bad faith, or names that clash with moral perceptions or Greek public policy.[41]

Policing the Internet

The Greek Police Division for Electronic Crime monitors Internet content in the sense of detecting criminal offenses such as fraud, child pornography, hacking, software piracy, credit card fraud, chat rooms crimes etc., but also in the sense of preventing harmful actions; it offers vital help in emergency cases of illegal content (e.g. cases of blackmail, suicide attempts etc.). This of course has nothing to do with blocking of content or general monitoring and specifically focusses on crime prevention. Police do not have the right to violate constitutional rights like freedom of expression or secrecy of communications, or to obtain unlawfully the personal data of citizens. Exceptions to reveal personal data or communication data or content are prescribed by law only if an investigation has been instigated and only subject to the safeguard of a judicial authority.

---

[40] See the proposal in Germany (Steegmann 2010) to 'cut' the basic infrastructure (*basisinfrastruktur*) which facilitates illegal gambling.

[41] Art. 10 par. 2 of NTPC Domain Name Regulation No.750/2/Official Gazette B-412/24-3-2015.

Secrecy of Communications and Disclosure of External Communication Data Under Greek Law

In Greece, both protection of personal data and protection of secrecy of communications are backed by Constitutional provisions (arts. 9A and 19). Two independent Authorities supervise each field: respectively, The Hellenic Data Protection Authority (HDPA) and the Hellenic Authority for Communication Security and Privacy (HACSP). In particular, for communication data to be revealed, a criminal investigation must be initiated for a particular list of serious crimes, contained in a catalogue described by law (L. 2225/1994). Under current legislation common Internet crimes resulting in harmful or illegal content, such as copyright infringements or defamation, are not included in this list of serious crimes.

Therefore, a major issue has arisen in Greece concerning the disclosure of personal data, and especially external communication data (including the IP address), by the Intermediaries to the Authorities. In an initial Guideline,[42] the Public Prosecutor of the Supreme Court (Areios Pagos) has adopted the view that no matter whether a Prosecutor's Order or a Judicial Decision exists, police and investigation authorities are entitled to request Internet Intermediaries to disclose external communication data and that HACSP has no jurisdiction over the matter. Subsequently, and in view of Directive 2006/24 (Greek Law 3917/2011), two newer Guidelines[43] have introduced a milder approach, limiting the disclosure of data to cases of malicious or threatening calls and only if a Preliminary Examination, Preliminary Criminal Investigation, or Criminal Investigation has been ordered by a Public Prosecutor.

Still, the dilemma for Internet Intermediaries whether to reveal or not the identity of users has not been resolved. CJEU case law, such as *Promusicae* and *Tele2*,[44] has been ambiguous, and the ECHR has not imposed an obligation to disclose the IP address.[45]

In practice, providers of mobile services in Greece have been asked by authorities (such as the police or public prosecutor) to disclose customers' data as per the above Guidelines. In most cases, they have refused to submit the data claiming that, since the crimes under investigation are not included in the list of serious crimes (Law 2225/1994), secrecy of communications may not be lifted. In some instances the Boards of Directors have been charged with disobedience and harbouring a

---

[42] Guideline 9/2009 of Prosecutor G. Sanidas.

[43] Guideline 12/2009 of Prosecutor I. Tentes and Guideline 9/2011 of Prosecutor Ath.Katsirodis. Furthermore, Decision 91/2012 of the Appeal Court of Thrace has clarified that these Guidelines may only be applied to criminal procedures and not to civil litigation.

[44] C-275/06 *Productores de Música de España v Telefónica de España SAU*, Reports of Cases 2008 I-00271and C-557/07 *LSG v Tele2* (see *supra* fn 6)

[45] See in favour of not disclosing the IP address *KU v Finland*, ECHR, 2.3.2009 (Application no. 2872/02); in favour of protecting external communication data, see *Kopland v UK*, ECHR 3.4.2007 (Application no. 62617/00).

criminal, but have been later acquitted. Currently, there exists an on-going investigation against providers in relation to illegal online gambling.

Case Law

In light of the above analysis, case-law on blocking and filtering has been limited to copyright cases and has dealt with the issue of imposing injunctions on ISPs to filter access to illegal sites (mainly, by blocking DNS names). This Section discusses the most significant of these decisions:[46]

Injunction Granted for DNS Blocking for IP Infringement [Decision 4658/2012 of the First Instance Court of Athens]

In an injunction case, following a petition by collecting societies, the Court ordered major access providers in Greece to cut off access to particular DNS addresses infringing copyright. Apart from the copyright rules, the Court mentioned articles 12 and 15 of the ECD, as well as CJEU decisions C-70/10 *Scarlet v Sabam* and C-360/10 *Sabam v Netlog*.[47] The Court accepted that a general blocking of access, in order to protect IP rights, would be disproportionate and incompatible with article 5A paragraph 2 (Right to Information Society) of the Greek Constitution. The Court denied the direct applicability of Directive 2009/140 (then not harmonised in Greece, see *supra*), but accepted that proportionate and necessary measures may be imposed in order to protect another right. Therefore, blocking of a particular webpage infringing copyright has been permitted and the court granted the injunction, ordering the blocking of specific URL and IP addresses. It is interesting that the blocked sites contained only hyperlinks to copies of protected works hosted in international file-sharing websites. The Court tried controversially (see *infra* case-law on hyperlinks) to associate hyperlinking with the hosting of copies of the works.

The particular injunction had been considered as a technological method in the area of copyright, which could have an educational effect on both providers and users, without charging them directly with liability. This has resulted in a more 'rigid' duty of care of Internet Intermediaries towards users, in addition to the standard duty of care requirements under contract or tort.

---

[46] See also Maniotis et.al., 2013, 105 et. seq.
[47] Case C-70/2010 Scarlet Extended SA v Société belge des auteurs, compositeurs et éditeurs SCRL (SABAM), Reports of Cases 2011 I–11959; Case C-.360/10 Belgische Vereniging van Auteurs, Componisten en Uitgevers CVBA (SABAM) v Netlog NV.

Injunction Not Granted for IP Infringement [Decision 13478/2014 of the First Instance Court of Athens]

In another case, the injunction was not granted. Five collective organisations filed a petition asking several access and host providers to block access to sites dispensing illegal copies of creative works (mainly music and movies) protected under the Greek IP legislation. The Decision initially clarified the legal position of Internet Intermediaries and their liability, stating that by a combination of the Greek IP law (Law 2121/1993), the Presidential Decree 131/2003 harmonising the ECD and Greek legislation concerning the secrecy of communications (L. 3471/2006, PD 47/2007 and L. 2225/1994), access providers are not allowed to disclose personal data of users connecting to the Internet.

The Court stated first that access providers do not play an active role and they neither initiate nor select the receiver of the transmission (and thus came within the scope of art. 12 of the ECD). Furthermore, it held that that the prerequisite of Recital 44 of the ECD i.e. "deliberate collaboration with one of the recipients of the service in order to undertake illegal acts", was not fulfilled. Therefore, the defendant access providers fell under the immunity of article 11 of PD 131/2003 (i.e. art.12 of ECD). The Court also concluded that the hosting defendants fell under the immunity of article 13 of PD 131/2003 (art. 14 of ECD), because as host providers "they [did] not have actual knowledge of illegal activity or information". The Court dismissed the claim of the Plaintiffs that they had informed the Defendants about the illegal activity. The Court required a particular degree of certainty in the 'knowledge' obtained, which should be equivalent to the notion of 'Direct Intent' of article 27 paragraph 2[48] of the Greek Criminal Code. The claimed 'notification' by the Plaintiffs to the Defendants did not meet the above criteria and, in particular, did not provide the Defendants with the required 'knowledge' of illegality of their actions. The Court emphasised that such 'notification' only referred to part of the stored (hosted) information, which otherwise was legal.

The Court noted recital 45 of ECD, which states that an injunction "can in particular consist of orders by courts or administrative authorities requiring the termination or prevention of any infringement, including the removal of illegal information or the disabling of access to it". The injunction sought would, however, have contradicted article 14 of PD 131/2003 (art. 12 ECD), since the Defendants did not host the allegedly infringed works themselves. These works were transmitted either through a forum, via hyperlinks or via p2p networks, and, therefore, the requested blockage of information could not be limited to the allegedly illegal content: it would extend to any kind of information connected or similar to the initial 'illegal' content. In that sense, the Defendants, as intermediaries, would end up with the burden of a general obligation to monitor any kind of transmitted information, since, according to the Court, other methods of control would not be effective. Furthermore, the Court

---

[48] Art. 27 par. 2 sec. 1 GPC: "If a statute requires knowledge of a certain particular as an element, conditional intent shall not suffice".

discovered that proposed technical means of blocking operated automatically and thus the software could not distinguish between 'legal' and 'illegal' uses of the protected works; the application of such automated 'filtering' system would therefore definitely block legal content as well.

The Court concluded that such methods, apart from the ECD considerations, contradict the principle of proportionality, the right to freedom of information (art. 5A par. 1 Greek Constitution), the right of participation to the Information Society (art. 5A par. 2 Greek Constitution), the right to protect personal data (art. 9A) and the right of secrecy of communications (art. 19). The Court remarked that by limiting access, legal actions of the users would be affected together with the 'illegal' ones and this was an unauthorised intervention that did not meet the terms of necessity and proportionality, which were required for an injunction. The argument of the Court was enhanced by the fact that several of the 'illegal' sites had, in the meantime, changed their IP address. The Court held, finally, that such 'filtering' infringes the right of the providers to conduct business (art. 16 of the EU Charter of Fundamental Rights), and also contradicts the basic principle of net neutrality, with the cost imposed on providers being disproportional to the envisaged gain. Interestingly, the decision, issued in 2014, did not take into account decision C-314/12 *UPC*[49] of the CJEU, which had considered the same arguments.

Decisions Involving Hyperlinks and Search Engines

The legal basis for injunctive relief derives from an identical paragraph 3 in articles 11, 12 and 13 of Greek Presidential Decree 131/2003 (i.e. arts. 12 par. 3, 13 par. 2 and 14 par. 3 of the ECD). In the hypothetical event of some kind of non-regulated quasi 'secondary' liability, the remedy could be outstreched and the Court could order any suitable measure against the intermediary. In this sense, case-law is more fascinating in the non-regulated areas of hyperlinks and search engines, which are not covered under the wording of the Greek PD 131/2003 (cf. art. 21 par. 2 ECD). Principally, courts have in several cases tried to establish liability by expanding the term "Information Society Services".[50]

In a recent case, the Multimember First Instance Court of Athens decided that 'deep linking' leading to downloading or streaming of videos and movies did not constitute infringement, as long as the right-holder had not introduced any measures to control access (e.g. some payment method, or creation of an account or even by requesting the simple registration of users).[51] The Court referred to CJEU Case

---

[49] C-314/12, UPC Telekabel Wien GmbH v Constantin Film Verleih GmbH and Wega Filmproduktionsgesellschaft mbH.

[50] See Yannopoulos 2013a, 178 fn 725.

[51] Decision MFIC Athens No. 5249/2014 (published in the Nomos database).

C-466/12 *Svensson* and CJEU Order C-348/13 *Bestwater*,[52] which support the view that an Intermediary who posts hyperlinks to freely accessible works does not host copies of those works on its own servers and, therefore, it has not communicated the audio-visual works to a 'new' audience as required to establish liability under copyright law. Earlier, in a criminal case known as the *Greek-movies* case,[53] a court decided that the inclusion of hyperlinks pointing to already published works on other websites does not fall within the notion of reproduction or public performance under Greek copyright law and, therefore, does not constitute infringement. In order to be immune on this basis, the Court held that the copyright holder must not have introduced any technological measures or other licensing limitations. Contrarily, in another case, the same Court decided that the owner of a radio station website infringed copyright by posting links to another site because the copyrighted music to which it linked was made available to the public without license; the court thus ordered the removal of the hyperlinks.[54]

Finally, in an injunction case, the chairing Judge of the Court of First Instance of Athens initially issued a Provisional Order and then the Court issued an Injunction Decision ordering Google to stop the auto-complete function, which had been producing insulting and defamatory results when typing the name of a known journalist.[55] With the same Provisional Order and Decision the Court ordered a known blog to stop reproducing the insulting information. The Court had identified a substantive danger of insult to the personality under article 57 of the Greek Civil Code, recognising also the ability of the search engine to institute preventive measures. As long as the search engine operator controls the specific algorithm of the auto-complete function, Google was, according to the Court, able to delete the insulting comments. The Court considered itself competent to order an injunction to be applied in the territory of Greece, dismissing the objection of Google that its legal seat is in California. The Court referred to the jurisdictional term of Google's Standard Terms of Service and to the fact that Google keeps a local branch in Greece.

It should be noted, however, that the relevance of the General Terms and Conditions of search engines and social networks, selecting the laws of the United States (mostly California) as governing law, has not yet been addressed. Such clauses have been characterised as illegal by the Greek Courts in cases of consumer protection, but the Courts have not yet produced any domestic judgements and case law for the jurisdictional dimension regarding the removal of content. In tort cases, Courts have to be guided by the existing case law of CJEU, such as joint Cases

---

[52] CJEU Case C-466/12, Nils Svensson and Others v Retriever Sverige AB; Order C-348//13, BestWater International GmbH v Michael Mebes and Stefan Potsch.

[53] See Decision of Magistrate's Court of Kilkis 965/2010 in Kalavrouzioti 2011a,194. The Court stated that adding a hyperlink does not constitute infringement but did not examine the particular details of the case i.e. who has uploaded the illegal content, if there was any financial profit in relation to advertisement etc.).

[54] See Decision of First Instance Court of Athens No. 4042/2010 in Kalavrouzioti 2011b, 195

[55] Decision 11339/2012 of the First Instance Court of Athens, commented by Yannopoulos 2013b, 168.

C-509/09 *eDate Advertising* and C-161/10 Olivier Martinez[56] and Case C-292/10 *G v Cornelius de Visser.*[57]

Soft Law

Codes of Practice

As regards Codes of Conduct, the Greek PD 131/2003 has recognised in article 15 the "wishful thinking" of article 16 of ECD for the introduction of 'soft law'. Such codes must, under the Greek PD, be ratified by the Minister of Development, a formality which causes bureaucratic burdens. In any event, self-regulation in Greece has a limited scope of application, although article 17(1) of PD 131/2003 provides for alternative dispute resolution by referral to the domestic rules for consumer protection.

E-Business Forum, a public consultation initiative of the Ministry of Development has drafted a Code of Practice and Ethics of ISPs in co-operation with Safenet, a non-profit organisation. The Code provides for the introduction of a hotline (www.safe-line.gr) that accepts complaints regarding illegal or harmful content, especially child pornography, racist and xenophobic postings. In a similar manner, the Ministry for Education has introduced blacklists of sites that should not to be visited by pupils and students. There is no official acceptance of such soft law and the Courts have not yet produced any decisions addressing these issues. The Gaming Commission's blacklist is prescribed by law and does not fall within the category of soft law.

Notice-and-Take-Down

Greek PD 131/3003 has not included the wording of art. 14 par. 3 of the ECD, which provides for notice and take-down procedures. Therefore, any such procedure in Greece requires either a change in legislation or may be introduced by contract or by a voluntary code of practice.

Scholars have argued that the introduction of a formal notice and take-down procedure, based on the similar procedure under the US Digital Millennium Copyright Act, may solve a number of problems, as has happened in several EU countries.[58] Such procedure may, to a certain degree, be extended to cover also other aspects of illegal or harmful content.

---

[56] C-509/09, *e*Date Advertising GmbH v X and C-161/10 Olivier Martinez and Robert Martinez v MGN Limited, Reports of Cases 2011 I–10269.

[57] C-292/10 *G v Cornelius de Visser.*

[58] See Yannopoulos 2013a, 298.

Procedural Aspects

As demonstrated above, apart from the Gambling Law, requiring blocking of black-listed sites and the not-yet-tested Orders of the Prosecutor for Child Pornography, in Greece there is no general law on blocking, filtering or taking down illegal content. Therefore, the only way to order an access provider to block/filter or a host provider to take-down/remove illegal content, is to obtain a court decision for injunctive relief.[59] The chairing Judge has the right *ex officio* to issue a 'Provisional Order',[60] prescribing the exact measures to be taken until the Injunction Decision. Normally, petitioners seek to obtain such an 'Order' because it can be issued within 1–2 days.

Article 17 of PD 131/2003 (corresponding to art. 18 of the ECD) allows for such remedy to be taken by means of an injunction that can be obtained if information society rights, as defined by article 1(a) of PD (art. 2(a) ECD) seem to be under threat of infringement. The Court may order any "adequate measure" and may even issue a provisional decision for immediate action against an Intermediary. Greek legislation, having in mind copyright and trademark infringements, has added a section in article 17 pf PD 131/2003, introducing the ability "to seize/confiscate the means for the illegal or harmful activity". In such cases, the Court must issue a Provisional Order and the case may proceed *in absentia* of the defendants. Nonetheless, such power of confiscation seems ineffective in the digital world. In cases where the Intermediaries do not comply with the court decision or provisional order, they face severe criminal sanctions (Greek Penal Code art. 232A).

As regards Domain Name abuse, the NTPC may examine complaints of 'abusive' or unlawful names and it may decide as a first instance *quasi* tribunal.[61] Following the NTPC's decision, the parties may appeal to the administrative courts.

Proportionality for Intermediaries

Host intermediaries have a duty of care to ensure that any restrictions regarding freedom of expression are reasonable and understandable by the users, with the content of the duty depending on the nature of each individual case. The Greek Courts have tried to establish a higher standard for service providers, because they are considered the Gatekeepers of modern era. This corresponds to the

---

[59] Articles 682 et seq. of the Greek Code of Civil Procedure (substantial changes to GCCP, to be inaugurated on 1.1.2016, have taken place in July 2015 by Law 4335/2015).

[60] Currently art 691 and as of 1.1.2016 new art. 691A of Greek Code of Civil Procedure as inserted by law 4335/2015.

[61] Article 10 par. 12 of NTPC Domain Name Regulation No.750/2/Official Gazette B-412/24-3-2015

attitude of the European Court of Human Rights in its recent *Delfi v Estonia* decision.[62]

As regards general injunctions for copyright infringement, trademark violations, defamation and distribution of other harmful content, the justification for any restriction may be found in article 10(2) of the ECHR, regarding the protection of "reputation or rights of others". As regards proportionality, the fact that the restriction is decided by a Court, even in an injunction case, means that the Court in each individual case will *ad hoc* weigh the two conflicting rights, in order to decide whether the freedom of expression of the Intermediary outweighs the right of the claimant and vice versa. The Courts in Greece have used that principle in order to establish, for example, whether personal data of users must be disclosed in case of infringements and whether blogs and bloggers fall under the rigid liability legislation for traditional editors.[63]

Concluding Remarks

Bearing in mind the problems of law enforcement on the Internet, the judiciary is best equipped to balance conflicting interests proportionately and to impose measures restricting fundamental freedoms. Otherwise, Internet Intermediaries would be charged with judicial duties: they would be obliged to decide the legality of content being transferred or hosted in their systems and to take action by blocking, filtering or taking down such content. Such a role is not appropriate and would create uncertainty as to what is the legitimate course of action.

Following review of Greek legislation and case-law, three suggestions can be made: (a) for activities understood to be illegal internationally, such as hacking, illegal intrusion to systems, data theft, fraud, child pornography and infringements of IP, as these crimes are defined in the Budapest Convention on Cybercrime, Intermediaries should be entitled to hinder any further spread of such activities, as soon as they are informed and subject to their mentality either in the case of 'intent with purpose' or 'negligence'; (b) for harmful content of international ethical demerit, such as hate speech, xenophobia, racism etc. Intermediaries must be allowed to act following notification, and (c) Immediate action for blocking/removal of content must be tolerated in case of threats to the integrity and security of networks.

However, we must realise that the role of intermediaries is changing from simple conduits to Gatekeepers of our modern Information Society. In a digital world, users

---

[62] ECHR Decision of 10.10.2013, *Delfi* AS v *Estonia,* appl. no. 64569/2009. Commented by Yannopoulos 2013c, p. 459.

[63] Attempts in Greece to introduce a specific law to control blogs and bloggers have failed. However, case-law varies when deciding whether a blog falls under Law 1178/1991 ("regarding press liability"), which imposes direct liability on the editors of traditional media. For case law regarding freedom of expression when republishing material over the Internet see ECHR Decision Board of pravoye delo and Shtekel v Ukraine, 5.5.2011, appl. no. 33014/2005.

must be convinced that they do not jeopardize something more than what they anticipate in a similar transaction in the analogue universe. In that sense, the demand for freedom to enjoy Information Society rights must be proportionally balanced with the demand for privacy, data protection and security. It is evident that oppressive enforcement of the rule of law over the Internet would be doomed to fail. To cover this vacuum, Internet Intermediaries are instrumental in finding an equilibrium and they remain the only reliable partners of the judiciary in order to enforce the law. Those who hold the keys of electronic transactions and who decide about actions such as blocking/filtering/taking-down, affect fundamental rights and, therefore, it is crucial that they develop a sense of responsibility. Such duty is not only a matter of statute or case-law, but rather a matter of attitude of the key Internet players who must seek, in the first place, to create confidence among users.

References

Fox, M. 2003. Controlling unlawful Internet gambling through the prohibition of bank instruments. *International Company and Commercial Law Review* 14 (5): 187.

Kalavrouzioti, D. 2011a. Comment on decision of Magistrate's Court of Kilkis 965/2010. *Journal for Media Law (DIMEE)* 2: 194. in Greek.

———. 2011b. Comment on decision of first instance court of Athens No. 4042/2010. *Journal for Media Law (DIMEE)* 2: 195. in Greek.

MacCarthy, M. 2010. What payment intermediaries are doing about online liability and why it matters. *Berkeley Technology Law Journal* 25: 1037.

Maniotis, D., M.-Th. Marinos, A. Anthimos, I. Iglezakis, G. Nouskalis. 2015. Chapter on Greece. J. Dumortier (Volume Editor), R. Blanpain (General Editor), M. Colucci (Associate General Editor), International encyclopaedia of laws for Cyber Law. Alphen aan den Rijn: Kluwer Law International BV.

Morse, E. 2007. The internet gambling conundrum: Extraterritorial impacts of US laws on internet businesses. *Computer Law and Security Report* 23: 529.

Schmoll, A. 2001. *Die deliktische haftung der Internet-Service-Provider, Peter Lang.* Frankfurt am Main: Peter Lang.

Steegmann, M. 2010. Die Haftung der Basisinfrastruktur bei rechtswidrigen Internetangeboten: Verantwortlichkeit von Internet- und Finanzdienstleistern im Rahmen des illegalen Online-Glücksspiels, Nomos Verlag.

Ufer, F. 2007. Die Haftung der Internet Provider nach dem Telemediengesetz, Recht der Neuen Medien, Hamburg.

Yannopoulos, G.N. 2013a. *The liability of internet intermediaries [in greek].* Athens: Nomiki Vivliothiki.

———. 2013b. The liability of search engines for suggest and autocomplete functions. Comment on FIC Athens Decision No. 11339/2012, Media Law (DIMEE) 2:168 [in Greek].

———. 2013c. Comment on ECHR decision Delfi v. Estonia, Media Law (DIMEE), 4, 2013, p.459 [in Greek].

———. 2013d. The liability of providers for advertisement of gambling sites. Is Real Madrid under threat? *Sinigoros* 96: 46 [in Greek].

———. 2014. Secondary liability of service providers, report for greece. *Revue Hellénique de Droit International* 1: 787.

Chapter 14
Internet Service Provider Copyright Infringement in Taiwan

Lung-Sheng Chen

Introduction

As the development of the Internet and digital technology continues, many of the largest economies across the globe are increasingly relying on this functional platform. Services provided through the Internet are vital for the integration of the global economy. Accordingly, the way that the law interacts with online activities profoundly impacts the way business is practised.

Internet legal issues arise in various contexts, such as pornography, white collar crime, consumer protection, free speech, privacy, and intellectual property (Zhang 2008). Among them, the question of whether or not an Internet Service Provider (hereinafter "ISP") should be responsible for its users' conduct has triggered fierce debate around the globe. While some countries adopt a horizontal approach to regulating the issue of ISP liability,[1] others provide different liability standards for various causes of action.[2] In 2006, the State Council of China passed the "Information Internet Distribution Protection Act," a statute that governs the liability of ISPs in a way that is similar to the US Digital Millennium Copyright Act of 1998 (hereinafter

---

[1] For instance, the European Union passed Directive 2000/31/EC of the European Parliament and of the Council on certain legal aspects of information society services, in particular electronic commerce, in the Internal Market in 2000. The Directive applies to all types of illegal activity and provides safe harbours for mere conduit, caching, and hosting service providers under specific circumstances listed in articles 12 through 15.

[2] Such as the United States, Australia, and Singapore. In the United States, the statutes that apply to internet service providers include "the Digital Millennium Copyright Act" and "Internet Gambling Prohibition Act of 2000."

L.-S. Chen (✉)
Department of Law, National Chung Hsing University,
No. 145 Hsing Da Rd., South Dist., Taichung City, Taiwan
e-mail: cls0910@gmail.com; lawchen@nchu.edu.tw

© Springer International Publishing AG 2017   339
G.B. Dinwoodie (ed.), *Secondary Liability of Internet Service Providers*,
Ius Comparatum – Global Studies in Comparative Law 25,
DOI 10.1007/978-3-319-55030-5_14

"DMCA").[3] In Taiwan, no statute has been enacted providing a general legal basis regulating the liability of ISPs for the conduct of their users. Taiwan's Copyright Act, however, was amended in 2009 to include provisions that regulate ISPs' liability for their users' conduct that infringes the copyright of others.[4] This amendment included a new chapter in the Copyright Act entitled "Internet Service Provider Liability Limitation" (hereinafter "the 2009 legislation" (Chapter VI-1 of Copyright Act)). The new legislation came into effect in May 2009. Under the new legislation, ISPs are free from copyright liability if they can establish that they complied with all the requirements set forth by the new legislation.[5] It is noteworthy, however, that the provisions set forth in that Chapter apply merely to copyright liability, not to any other liabilities, such as trademark infringements, privacy, and defamation.

The increase in online piracy has drawn the attention of many industries (Mag Chang 2012). According to research by the International Intellectual Property Alliance, every year the music industry, for instance, suffers a loss of at least US $ 130 million because of online piracy (International Intellectual Property Alliance 2013). Copyright groups have urged governmental authorities to take effective measures to prevent ISP users from illegally downloading, copying, or sharing copyrighted works. At the same time, however, many have objected to a legal regime that overly protects the rights of copyright holders. The 2009 legislation reflects the Taiwan government's attempt to reconcile and balance competing interests of copyright holders, users, and ISPs.

Several years have passed since Taiwan's enactment of the 2009 legislation. This legislation was the Taiwan Legislation Yuan's first attempt to update copyright law. The Act establishes four safe harbours, which limit the copyright liability of ISPs for a third party's infringing use of its services.[6] At the same time, the Act also provides a means by which content owners can request ISPs to remove the infringing materials from their web pages.[7] Many people believe that the legislation provides a creative solution to the issue of ISPs' responsibility for infringing activities that occur on their platforms. Nonetheless, like the DMCA, the 2009 legislation's counterpart in the United States, it is far beyond the legislation's ability to address all issues in relation to online copyright infringements. Disputes and lawsuits are on their way. This is especially true after considering the post-DMCA cases in the U.S., where copyright owners have pursued claims against ISPs.[8]

---

[3] Digital Millennium Copyright Act of 1998, 17 U.S.C. § 512 (2000).

[4] The English version of Copyright Act of Taiwan is provided by the Taiwan Intellectual Property Office, Ministry of Economic Affairs, Taiwan, *available at* http://www.tipo.gov.tw/dl.asp?fileName=332914394849.pdf

[5] *See* Copyright Act of Taiwan, arts. 90*quinquies* through 90*terdecies* (2010).

[6] See Copyright Act of Taiwan, art. 90*quinquies*.

[7] See Copyright Act of Taiwan, arts. 90*septies*, 90*octies*, and 90*novies*.

[8] See, generally, *Metro-Goldwyn-Mayer Studios Inc.v.Grokster, Ltd.*, 545 U.S. 913 (2005)[hereinafter *Grokster*]; *In re Aimster Copyright Litig.*, 334 F.3d 643 (7th Cir. 2003); *A&M Records, Inc. v. Napster, Inc.*, 239 F.3d 1004 (9th Cir. 2001); *IO Group, Inc. v. Veoh Networks, Inc.*, 586 F. Supp. 2d 1132 (N.D. Cal. 2008).

In order to give authors an incentive to engage in innovative creation, Taiwan's copyright law gives authors certain exclusive rights with respect to their works. Nevertheless, in the digital era, with the establishment of safe harbour provisions under the 2009 legislation, ISPs are immune from liability in certain circumstances. Consequently, the safe harbour provisions, which serve as an important legal means to balance and to protect ISPs from copyright infringement liability, might become so expansive that it jeopardises the goal of the Copyright Act—to promote scientific development and artistic creation. This balanced result, in the language of economics, promotes optimal production, and thus allocates efficiency, of copyrighted works in cyberspace.

This chapter proceeds in five parts. The introductory section, "Secondary liability theories" is a brief introduction to the secondary liability theories rooted in U.S. law. In the section entitled "Legal standards for secondary liability in Taiwan", I focus on the legislative basis for secondary liability in Taiwan. In the section "ISP copyright infringement liability and safe harbour in Taiwan", I address ISPs' copyright infringement liability and safe harbour provisions under Taiwan's Copyright Act. And in section "ISP copyright infringement liability in practice", I consider two cases relating to ISPs' copyright infringement liability in Taiwan and examine the specific problem of the safe harbour's scope.

Secondary Liability Theories

U.S. law, and in particular the DMCA and secondary liability theories in US tort law, had a great impact on Taiwan's ISP Liability Limitation legislation. As such, to better understand this legislation, it will be helpful to take a brief look at US law on these issues.

In the US, Section 106 of the Copyright Act of 1976 grants numerous exclusive rights to authors. Liability for infringement of those rights may be founded on theories of direct infringement or secondary liability.[9] Secondary liability doctrines are not expressly articulated in the Copyright law of 1976. Nonetheless, two types of secondary liability are generally recognized by common law doctrines: contributory liability and vicarious liability.[10]

---

[9] *Ellison v. Robertson,* 357 F.3d 1072, 1076 (9th Cir. 2004) (holding that "[w]e recognize three doctrines of copyright liability: direct copyright infringement, contributory copyright infringement, and vicarious copyright infringement").

[10] *Id.* See, generally, Gorman and Ginsburg 2001, 782–806.

Contributory Liability

Contributory liability arises when a person, "with knowledge of the infringement activity, induces, causes, or materially contributes to the infringing conduct of another."[11] A finding of knowledge of the direct infringement is necessary to establish contributory infringement liability.[12] The knowledge requirement is met "where a party has been notified of specific infringing uses of its technology and fails to act to prevent such infringing uses, or willfully blinds itself to such infringing uses."[13] The theory of contributory liability has significantly impacted technologies which can be used for infringing and non-infringing purpose. The Supreme Court's 1984 decision in *Sony Corp. of America v. Universal City Studios*[14] is one of the most important cases addressing issues of secondary liability.

The *Sony Corp.* case illustrates the economics of secondary liability. The Court analysed whether Sony was liable for contributing to the infringement of consumers who purchased VCRs manufactured and sold by Sony and used them to tape television broadcasts. Holding Sony liable would have profound economic consequences. First, if the copyright holder, Universal, has a viable claim against Sony, it is much easier for Universal to enforce its copyrights. Undoubtedly, it is very difficult for a copyright holder to locate each individual purchaser. Even if each purchaser is able to compensate for the harm he causes, the costs of identifying each purchaser, gathering evidence, and initiating separate lawsuits would likely make it uneconomical for Universal to enforce its copyright. The theory of secondary liability, by contrast, is a promising, and cheaper, avenue for copyright holders to enforce their rights.

Second, as the Supreme Court held in *Sony Corp.*, a contributory liability claim is unlikely to prevail if the infringing products were found to be capable of "substantial noninfringing uses." By borrowing patent law's "staple article of commerce" doctrine, the Supreme Court declined to impose secondary liability on Sony by concluding, "[T]he sale of copying equipment, like the sale of other articles of commerce, does not constitute contributory infringement if the product is widely used for legitimate, unobjectionable purposes."[15] Thus, under the "substantial noninfringing uses" test, a manufacturer like Sony is not liable for purchasers' infringement of copyrighted works. Conversely, if a product is distributed with the principal, if not exclusive, objective of promoting its use to infringe copyright, the distributors

---

[11] *Gershwin Publ'g Corp. v. Columbia Artists Mgmt., Inc.*, 443 F.2d 1159, 1162 (2d Cir. 1971) [hereinafter *Gershwin*]. Contributory liability originates in tort law and stems from the notion that one who directly contributes to another's infringement should be held accountable. See *Fonovisa, Inc. v. Cherry Auction, Inc.*, 76 F.3d 259, 264 (9th Cir. 1996).

[12] *Gershwin* id 1162.

[13] *Monotype Imaging, Inc. v. Bitstream, Inc.*, 376 F. Supp. 2d 877, 886 (N.D. Ill. 2005); *Newborn v. Yahoo!, Inc.*, 391 F. Supp. 2d 181, 186 (D.D.C. 2005)

[14] *Sony Corp. of America v. Universal City Studios*, 464 U.S. 417 (1984)[hereinafter *Sony Corp.*].

[15] *Id.* at 442.

of the product could be held liable for contributory infringement, regardless of the product's lawful uses.[16]

The U.S. Supreme Court in *MGM Studios Inc. v. Grokster, Ltd.* established an alternative basis on which copyright holders could found a secondary liability claim—the inducement theory. In *Grokster*, the Court explained that the inducement theory was built not to penalise ordinary commerce in the form of legitimate product distribution, but to impose liability on "purposeful, culpable expression and conduct, and thus does nothing to compromise legitimate commerce or discourage innovation having a lawful promise."[17] This theory will have an impact on ISPs though they have little control over the conduct of end user. In order to avoid contributory liability under the inducement theory, ISPs might have to reconsider their practices and policies or restructure their business models (Hancock 2006). As such, the costs of ISPs in altering their business operation will significantly increase (*id*, 212).

Vicarious Liability

Vicarious liability is derived from the principle of *respondeat superior*.[18] This theory applies when (1) a direct infringement caused by a third party and this third party's infringing conduct is subject to the defendant's supervision and control and (2) the defendant acquires a direct financial benefit from the infringement.[19]

One reason for imposing liability in this instance is to encourage employers to exercise due care in hiring, supervising, controlling, and monitoring their employees so as to avoid copyright infringement. Another reason is that, like contributory liability, the cost of enforcing rights will be substantially lower if copyright holders are able to go after the employers of actual infringers (Litchman and Landes 2003). In cases of this kind, a court will usually consider how much the presenting case looks like a landlord-tenant case, in which landlords are not liable if they had no knowledge of their tenants' infringing acts and exercised no control over leased premises, or a "dance hall" case, in which dance hall operators who had control over the infringing performances and directly received a financial benefit from the performances are held liable for infringing performances on premises.[20]

---

[16] *Grokster*, at 933–35.

[17] *Id.* 936–37.

[18] *A&M Records, Inc. v. Napster, Inc.*, 239 F.3d 1004, 1022 (9th Cir. 2001). The principle of *respondeat superior* holds an employer responsible for its employees' actions.

[19] *Gordon v. Nextel Communs.*, 345 F.3d 922, 925 (6th Cir. 2003).

[20] *Fonovisa, Inc. v. Cherry Auction, Inc.*, 76 F.3d 259, 264 (9th Cir. 1996). The example of a dance hall operator shows that secondary liability is attractive under these circumstances. While performers hired by the venue operators are often without sufficient resources needed for compensating the associated harm, vicarious liability in this instance prevents the externalisation of copyright harm. From an economic perspective, allowing a copyright holder to sue the venue operator rather than suing each performer individually is on the one hand likely to reduce enforcement costs, such as

Legal Standards for Secondary Liability in Taiwan

Taiwanese law does not expressly recognise concepts of contributory or vicarious liability (Tsai 2005; Sun 2007). Nonetheless, this does not necessarily mean that an ISP will never be held liable for its users' conduct. And ISPs can be held liable for indirect infringement under the Civil Code and Copyright Act.

General Standard Under Taiwan Civil Code

Under paragraph 1, Article 185 of Taiwan's Civil Code, "[i]f several persons jointly conducted a tort to another person, they are jointly liable for the damage arising therefrom.[21] The provision applies even if the person harmed cannot indicate who in fact committed the tort."[22] Paragraph 2 of the same Article provides that "[i]nstigators and accomplices are deemed to be joint tortfeasors."[23] For an ISP to be held as an accomplice, the plaintiff must prove: (1) a tort conducted by a third party; (2) the defendant has intent to aid; and (3) the defendant provides aid to the third party. According to one commentator, the liabilities under Article 185(2) might be close to the concept of contributory infringement (Tsai 2005, 78).

Moreover, paragraph 1, Article 188 of the Civil Code holds an employer liability for his/her employees' conduct. The concept of employer liability under the Civil Code is similar to the concept of vicarious infringement (*id*, 78–9). Article 188 (1) of Taiwan's Civil Code provides, "[t]he employer shall be jointly liable to make compensation for any injury which the employee has wrongfully caused to the rights of another in the performance of his duties. However, the employer is not liable for the injury if he has exercised reasonable care in the selection of the employee, and in the supervision of the performance of his duties, or if the injury would have been occasioned notwithstanding the exercise of such reasonable

---

identifying each infringer, gathering evidence and litigating in several lawsuits. On the other hand, the costs for controlling the acts of performers of the venue operators are low because the operator is probably already monitoring the dance hall quite carefully to ensure that patrons are being well treated. In order to avoid vicarious liability, entities like concert halls, stadiums, radio stations, restaurants look for an inexpensive way to acquire performance rights by purchasing blanket licenses from performing rights societies. This saves enormous transaction costs by excluding the need for licenses from each individual copyright holder on the one hand, and by excluding performers from notifying copyright holders in advance every time they intend to perform the copyright work on the other. Additionally, the marginal use problem will be solved because "each licensee will act as if the cost of an additional performance is zero—which is, in fact, the social cost for music already created." See Lichtman and Landes 2003, id, at 399.

[21] Civil Code of Taiwan, article 185 paragraph 1(2012). The English version of the Code provided by Ministry of Justice can be found at "Laws & Regulations Database of the Republic of China" website, *available at* http://law.moj.gov.tw/Eng/LawClass/LawAll.aspx?PCode=B0000001

[22] Civil Code of Taiwan, article 185 paragraph 2.

[23] Civil Code of Taiwan, article 185.

care."[24] Therefore, if an ISP falls within the category of an employer, it is likely to be held liable for the conduct of its user (employee) if it is proved that: (1) there is a tort conducted by an user (employee); (2) the conduct of the user (employee) is part of the work for the usage (employment); and (3) the ISP should have supervised the conduct of the user (employee).

Secondary Liability Standard Under Taiwan Copyright Act

In addition to the Civil Code, Taiwan's Copyright Act provides a cause of action for indirect copyright infringement. Under subparagraph 7, paragraph 1, Article 87 of the Copyright Act, except as otherwise provided under the Act, it shall be deemed a copyright infringement when "[p]rovid[ing] to the public computer programs or other technology that can be used to publicly transmit or reproduce works, with the intent to allow the public to infringe on others' economic rights by means of public transmission or reproduction by means of the Internet of the copyrighted works of another, without the consent of or a license from the economic right owners, and to receive benefits therefrom." This provision was added to the Copyright Act in 2007 after having regard to the U.S. *Metro-Goldwyn-Mayer Studios, Inc. v. Grokster, Ltd.* decision.[25] The theory underlying this Amendment was that a person, who distributes the means with the purpose of promoting its use to infringe copyrighted works, as evidenced by their clear expression or other affirmative steps taken to foster the infringing actions, should be held liable for the resulting infringing acts by third parties. There is a difference, however, between the U.S. and Taiwan positions when it comes to the requirements for establishing secondary liability.

The *Grokster* Court held that "one who distributes a device with the object of promoting its use to infringe copyright, as shown by clear expression or other affirmative steps taken to foster infringement, is liable for the resulting acts of infringement by third parties."[26] Under this rule, in the case of online copyright infringement, to prevail in a claim for inducement, the plaintiff must prove that: (1) the ISP intends to enable infringement; (2) the ISP provides a device (website) suitable for infringing use; and (3) an actual infringement by recipients of the device (Ballon 2009). As the Court noted, secondary liability will not be found "on the basis of presuming or imputing fault, but from inferring a patently illegal objective from statements and actions showing what that objective was."[27]

Unlike the U.S., in Taiwan an ISP is liable for its users' copyright infringements under the aforementioned provision of the Copyright Act if the following four requirements are satisfied. First, the ISP provides its users with computer programs or other technologies that can be used to publicly transmit or reproduce copyrighted

---

[24] Civil Code of Taiwan, article 188.
[25] 545 U.S. 913 (2005).
[26] *Id.*, 918.
[27] *Grokster*, 941.

works. Second, the ISP has the intent to allow its users to infringe on the copyrights of another. Third, the ISP does not obtain the consent or a license from the copyright owner. Lastly, the ISP receives benefits from the infringement. This provision was codified in 2007 by referring to the *Grokster* decision. Pursuant to the legislative documents, the purpose of this provision was to impose liability on means providers and it is the "providing act" itself that was regulated.[28] Therefore, proof of a resulting copyright infringement by a third party is not a necessary requirement to establish inducement liability. Instead, an individual's conduct of providing means that help others to infringe copyrighted works is deemed infringing conduct under this provision. In addition, under paragraph 2, Article 87 of the Copyright Act, the second requirement (an intent allowing its users to infringe) is satisfied if the ISP instigates, solicits, incites, or persuades its users to make use of the computer program or other technology the ISP provided for the purpose of infringing copyrights of others by advertising or other active measures.

Comparison

A comparison of the general standard for establishing secondary liability under the Civil Code and that under the Copyright Act reveals differences between them. For an ISP to be held liable for another's conduct under the Civil Code, there must be a direct infringement. However, under the Copyright Act, an ISP is liable when the aforementioned four elements are satisfied, regardless of whether a direct infringement is found. In other words, no direct infringement of others is required to establish secondary liability of an ISP under the Copyright Act.

Additionally, both direct and secondary liability involve the requirement of "intent". Generally, the elements of the direct infringement are: (1) direct intent (*mens rea* element); and (2) infringing conduct (*actus reus* element) (Tsai 2005, 85). The establishment of secondary liability also requires a proof of "an intent to aid or induce." To prove the defendant's intent in a lawsuit has never been easy. When deciding whether an ISP has the "intent" of infringement, courts take the role and function of technologies into account.[29] For instance, in the case of peer-to-peer file sharing, courts may look into whether the structure of the ISP is "server-centered or decentralized" to determine the extent to which the ISP is involved in the reproduction and transmission of copyrighted works when proving the ISP's intent of infringement (*id*). If the service/technology provided by ISPs is not designed for an infringing purpose, it is unlikely to prove there is intent.

When an ISP is found secondarily liable for the conduct of others under Article 185 or paragraphs 1, Article 188 of the Civil Code, the ISP has to compensate the

---

[28] See Legislative Yuan of Taiwan, *Copyright Act Amendment Documents*, *available at* http://lis. ly.gov.tw/lgcgi/lgmeetimage?cfc9cfcdcec6cfcec5cec8cad2cec8c6
[29] See infra "ISP Copyright Infringement Liability in Practice" (discussing the ezPeer case and KURO case).

victim for any damage arising from the direct infringement.[30] The ISP's responsibility for this compensation is not different from that available against the third party who committed the infringing conduct and is thus primarily liable. On the other hand, under Taiwan's Copyright Act, both primary infringers and those with the secondary liability are subject to civil liability and criminal penalties.[31] The copyright owner may request the court to remove the infringing material and to prevent the infringing parties from continuing to infringe.[32] Moreover, the direct infringer and the indirect infringer are both liable for damages arising from the infringement.[33] If the copyright owner's moral rights are infringed, the injured parties may also "claim a commensurate amount of compensation" as well as appropriate measures necessary for the restoration of their reputation.[34] For instance, the copyright owner may request the court to order the infringing party to make a public statement that indicates "the name of the authors or appellation, correction of content, or adoption other appropriate measures necessary for the restoration of its reputation."[35] If the copyright owner's economic rights are injured, the owner may claim damages calculated based on the standards provided in paragraph 2 and 3, Article 88.[36] In addition, the copyright owner may also "request the destruction or other necessary disposition of goods produced as a result of the infringing act, or of articles used predominantly for the commission of infringing acts."[37] When the court delivers its

---

[30] See also article 184 of the same statute (providing that "[a] person who, intentionally or negligently, has wrongfully damaged the rights of another is bound to compensate him for any injury arising therefrom. The same rule shall be applied when the injury is done intentionally in a manner against the rules of morals.")

[31] Copyright Act of Taiwan, articles 84–103.

[32] Copyright Act of Taiwan, article 84 (providing that "[t]he copyright holder or the plate rights holder may demand removal of infringement of its rights. Where there is likelihood of infringement, a demand may be made to prevent such infringement.")

[33] See Copyright Act of Taiwan, article 85, paragraph 1 and article 88.

[34] Copyright Act of Taiwan, article 85.

[35] Copyright Act of Taiwan, article 85, paragraph 2.

[36] Copyright Act of Taiwan, article 88, paragraph 2 (providing that:With regard to the damages referred to in the preceding paragraph, the injured party may make claim in any of the following manners:

1. In accordance with the provisions of Article 216 of the Civil Code; provided, when the injured party is unable to prove damages, it may base the damages on the difference between the amount of expected benefit from the exercise of such rights under normal circumstances and the amount of benefit from the exercise of the same rights after the infringement.

2. Based on the amount of benefit obtained by the infringer on account of the infringing activity; provided, where the infringer is unable to establish costs or necessary expenses [of the infringing act or articles], the total revenue derived from the infringement shall be deemed to be its benefit.

If it is difficult for the injured party to prove actual damages in accordance with the provisions of the preceding paragraph, it may request that the court, based on the seriousness of the matter, set compensation at an amount of not less than ten thousand and not more than one million New Taiwan Dollars. If the damaging activity was intentional and the matter serious, the compensation may be increased to five million New Taiwan Dollars.)

[37] Copyright Act of Taiwan, article 88*bis*.

decision, the plaintiff "may request the court to order the infringing parties to, at its own expense, make a public statement in the media all or part of the court's judgment concerning said infringement."[38]

ISP Copyright Infringement Liability and Safe Harbour in Taiwan

The Copyright Act of Taiwan provides a so-called "safe harbor" for ISPs. The safe harbour provisions were incorporated into the Copyright Act in 2009 and based upon the DMCA. It attempts to immunise from civil liability connection service providers, caching service providers, information storage service providers, and search service providers.[39]

Applicable Service Provider

The 2009 legislation was enacted specifically to address the issue of ISP secondary liability. The legislation provides a definition of ISPs. A party may claim immunity by proving that they fall under the definition of an ISP. Under subparagraph 19, paragraph 1, Article 3 of Taiwan's Copyright Act, the term "Internet service providers" means those who provide any of the following services:

(1) Connection service provider: those who provide services, by wire or wireless means, of transmitting, routing, or receiving, information through a system or network controlled or operated by the service provider, or of the intermediate and transient storage of information in the course of such transmitting, routing, or receiving.
(2) Caching service provider: those who, after information has been transmitted at the request of a user, provide services of intermediate and temporary storage of the information through a system or network controlled or operated by the service provider, for purposes of providing accelerated access to the information by users who subsequently request transmission of the information.
(3) Information storage service provider: those who provide information storage services at the request of a user through a system or network controlled or operated by the service provider.

---

[38] Copyright Act of Taiwan, article 89.

[39] In Taiwan, copyright infringement can give rise not only to compensation of damages but also to criminal liability. Articles 84-90*quarter* of Copyright Act of Taiwan provide for civil liability, and articles 91–103 provides for criminal liability.

(4) Search service provider: those who provide users with services, including an index, reference, or hyperlink, to search or hyperlink to online information.[40]

According to this definition, therefore, ISPs can be divided into four categories: (1) Connection Service Providers, such as Chunghwa Telecom Co., Ltd.,[41] or AT&T which are mere conduits; (2) Caching Service Providers, such as Facebook; (3) Information Storage Service Providers, such as Gmail and Blogs; and (4) Search Service Providers, such as Google, Yahoo!, and Bing. Operators of online markets provide information storage services upon the request of a user. ISPs could qualify in two or more categories, depending upon the services provided. For instance, while Yahoo! falls within the fourth category of ISPs in light of its search service, its online auction service, like eBay that provides users with a platform for online transactions, falls within the third category of ISPs as provided in the Copyright Act.

Requirements of Safe Harbour

Under the 2009 legislation, if ISPs comply with both the general and additional requirements set out in the statute, they will be exempted from secondary liability for copyright infringement conducted by a third party.

In order for ISPs to be eligible for safe harbour immunity, they must implement all four general requirements as provided for in Article 90 *quinquies*. These general requirements include: (1) adopting a set of copyright protection measures and policy and informing users of such measure and policy; (2) adopting a "graduated response system" (also called "three-strikes system") that in the event of repeated (three) alleged infringements the ISP shall terminate the service in whole or in part, and inform its users of such system; (3) publicly announcing information regarding its contact window[42] for receipt of notification documents; and (4) accommodating and implementing technical measures to identify or protect copyrighted works. It should be noted that ISPs do not have the obligation to monitor content on their webpages.

Some of these general requirements impose on ISPs the obligation to inform users of certain information. ISPs can do so in many forms. For instance, ISPs may

---

[40] Copyright Act of Taiwan, article 3, paragraph 1, subparagraph 19.

[41] Chunghwa Telecom Co., Ltd. is a Taiwan Company which provides services of internet network, GPRS, 3G and mobile Virtual Private Network (MVPN). See Chunghwa Telecom, *About CHT*, *available at* http://www.cht.com.tw/en/aboutus/aboutcht.html

[42] The "contact window information" must include: (1) the name of the individual or institution, address, contact telephone, fax number and electronic mail address of the contact window, and (2) the format of electronic signatures accepted, or the information on willing to accept the notification document without electronic signature. See Regulations Governing Implementation of ISP Civil Liability Exemption, article 2 (2009). These Regulations are adopted pursuant to Article 90terdecies of the Copyright Act by TIPO.

put the relevant information in the contract with its users, or in an electronic transmission, automatic detective system, or other means. If a connection service provider forwards to its users a notification by a copyright owner regarding infringement by a user immediately after the service provider receives the notification, the service provider is deemed to have informed users of the relevant measures and policies mandated by the first requirement.[43]

In addition to these general requirements, the 2009 legislation sets forth additional requirements that ISPs must satisfy in order to acquire immunity. The legislation imposes different additional requirements upon different types of ISPs.

For connection service providers to be immune from secondary liability, the transmission of the information must be initiated by or at the request of the user, and must be carried out through an automatic technical process, without any selection of the material or modification of its content at the end of connection service providers.[44] With regard to caching service providers, the cached information must not be modified by service providers. When the person who made the original information available subsequently updates, deletes, or blocks access to it, the cached information is altered in the same way as a result of an automatic technical process. In addition, caching service providers must respond expeditiously to remove, or disable access to, the allegedly infringing content or related information upon notification by a copyright holder of the alleged infringement by the user of the service provider.[45]

Additionally, information storage service providers must not have knowledge of the allegedly infringing activity of the user, and must not receive a financial benefit directly attributable to the infringing activity of the user. Information storage service providers must also respond expeditiously to remove, or disable access to, the allegedly infringing content or related information upon notification by a copyright holder of the alleged infringement by the user of the service provider.[46]

Finally, search service providers must not have knowledge that the searched or linked information may be infringing others' copyrights, and must not receive a financial benefit directly attributable to the infringing activity of the user. The ISPs must respond expeditiously to remove, or disable access to, the allegedly infringing content or related information upon notification by a copyright holder or plate rights holder of the alleged infringement by the user of the service provider.[47]

---

[43] Copyright Act of Taiwan, article 90*quinquies*, paragraph 2.

[44] Copyright Act of Taiwan, article 90*sexies*.

[45] Copyright Act of Taiwan, article 90*septies*.

[46] Copyright Act of Taiwan, article 90*octies*.

[47] Copyright Act of Taiwan, article 90*novies*.

Notice and Takedown Regime

The 2009 legislation has two components. In addition to providing a "safe harbor" for an ISP, it provides a copyright owner a mechanism to give an ISP a notice claiming that the ISP is hosting infringing material.[48] Upon receipt of such a notice, the ISP has a duty to respond expeditiously to remove, or disable access to, the allegedly infringing content or related information. This is known as the "notice and takedown" procedure, which was adopted by the DMCA.[49] In particular, an information storage service provider must notify the owner of the allegedly infringing content that the content has been removed or rendered inaccessible.[50]

The 2009 legislation (Articles 90*septies* through Article 90*novies*) requires the notice to be made by the copyright holder. More importantly, the copyright holders may not make such a notice anonymously. They should state their name in the notice. In addition, such notice shall contain information including: (1) the name, address, and telephone number or fax number or electronic mail address or description of other automatic communication of the rights holder or agent thereof; (2) the name of the copyrights infringed; (3) a statement requesting the removal of, or disabling of access to, the content that allegedly infringes copyrights; (4) access or relevant information sufficient to enable the ISP to identify the allegedly infringing content; (5) a statement that the rights holder or the agent thereof is acting in good faith and in the belief that the allegedly infringing content lacks lawful licensing or is otherwise in violation of the Copyright Act; and (6) a declaration that the rights holder is willing to bear legal liability in the event there is misrepresentation with resultant injury to another.[51]

Counter Notification Regime

Additionally, the 2009 legislation provides the accused infringer with an opportunity to "counter notify" the ISP if he believes that the materials are not infringing.[52] Upon receipt of the accused infringer's "counter notification," the information storage service provider "shall expeditiously forward such documents to the copyright holder."[53] A "counter notification" shall be signed or sealed by the user or agent

---

[48] See Copyright Act of Taiwan, articles. 90*quinquies*, 90*septies*, 90*octies* and 90*novies*.

[49] See 17 U.S.C. § 512 (c).

[50] Copyright Act of Taiwan, article 90*decies*, paragraph 1 (providing that "[a]n information storage service provider shall forward notice to the allegedly infringing user of any measures taken under Article 90octies, subparagraph 3, by the contact method stipulated between the service provider and the user or by the contact information left by the user. However, this requirement shall not apply if the nature of the service provided makes such notice impossible.")

[51] Regulations Governing Implementation of ISP Civil Liability Exemption, article 3.

[52] Copyright Act of Taiwan, article 90*decies*, paragraphs. 2 and 3.

[53] Copyright Act of Taiwan, article 90*decies*, paragraph 3.

thereof and contain information including: (1) the name, address, and telephone number or electronic mail address of the user or agent thereof; (2) a statement of the request to replace the content that has been removed or to restore access to the content; (3) relevant information sufficient to enable the ISP to identify the content; (4) a statement that the user is acting in good faith and in the belief that the user has a lawful right to exploit the content, and that the removal or disabling of access to the content is the result of a misrepresentation or error on the part of the rights holders or agent thereof; (5) a statement giving consent for the information storage service provider to forward the contents of the counter notification and the user's personal information to the rights holder or agent thereof; and (6) a declaration that the user is willing to bear legal liability in the event there is misrepresentation with resultant injury to another.[54]

The information storage service provider does not have the obligation to restore the content or related information if, within 10 business days of receiving counter notification, the copyright holder provides the ISP with proof that a lawsuit has been duly filed by it.[55] However, if the copyright holder fails to do so, the information storage service provider "shall, within no more than 14 business days since one day after the date of forwarding the counter-notification documents, restore the removed content or related information or restore the access to it."[56] If restoration is impossible, the service provider shall notify the user in advance, or provide another appropriate method by which the user may restore it.[57] If a person makes a misrepresentation to an ISP, by means of a notification or counter notification, either intentionally or negligently, he shall be liable for damages for any injury incurred by the user, copyright holder, plate right holder or ISP.[58]

Consequently, an ISP is not liable for damages to the allegedly infringing user if the ISP: (1) removes, or disables access to, the allegedly infringing content or related information in accordance with Articles 90*septies* to 90*novies*; or (2) upon obtaining knowledge of suspected infringement by the user, in good faith removes, or disables access to, the allegedly infringing content or related information.[59] The damages awarded against service providers are used to compensate the copyright holder, and there is no legal basis for a court to grant remedies to help restrain wrongful conduct by others. Some commentators argue that the safe harbor provisions seem to impose a newly created duty of care on ISPs rather than limit ISPs' copyright infringement liability (Lee 2014, 159; Wang 2009, 37–39). Before the 2009 legislation, an ISP does not necessarily need the safe harbor provision to exempt itself from liability. Nevertheless, after the 2009 legislation, the pressure of complying with the requirements of safe harbor on an ISP and the increasing cost of compliance cause an ISP take the safe harbor provisions as a legal duty (*id*).

---

[54] Regulations Governing Implementation of ISP Civil Liability Exemption, article 5.

[55] Copyright Act of Taiwan, article 90*decies*, paragraph 4.

[56] Copyright Act of Taiwan, article 90*decies*, paragraph 5.

[57] *Id.*

[58] Copyright Act of Taiwan, article 90*duodecies*.

[59] Copyright Act of Taiwan, article 90*undecies*.

Consequently, if an ISP does not satisfy the requirements of immunity, it is likely to be held liable for the damages suffered by the copyright holder. This contradicts the purpose of the safe harbor provisions, which are designed to provide protection for ISPs rather than copyright holders.[60]

ISP Copyright Infringement Liability in Practice

In this part, I consider two famous cases dealing with secondary liability of ISPs. To be clear, these two cases occurred prior to the 2009 legislation. As such, the courts in these two cases had no basis to consider the new legislation. Nonetheless, the cases are still good examples of how Taiwanese courts consider secondary liability for ISPs.

The ezPeer and KURO Cases

In most cases, it is difficult and costly for a copyright holder to detect individual wrongdoers (Hamdani 2002). This is because many online users tend to remain anonymous. Copyright holders' efforts to go after individual online wrongdoers for the purpose of deterring online infringements are unlikely to succeed (Wan 2011, 384). But making a direct liability claim against individual wrongdoers has never been cost-effective (*id* and Becker 1968, 179–80). At the same time, in addition to having to establish the alleged direct liabilities, the copyright holder needs to produce evidence showing the damages resulting from the infringements. Expanding liability to ISPs thus becomes a solution to online piracy cases.

In 2003, two operators of different websites that helped its member users to share and to download digital music files were indicted and accused of violating the Copyright Act. One of the websites was KURO and the other was ezPeer. Both websites provided peer to peer (P2P) software that enabled its users to engage in online file sharing (Tsai 2005, 61–4). The operator of the KURO website and one of its members were indicted by the Prosecutor's Office, Taipei District Court,[61] while the operator of ezPeer website and four ezPeer members were indicted by the Prosecutor's Office, Shi Lin District Court.[62] The defendants were accused of direct

---

[60] *Online Policy Group v. Diebold, Inc.*, 337 F. Supp. 2d 1195, 1204–05 (N.D. Cal. 2004); see Lee 2014, 160.

[61] The Iindictment of the Prosecutor's Office of the Taipei District Court, Taiwan, Case Number: 92 Jen Tzyh No.16389/21865 (Dec. 1, 2003).

[62] The indictment of the Prosecutor's Office of the Shih Lin District Court, Taiwan, Case Number: 91 Jen Tzyh No.10786, 92 Jen Tzyh No.4559 (Dec. 4, 2003).

copyright infringement in violation paragraph 1 of Article 91, paragraph 1 of Article 92 and Article 94 of Copyright Act.[63]

The Taipei District Court found the operator of KURO guilty. The court in its reasoning explained that the service provider had actual knowledge about its users' infringing conduct but the operator did nothing to stop such infringing activities. Relying upon the conspiracy theory, the court held the operator (service provider) guilty because the court found the operator was committing the criminal offense jointly with its users. According to the court, whether or not the operator was actually engaging in the reproduction or transmission of copyrighted works made no differences to this finding.[64] On the other hand, the Shihlin District Court found the operator of ezPeer not guilty.[65] The court in the ezPeer case held that the service provider did not engage in the commission of the infringing activities and there was no evidence showing that the service provider had knowledge about the infringement conducted by its users.[66] Therefore, the operator of ezPeer did not commit any direct copyright infringement offense.[67] Nevertheless, the Supreme Court remanded the case and the appellate court (the Intellectual Property Court) reversed the district court's decision and held the operator of ezPeer guilty.[68]

After the Copyright Act codified the secondary liability of service provider and the safe harbor provisions in 2009, courts have taken the concept of media neutrality[69] into consideration when dealing with cases related to service providers. The Intellectual Property Court held in the ezPeer case that technology advances distribution of information and data with features of "fidelity," "facility," and "ubiquity."[70] The P2P technology helps the public easily access more copyrighted works, and thus stimulates the creation of works.[71] Requiring the intermediary to monitor all the contents transmitted by its user would be impossible for ISPs and would highly increase transaction costs (Lee 2014, 155). Therefore, ISPs may raise media neutrality as a defense if the services/technologies remains neutral (Tasi 2005, 85). The more the technology is neutral, the more possibility that the service providers will be deemed as mere conduits and thus is not liable for the conduct of others. The neutrality of the technology provides a clue in proving the intent of ISPs.

---

[63] *Id.*

[64] Decision of Taipei District Court, Taiwan, Case Number: 92 Su Zi No. 2146 (2005); *see also* Tsai, *supra* note 32, at 85.

[65] Decision of Shih Lin District Court, Taiwan, Case Number: 92 Su Zi No. 728 (2005).

[66] *Id.*

[67] *Id.*

[68] Decision of Intellectual Property Court, Taiwan, Case Number: 99 Xing Shang Geng Er Zi 24 (2012).

[69] The concept of media neutrality arose from the U.S. Supreme Court *White-Smith Pub. Co. v. Appollo Co.* decision. See *White-Smith Pub. Co. v. Appollo Co.*, 209 U.S. 1 (1908).

[70] Decision of Intellectual Property Court, Taiwan, Case Number: 98 Xing Zhi Shang Geng Yi No. 16 (2009); *see also* Goldstein 2003 and Tsai 2005, 85–6.

[71] Decision of Intellectual Property Court, Taiwan, Case Number: 98 Xing Zhi Shang Geng Yi No. 16 (2009); see Tsai 2005, 86.

In order to avoid secondary liability for conduct by others, ISPs have developed policies and increased their responsiveness to complaints of infringement.[72] For instance, eBay developed the Verified Rights Owner program (VeRO), which is similar to the "notice and take down" procedure in the DMCA and Taiwan's 2009 legislation.[73] Under the Copyright Act, both the incentive to distribute and the incentive to create are compelling public interests that the legislature wanted to ensure. A proper balance between the two competing interests is a crucial issue for courts when determining a case that involves secondary liability. The 2009 legislation requires courts to take into consideration the concern that holding service providers liable for users' conduct will impede the development of new technology (Lee 2014, 155). Eventually, the increased cost of distribution will be shifted to customers (Lee 2014, 160; Tsai 2005, 86). Consequently, society as a whole will have to assume the negative effects, if any.

Concerns About ISP Liability

These cases best exemplify what is actually happening in cyberspace. For policy makers, whether ISPs are liable for secondary liability is not just a legal question. The policy maker also needs to take into consideration the economic effects of providing safe harbours. If a copyright holder is allowed to sue an ISP for secondary liability, substantial enforcement and administrative savings can be achieved. Those who argue for ISPs' secondary liability argue that imposing secondary liability on ISPs will significantly increase ISPs' incentives to identify wrongdoers and to block infringing or other offensive materials (Lichtman and Landes 2003; Freiwald 2001). Nonetheless, one of the obvious problems is that making ISPs liable for their users' infringements will require ISPs to bear transaction costs (Kim 2007, 165). This is especially true when considering that current information technology does not provide ISPs with a simple/cheap method to filter billions of links and messages they process every day and identify all the information on their webpages that may constitute infringements.

In addition to the technical issue, imposing secondary liability on ISPs leads to ISPs having to monitor their users' online activities. This means that ISPs will have to determine whether there are any infringements occurring on the platform they provide to their users (*id*). It is difficult to automate the process of determining legal liability due in part to the vagueness of the fair use doctrine. For instance, it is far beyond an ISP's ability to determine whether pictures posted by its users are fair use, or whether they are legitimate copies or displays that qualify for one of the exceptions of copyright law.

---

[72] Stone 2007, C9.

[73] See eBay, *What is VeRO and why was my listing removed because of it?*, available at http://pages.ebay.com/help/policies/questions/vero-ended-item.html

Another problem that will arise from imposing secondary liability on ISPs is the collapse of some online services as a result of the insurmountable transaction costs (*id*). Although advocates of liability for ISPs argue that this can be overcome by transferring the costs to users (Lichtman and Landes 2003, 404–7), this argument lacks merit. Internalising costs is not possible for ISPs under the copyright regime because courts often order damages that exceed the statutory damages set forth in Copyright Act (Lemley 2007, 111). Thus, the uncertainty of expected damages that need to be paid for infringing content on ISPs' websites means that ISPs are unable to estimate the costs that they would likely have to internalise.

Additionally, a court granting injunctive relief prohibiting the display of infringing content may curtail the operation of ISPs altogether because there is no effective way for ISPs to preclude the infringing content from every source, without blocking the other non-infringing material as well (*id*). Thus, the argument that ISPs can internalise the costs of compensation by passing the costs on to their users would require elimination of statutory damages rules, punitive damages in tort, and all injunctive relief (*id*).

Lastly, imposing a strict liability approach without providing safe harbor for ISPs is likely to be inefficient (*id*, 112). Because ISPs can hardly reasonably capture anything like the full social value of the use of their system, imposing the full social costs of harm caused by end users' posting on the ISPs' websites will cause ISPs to respond by inefficiently restricting third parties' use of the Internet (*id*; Frischmann and Lemley 2007). Thus, a strict liability approach, without considering any safe harbour, is likely to be inefficient. (Frischmann and Lemley 2007, 112; Hylton 2007). As a result, technological innovation will be dampened and the diversity of the Internet, with the dissemination of user-generated content, will be undermined (Frischmann and Lemley 2007).

The 2009 legislation attempts to apply the copyright laws to cyberspace, particularly to provide statutory limitations on the liability of companies providing online services. Under the safe harbour provisions, an ISP as defined under the statute, subject to varying conditions, may be exempted from liability for copyright infringements with respect to four categories of activities, if it meets two basic requirements as required by the statute. Overall, like the DMCA, the safe harbour provisions are a balance among the need to protect copyright holders, users, and ISPs (Elkin-Koren 2006).

Conclusion

The Internet makes it possible for an individual to exchange information at virtually no cost. When a copyright infringement is committed by an ISP user, it is likely that the copyright holder will turn to the ISP and request the ISP to take responsibilities for the infringement. Those who argue that ISPs should be responsible for their users' online infringements ground their arguments on the fact that ISPs are typically in a good position to either prevent copyright infringement or pay for the harm it causes. However, secondary liability has a significant drawback, in that it inevitably interferes with the legitimate use of the tools, services, and venues involved. For instance, while ISPs can reduce copyright infringement caused by their users through aggressive monitoring and filtering, it would also raise the transaction costs of operating the internet services and as a result it would force (some) ISPs out of Internet-related business. Not to mention, an aggressive monitoring approach raises concerns about users' freedom of speech. In response to the emergence of these tensions, the U.S. Congress amended the Copyright Act with the DMCA. Section 512 of the DMCA represents a reasonable balancing of the interests of the ISPs and copyright holders.

Taiwan law does not expressly recognize concepts of indirect liability. However, both the Civil Code and the Copyright Act provide statutory bases for establishing liability for the conduct of others. Service providers might be held liable for the conduct of others under Articles 185 and 188 of the Civil Code of Taiwan. They are subject to secondary liability for copyright infringement under subparagraph 7, paragraph 1, Article 87 of Copyright Act. The standards of secondary liability created by law vary according to the cause of action. Before the Copyright Act incorporated the safe harbor provisions, courts in Taiwan adopted the media neutrality principle as a defense in deciding whether an ISP is liable for its users conducts of copyright infringement. Although commentators have argued, that "[u]nder the media neutrality principle, there is no need to enact another statute to regulate copyright disputes regarding the Internet" (*id*, 88), uniform and definite rules are advisable. In order to avoid secondary liability, service providers will take down alleged infringement content as soon as they receive notification of infringing activity. Therefore, uncertainty and lack of uniformity create the risk of online bullying (Levin 2009, 521–2). In 2009, Taiwan's legislature added to its Copyright Act the safe harbour provisions to shield ISPs from being held liable for its user's direct infringement. Nevertheless, because there is no single statute regulating the liability of service providers in Taiwan, commentators have argued that Taiwan should consider amending the Trademark Act to encompass the safe harbour provisions similar to those in Copyright Act. The next reform of Taiwan's current law would be the amendment of Trademark Act, by adding such standards for secondary liability of service providers and its requirements for immunity in the context of trademark infringement.

References

Journal Articles

Ballon, Ian C. 2009. Secondary Copyright Liability. *American Law Institute – American Bar Association Continuing Legal Education* SP016: 1257–1315.
Becker, Gary S. 1968. Criminal and punishment: An economic approach. *Journal of Political Economy* 76: 169–224.
Elkin-Koren, Niva. 2006. Making technology visible: Liability of internet service providers for Peer-to-Peer traffic. *New York University Journal of Legislation and Public Policy* 9: 15–76.

Freiwald, Susan. 2001. Comparative institutional analysis in cyberspace: The case of intermediary liability for defamation. *Harvard Journal of Law and Technology* 14: 569–655.

Frischmann, Brett M., and Mark A. Lemley. 2007. Spillovers. *Columbia Law Review* 107: 257–301.

Hamdani, Assaf. 2002. Who's liable for cyber wrongs? *Cornell Law Review* 87: 901–957.

Hylton, Keith N. 2007. Property rules, liability rules, and immunity: An application to cyberspace. *Boston University Law Review* 87: 1–39.

Kim, Eugene C. 2007. Youtube: Testing the safe harbors of digital copyright law. *Southern California Interdisciplinary Law Journal* 17: 139–171.

Lee, Jyh-An. 2014. Policy implications of the ISP safe harbor in copyright law. *National Taiwan University Law Review* 43 (1): 143–207.

Lemley, Mark A. 2007. Rationalizing internet safe harbors. *Journal on Telecommunications and High Technology Law* 6: 101–119.

Levin, Elizabath K. 2009. A Safe harbor for trademark: Reevaluating secondary trademark liability after Tiffany v eBay. *Berkeley Technology Law Journal* 24: 491–527.

Lichtman, Douglas, and William Landes. 2003. Indirect liability for copyright infringement: An economic perspective. *Harvard Journal of Law and Technology* 16: 395–410.

Tsai, Huei-ju. 2005. Media neutrality in the digital Era a study of the Peer-To-Peer file sharing issues. *Chicago-Kent Journal of Intellectual Property* 5: 46–89.

Wan, Ke Steven. 2011. Internet service providers' vicarious liability versus regulation of copyright infringement in China. *Journal of Law, Technology and Policy* 2011 (2): 375–412.

Wang, Yi-Ping. 2009. Internet service provider civil liability immunity provisions in copyright act. *The Taiwan Law Review* 173: 25–41.

Zhang, Zhong-Xin. 2008. The legislative trend of internet service provider copyright infringement liability and its limitation (Wang Lu Fu Wu Ti Gong Zhe Zhe Zuo Quan Qin Hai Ze Ren Xian Zhi Zhi Li Fa Si Kao Yu Fang Xiang). *Taipei Bar Journal* 347: 28–44.

Book

Goldstein, Paul. 2003. *Copyright's highway: From Gutenberg to the celestial jukebox*. California: Stanford University Press.

Gorman, Robert A., and Jane C. Ginsburg. 2001. *Copyright: Cases and materials*. New York: Foundation Press.

Book Chapter

Sun, Andy Y. 2007. Contributory and vicarious liability for copyright infringement. In *Copyright law and the information society in Asia*, ed. Cristopher Heath and Kung-Chung Liu, 227–268. Oxford: Hart Publishing.

Online Document

Chang, Mag. 2012. Internet Use Hits All-Time High in Taiwan. *Taiwan Today* http://taiwantoday.tw/ct.asp?xItem=193241&ctNode=413. Accessed 14 Aug 2015.

International Intellectual Property Alliance. 2013. 2013 Special 301 Report on Copyright Protection and Enforcement: Taiwan. http://www.iipa.com/rbc/2013/2013SPEC301TAIWAN. PDF. Accessed 14 Aug 2015.
Stone, Brad. 2007. EBay says fraud crackdown has worked. *New York Times*. http://www.nytimes.com/2007/06/14/technology/14ebay.html?_r=0. Accessed 14 Aug 2015.

Organization Site

Taiwan Intellectual Property Office. 2015. *Copyright Law*. http://www.tipo.gov.tw/. Accessed 14 Aug 2015.

Online Database

Laws & Regulations Database of the Republic of China. 2015. *Civil Code*. http://law.moj.gov.tw/ENG/LawClass/LawAll.aspx?PCode=B0000001. Accessed 14 Aug 2015.

Chapter 15
Secondary Liability for Open Wireless Networks in Germany: Balancing Regulation and Innovation in the Digital Economy

Christoph Busch

Introduction

The rapidly developing digital economy creates a growing need for public access to the Internet everywhere and anytime. In many countries, wireless local area networks (WLAN) providing such access have become a standard in hotels, cafés, airports, train stations or public areas. Interestingly, in Germany the number of open wireless networks is much smaller than in many other developed economies.[1] One reason for this situation seems to be that it is rather unclear whether providers of an open WLAN can be held liable for unlawful conduct of network users, in particular for infringements of intellectual property rights (IPRs). As a consequence, potential providers of open wireless networks are often discouraged by legal uncertainty regarding the risk of liability.

This observation brings into focus the effect of liability rules on innovation and technological development. Apparently, German law has not yet found the right balance between the protection of IPRs and the development of innovative technologies that provide an essential infrastructure for the digital economy.[2] Following a

---

[1] According to the German Federal Ministry of Economic Affairs and Energy (BMWi) there are only 1.87 WLAN Hotspots per 10,000 inhabitants in Germany (as compared to 37.35 in South Korea, 28.67 in the United Kingdom and 9.94 in Sweden), see BMWi Website: <http://www.bmwi.de/DE/Themen/Digitale-Welt/Netzpolitik/rechtssicherheit-wlan.html> (all websites last visited 31 March 2017).

[2] Open wireless networks are not only an essential part of the infrastructure for the digital economy, they also play an important role as a communication infrastructure in case of natural disasters where they can facilitate emergency services. For examples see the open letter formulated by the Electronic Frontier Foundation regarding the *McFadden* case (C-484/14) available at <https://www.eff.org/files/2015/07/20/closedwifiasanobstacletolegitimatetrade-4.pdf>

C. Busch (✉)
European Legal Studies Institute, University of Osnabrück, Osnabrück, Germany
e-mail: christoph.busch@uos.de

© Springer International Publishing AG 2017   361
G.B. Dinwoodie (ed.), *Secondary Liability of Internet Service Providers*,
Ius Comparatum – Global Studies in Comparative Law 25,
DOI 10.1007/978-3-319-55030-5_15

lengthy and controversial political debate, the German government in July 2016 enacted a law reform which aims to create a new legislative framework offering legal certainty for providers of open wireless networks and thus establishing the necessary legal environment for better WLAN coverage.[3] It is doubtful, however, whether the new law will really change the situation for the better.

This Chapter analyses the recent changes in the regulatory environment for open wireless networks in Germany. The first part provides an overview of the general liability principles regarding the liability for online intermediaries and their application by the courts (Section "General principles and legal framework"). In a second step, the Chapter examines the recent reform of the Tele Media Act and assesses whether it enhances legal certainty (Section "The 2016 reform of the Tele Media Act"). Finally, the Chapter analyses the CJEU's decision in the *McFadden* case[4] and its impact on the new legal framework for WLAN providers in Germany (Section "Waiting for a clarification from Luxembourg—the *McFadden* case"). The Chapter concludes that the German legislator may have to revise again the legal framework in the light of the CJEU ruling in *McFadden*.

General Principles and Legal Framework

The Concept of Störerhaftung

The liability of WLAN providers is embedded in the broader context of online service providers.[5] Various kinds of service providers or intermediaries play a crucial role in the online world. Access providers enable users to enter the Internet and transmit information through online networks. Host providers store information for third parties. Other services providers create online platforms and social networks for communication among users. Search engines help users to navigate through the Internet and find the content they are looking for. All of these intermediaries make valuable contributions to the complex infrastructure of the Internet. At the same time they create a source of danger by enabling users to commit infringements of IPRs and to commit other unlawful acts.

Cases in which online intermediaries are held directly liable as tortfeasors (*Täter*) or infringers of copyright or trademark law are quite rare in Germany.[6] Similarly, service providers are only rarely found liable for contributory liability (or participant

---

[3] *Gesetzentwurf der Bundesregierung, Entwurf eines Zweiten Gesetzes zur Änderung des Telemediengesetzes* of 25 September 2015, Bundesrat-Drucksache 440/15. The original legislative proposal also provided for an amendment of the Tele Media Act to the effect that host providers whose business model is largely established on violations of IPRs are no longer able to rely on the liability privilege under § 10 Tele Media Act. In the final version adopted by the German Parliament in June 2016 this part of the proposal has been dropped. For details see Volkmann 2015, 289, 291.

[4] CJEU, Judgment of 15 September 2016, *McFadden*, C-484/14, ECLI:EU:C:2016:689.

[5] See Ohly 2015, 308–318; see also Ohly 2014.

[6] See Hoeren and Yankova 2012, 503.

liability). Liability for participation as accomplice or instigator in unlawful actions requires that the participant (*Teilnehmer*) is positively aware of the unlawful action of the direct tortfeasor or infringer. In most cases, however, online intermediaries have no actual knowledge of the concrete infringement committed by the individual user.

As the online intermediaries usually do not meet the subjective requirements of contributory liability, in most cases their liability is based on the concept of *Störerhaftung* (the liability of a *Störer*, meaning 'interferer').[7] This concept is not based on tort law but on analogy to § 1004 of the German Civil Code (BGB), a statutory provision offering injunctive relief against infringements of property. § 1004 BGB is directly applicable only to infringements of corporeal property, but by analogy case law has extended the doctrine of *Störerhaftung* to infringements of IPRs.[8]

The concept of *Störerhaftung* is a form of strict or objective liability. Consequently, no negligence of the intermediary need be established. However, in order to prevent limitless liability, injunctions against 'interferers' are granted only if the following three conditions are satisfied: (1) an adequate causal contribution to the infringing act of a third party; (2) the legal and factual possibility of preventing the resulting direct infringement; and (3) the violation of a reasonable 'duty of care' or 'monitoring duty' to prevent these infringements.[9]

Quite importantly, remedies available under the concept of *Störerhaftung* are limited. It can only serve as a legal basis for injunctive relief, including preventive injunctions, but not for damage claims. This follows from the fact that the 'interferer' is neither liable for primary nor secondary infringement according to the general rules of tort law.[10]

Limitations of service providers' liability are laid down in §§ 7 to 10 Tele Media Act (*Telemediengesetz*). These 'safe harbour' provisions contain an almost literal transposition of Articles 12 to 15 of the E-Commerce Directive (2000/31/EC).[11] According to established case law the liability privileges under §§ 7 to 10 Tele Media Act only provide immunity against claims for damages and criminal liability, but not from injunctive relief.[12] Hence, the 'safe harbour' provisions provide no

---

[7] See Kur 2014, 532–535; Leistner 2014, 78–82; Hoeren and Yankova 2012, 504–506; Busch 2014, 765–779. For a comprehensive overview see Neuhaus 2011.

[8] The application of the doctrine of *Störerhaftung* in the context of IPRs was well established before the age of the Internet. Based on this concept, injunctions could be issued, for example, against freight carriers who were unaware of transporting goods infringing third party trademarks, see e.g. BGH, Judgment of 15 January 1957, Case ref. I ZR 56/55, GRUR 1957, 352; see also Neuhaus 2011, 44 and Kur 2014, 532.

[9] Leistner 2014, 78 and Busch 2014, 768.

[10] Kur 2014, 533.

[11] For more details on §§ 7 to 10 Tele Media Act, see Hoeren and Yankova 2012, 507–509. For case law examples see Busch 2014, 769–774.

[12] See e.g. BGH, Judgment of 11 March 2004, Case ref. I ZR 304/01, MMR 2004, 668 at 670 (*Internetversteigerung I*); BGH, Judgment of 19 April 2007, Case ref. I ZR 35/04, MMR 2007, 507 at 508 (*Internetversteigerung II*); BGH, Judgment of 30 April 2008, Case ref. I ZR 73/05, MMR 2008, 531 at 532 (*Internetversteigerung III*); see also Kur 2014, 533.

protection against injunctions based on the concept of *Störerhaftung*. Whether this interpretation is compatible with EU law is a matter of controversial debate.[13]

Case Law

The question whether providers of wireless networks are liable for unlawful acts committed by network users has given rise to a number of court decisions in recent years. Several decisions by the German Federal Supreme Court, the *Bundesgerichtshof* (BGH), have clarified the liability of private owners of unsecured wireless networks for copyright infringements committed by third parties (Section "Private networks"). By contrast, the liability regime applicable to operators of commercial networks remains rather unclear. There are several decisions by first instance courts, but so far the case law has not revealed a clear position (Section "Commercial networks").[14]

Private Networks

In the decision *Sommer unseres Lebens*, a landmark case decided in 2010 which stirred much debate,[15] the BGH held that the private owner of an unprotected wireless network is liable for a copyright infringement committed by an unidentified third party.[16] According to the Court the private owner of a WLAN connection has a duty to apply adequate safety measures to prevent abuse of the network by unauthorised third parties. The adequacy of the safety measures is to be assessed by the technical standards applicable at the time of installation of the router. In other words, the subscriber is not required to continuously update the safety measures. However, as the subscriber in the case decided by the BGH had not replaced the password set by the producer of the router by a secure personal password, the Court held that he violated a specific duty of care.[17]

---

[13] See *infra*, Section "Scope of the liability privilege for access providers".

[14] For a brief overview of recent case law, see Hoeren and Jakopp 2014, 72–75. See also Hofmann 2014, 654–660; Borges 2014, 2305–10.

[15] See e.g. Spindler 2010, 592–600; Borges 2010, 2624–2627.

[16] Bundesgerichtshof, Judgment of 12 May 2010, Case ref. I ZR 121/08, GRUR 2010, 633 (*Sommer unseres Lebens*).

[17] In two more recent decisions local courts in Frankfurt am Main and Hamburg decided that the WLAN subscriber complies with his duty of care if he does not change the individual 13-digit password printed on the back of the router. This password usually meets the high safety standard by the Bundesgerichtshof, see Landgericht Frankfurt am Main, Judgment of 14 June 2013, Case ref. 30 C 3078/12 (75), MMR 2013, 605 at 607; Amtsgericht Hamburg, Judgment of 9 January 2015, Case ref. 36a C 40/14, BeckRS 2015, 08939.

Interestingly, the BGH did not apply nor even discuss the liability privilege for access providers (§ 8 Tele Media Act) in this case.[18] As the BGH only awarded injunctive relief and reimbursement of pre-trial costs for a warning letter but no damages, one could assume that this 'omission' is based on the implicit assumption that § 8 Tele Media Act provides no defence against injunctions.[19]

In two more recent decisions, the BGH applied the principles set out in *Sommer unseres Lebens* to cases in which the owner of a private wireless network had provided network access to family members who abused the connection for copyright infringements. In *Morpheus*, the court had to decide whether a network owner was liable for illegal file sharing by his 13-year old son.[20] The Court held that the network owner fulfils his duty of care if he has instructed the minor not to use the Internet connection for any unlawful act. In contrast, there is no general duty to monitor the Internet use of the minor or to block the Internet connection. Such measures are only required if there is concrete evidence that the child does not comply with the instruction.[21] With regard to the prevention of copyright infringements by adult family members, the duty of care imposed on the network owner is even less stringent. In *BearShare*, the BGH decided that the owner of a wireless network could not be held liable for illegal file sharing by his 21-year old stepson.[22] The Court held that there was no duty to instruct adult family members about the prohibition of illegal file sharing as long as there is no concrete evidence of any illegal activities. The argumentation of the BGH in *BearShare* also shows that the liability principles developed by the Court in cases concerning private networks cannot be transferred directly to wireless networks that are open to the public. When assessing the duties of care that can reasonably imposed on the owner of a private network in a family context, the court discussed in detail the relevance of fundamental rights, which protect the relationship of trust within the family.[23] In this context, Article 6 of the German constitution, which enshrines the protection of family life, sets a limit to the network owners' 'monitoring duties'.

Commercial Networks

For commercially used open wireless networks there is not yet a clear line of case law. So far only a number of decisions from lower instance courts have been reported. It appears that the courts distinguish between cases in which network access is granted to users usually known by name (e.g. hotels, holiday apartments

---

[18] Hoeren and Jakopp 2014, 73. The BGH only discusses whether § 10 Tele Media Act concerning host providers is applicable in the case, which is eventually answered in the negative.

[19] See *supra*, Section "The Concept of *Störerhaftung*".

[20] BGH, Judgment of 15 November 2012, Case ref. I ZR 74/12, NJW 2013, 1441 (*Morpheus*).

[21] BGH, Judgment of 11 June 2015, Case ref. I ZR 7/14, NJW 2016, 942 (*Tauschbörse II*); see also Obergfell 2016a, 910.

[22] BGH, Judgment of 8 January 2014, Case ref. I ZR 169/12, NJW 2014, 2360 (*BearShare*).

[23] Ibid. para. 27–29.

and hospitals) and cases in which network users are usually unknown (e.g. cafés, airports and at so-called 'free radio' hotspots).[24]

To the first category of cases belongs a decision of 18 August 2010 of the Regional Court of Frankfurt am Main. In its judgment the court held that a hotel owner is not responsible for copyright infringements committed via the hotel's WLAN by an unidentified third party, provided that the network is protected by industry standard encryption technology.[25] In a more recent ruling of 28 June 2013, the same court decided that the owner of a holiday apartment could not be held liable for an infringement committed by one of his guests if the landlord can prove that he has instructed the guests not to use the WLAN for any unlawful acts.[26] In a decision of 10 June 2014, the Hamburg district court held that the liability privilege under § 8 Tele Media Act applies to a wireless network operated by the owner of a hotel.[27] As a consequence, the court dismissed a claim for damages against the hotel owner for a copyright infringement by one of the hotel guests. The court also held that there was no claim based on the doctrine of *Störerhaftung*. According to the court, it is neither necessary to provide hotel guests with detailed instructions regarding the prohibition of copyright infringements nor to block certain ports of the router.

In summary, network operators who provide an Internet connection to users known by name are shielded from liability if they protect the network by encryption technology and instruct network users not to commit any unlawful acts. Such safety measures have been considered sufficient in the case of hotel operators,[28] owners of rental flats[29] and holiday apartments[30] as well as for hospital operators.[31]

The standards applicable to the second category of cases, in which wireless network access is provided to users who are unknown to the network operator, are less clear. In its decision of 25 November 2010 the Regional Court of Hamburg held that the owner of an Internet café was liable for a copyright infringement committed by

---

[24] Sesing 2015, 424; see also Borges 2014, 2308.

[25] Landgericht Frankfurt am Main, Judgment of 18 August 2010, Case ref. 2-6 S 19/09, MMR 2011, 401.

[26] Landgericht Frankfurt am Main, Judgment of 28 June 2013, Case ref. 2-06 O 304/12, GRUR-RR 2013, 507.

[27] Amtsgericht Hamburg, Judgment of 10 June 2014, Case ref. 25b C 431/13, CR 2014, 536; see also Amtsgericht Hamburg, Judgment of 24 June 2014, Case ref. 25b C 924/13, BeckRS 2014, 13884.

[28] Landgericht Frankfurt am Main, Judgment of 18 August 2010, Case ref. 2-6 S 19/09, MMR 2011, 401; Amtsgericht Hamburg, Judgment of 10 June 2014, Case ref. 25b C 431/13, CR 2014, 536; Amtsgericht Koblenz, Judgment of 18 June 2014, Case ref. 161 C 145/14, BeckRS 2014, 15122.

[29] Amtsgericht München, Judgment of 15 February 2012, Case ref. 142 C 10921/11, CR 2012, 340.

[30] Landgericht Frankfurt am Main, Judgment of 28 June 2013, Case ref. 2-06 O 304/12, GRUR-RR 2013, 507; Amtsgericht Hamburg, Judgment of 24 June 2014, Case ref. 25b C 924/13, BeckRS 2014, 13884.

[31] Amtsgericht Frankfurt am Main, Judgment of 16 December 2014, Case ref. 30 C 2801/14 (32), NJOZ 2015, 588.

a third party via the WLAN offered to the customers of the café.[32] The court argued that the owner of the Internet café had violated reasonable duties of care because he offered a WLAN connection without blocking the ports that are typically used for file sharing.

A different approach was taken by the district court of Berlin-Charlottenburg in its decision of 17 December 2014.[33] The court ruled out the liability of the operator of a 'free radio' (*Freifunk*) WLAN hotspot under the doctrine of *Störerhaftung*. Following the approach taken by the BGH regarding the liability of private operators of wireless networks, the Charlottenburg court discussed whether the network operator violated its duty to observe a reasonable 'duty of care' to prevent any infringements through the network. According to the court, particularly strict requirements have to be applied when assessing the reasonableness of safety measures taken by the 'free radio' operator. Such measures must not endanger the business model of the 'free radio' operator, as the court underlined. This would be the case if one required the blocking of certain ports or domain main servers (DNS) or a duty to register for the network users. Similarly, a duty of the operator to instruct all users not to commit any unlawful acts was considered impracticable by the court. The line of argument taken by the Charlottenburg district court is not entirely free from logical flaws. In assessing the reasonableness of the 'free radio' operator's duties of care, the court starts from the premise that such duties must not endanger the operator's legal business model. The legality of this business model depends, however, precisely on the extent of the duties of care owed by the 'free radio' operator. This is somewhat circular.[34]

At the time of writing there is not yet any decision by the BGH regarding the liability of commercially used open wireless networks. However, in the above cited leading case concerning private WLAN subscribers, the Court mentioned commercial networks in an *obiter dictum* when justifying the high standards for safety measures. The Court argued that in case of a private network used in a family context imposing a preventive duty of care does not endanger any 'business model'.[35] This could indicate *a contrario* that the BGH might apply a less stringent standard to commercially used networks.

---

[32] Landgericht Hamburg, Decision of 25 November 2010, Case ref. 310 O 433/10, MMR 2011, 475.

[33] Amtsgericht Charlottenburg, Judgment of 17 December 2014, Case ref. 217 C 121/14, CR 2015, 192.

[34] Sesing 2015, 424; cf. also CJEU, Judgment of 27 March 2014, *UPC Telekabel Wien*, C-314/12, ECLI:EU:C:2014:192, para 49, where the Court notes that 'the freedom to conduct a business includes, inter alia, the right for any business to be able to freely use, within the limits of its liability for its own acts, the economic, technical and financial resources available to it'.

[35] Bundesgerichtshof, Judgment of 12 May 2010, Case ref. I ZR 121/08, GRUR 2010, 633 (*Sommer unseres Lebens*), para. 24; see also BGH, Judgment of 11 March 2004, Case ref. I ZR 304/01, MMR 2004, 668 at 670 (*Internetversteigerung I*), para. 671.

The 2016 Reform of the Tele Media Act

Against the background of the inconsistent court decisions and the rather unclear legal situation of public wireless networks there has been a controversial debate about how to create a regulatory environment that balances the interests of IPR owners and legal certainty for WLAN operators. Several proposals for law reform have been submitted by political parties and various interest groups.[36] In March 2015 the Federal Ministry of Economic Affairs and Energy published a legislative draft for a reform of the Tele Media Act, which contains the liability privileges for online service providers. Following harsh criticism from stakeholders,[37] the Ministry tabled a revised legislative draft in June 2015,[38] which in September 2015 was introduced as a government bill in Parliament.[39] In a surprise turn—as a reaction to the publication of the Opinion[40] of Advocate General *Szpunar* in the *McFadden* case[41]—the German legislator changed the pending proposal again in June 2016. The reform act entered into force on 27 July 2016.[42]

In essence, the reform aimed at answering two crucial questions with regard to the liability of WLAN operators: (1) Does the 'safe harbour' provision of § 8 Tele Media Act, which transposes Art. 12(1) of the E-Commerce Directive into German law and exempts access providers from liability for infringements by third parties, also apply to operators of wireless networks? (2) Under which conditions does this exemption also shield the network operator against injunctive claims?

Application of the Mere Conduit Defence to WLAN Operators

Regarding the first question, the answer given by the reform act is a clear and simple 'yes'. The newly added paragraph 3 of § 8 Tele Media Act states that the liability privilege for access providers under § 8(1) Tele Media Act also applies to operators

---

[36] For an overview of the discussion and recent proposals see Mantz and Sassenberg 2015a, 298.

[37] See Sesing 2015, 423–427; Mantz and Sassenberg 2015a, 298–306; Volkmann 2015, 289–91.

[38] *Entwurf eines Zweiten Gesetzes zur Änderung des Telemediengesetzes* of 11 March 2015. A revised draft was published on 15 June 2015. Both drafts are available at the website of the German Ministry of Economic Affairs and Energy: <http://www.bmwi.de/BMWi/Redaktion/PDF/S-T/ telemedienaenderungsgesetz-aenderung>. Multiple language versions of the revised draft, which has been notified to the European Commission under the EU Technical Regulation Information System (TRIS) and is currently being scrutinized for compatibility with EU law, are available online: <http://ec.europa.eu/growth/tools-databases/tris/en/> (Notification Number: 2015/0305/D).

[39] *Gesetzentwurf der Bundesregierung, Entwurf eines Zweiten Gesetzes zur Änderung des Telemediengesetzes* of 25 September 2015, Bundesrat-Drucksache 440/15.

[40] Opinion of Advocate General *Szpunar*, 16 March 2016, C-484/14, ECLI:EU:C:2016:170.

[41] Case C-484 (*Tobias McFadden v. Sony Music Entertainment Germany GmbH*).

[42] Bundesgesetzblatt I 2016, 1766.

who provide users with Internet access via a wireless network. This is certainly the less controversial part of the reform act. In so far the new law only codifies what has been already the widely shared view in academic writings.[43]

Additional Duties of Care Imposed on WLAN Operators?

By far more controversial is the question under which conditions the application of the safe harbour provision also shields the network operator against injunctive claims based on the concept of *Störerhaftung*. With regard to this question, the legislative proposal has undergone an interesting development from the original draft to the final version. In order to fully understand the new legislative framework, it may be helpful to briefly outline the different stages of development of the legislative proposal.

Ministerial Draft (March 2015)

The first draft published in March 2015 followed a 'split-level approach' applying two different 'standards of care' for commercial WLAN operators and public entities on the one hand, and private network operators on the other hand.

According to the draft, commercial operators and public entities who provide network access to third parties should only benefit from the 'safe harbour' provision of § 8 Tele Media Act if they have taken 'reasonable measures to prevent infringements by users'. The draft further stipulated that this duty of care is only fulfilled if two conditions are satisfied: (1) the network operator has taken appropriate security measures against unauthorised access of WLAN by using 'industry standard encryption technology' for the router, and (2) the operator only grants internet access to users who have declared not to commit any unlawful acts in the context of the use.

For private WLAN operators a stricter standard of care was foreseen in the Ministerial draft. They should only benefit from the liability exemption if, in addition to the two conditions mentioned above, they knew the names of the users to which they have granted network access.

Revised Ministerial Draft (June 2015)

In response to harsh criticism from all sides,[44] the Federal Ministry of Economic Affairs and Energy presented a revised legislative draft in June 2015 which abandoned the 'split-level approach'. The new proposal, which was introduced into the

---

[43] See e.g. Spindler 2010, 595; Kaeding 2010, 168; Mantz 2013, 498; Spindler 2016a, 48, 50.
[44] Drücke 2015, 95; Mantz and Sassenberg 2015a, 298; Sesing 2015, 423; Solmecke 2015, 95; Volkmann 2015, 289.

parliamentary process in September 2015, applied the same standard of care to all kinds of WLANs, regardless of whether they are commercial or private networks. The additional requirement for private network operators to know the names of the users was dropped. Moreover, the revised draft no longer explicitly required the use of encryption technology, but only vaguely refers to 'appropriate security measures against unauthorised access'.

Appropriate Security Measures Against Unauthorised Access

However, also the revised draft still contained several questionable points. First, it remained unclear, what kind of security measures would be considered as 'appropriate'. The explanatory memorandum attached to the proposal explained that the open wording should make sure that the new regulation is technologically neutral. This is certainly a laudable approach. It comes, however, at the cost of reducing legal certainty and therefore endangers the very aim of the law reform.

Moreover, the requirement of an industry standard encryption (e.g. the secure WPA2 standard), which had been deleted from the text of the legislative proposal, was still mentioned in the explanatory memorandum as an example of 'appropriate security measures'. If, however, WLAN operators had to use encryption technology for their networks in order to be securely immune from liability, this would significantly hamper the legislative goal of increasing the free WLAN coverage in Germany. Advocates of the open wireless movement somewhat justifiably pointed out that the idea of an 'encrypted open network' is a contradiction in terms.[45]

As an alternative 'security measure' the explanatory memorandum attached to the draft law mentioned a 'voluntary registration' of network users. Yet, it still remained unclear what information would be necessary (e.g. name, address) in order to consider the registration process as equivalent to the encryption of the network. In addition, such a registration requirement would probably be incompatible with § 13(6) of the Tele Media Act, according to which 'the service provider must enable the use of telemedia [...] to occur *anonymously* or via a pseudonym where this is technically possible and reasonable'.[46]

In any case, similar to the encryption requirement a registration requirement would put up hurdles for users and prevent customers and passers-by from simply logging onto the network without the annoying task of enquiring about a password. Thus, both encryption and registration are hardly compatible with the idea of an *open* wireless network.

---

[45] Solmecke 2015, 95; Müller and Kipker 2016, 87, 89.

[46] See § 13(6) of the Tele Media Act (emphasis supplied). There is some disagreement whether § 13(6) Tele Media Act gives the user only a right to anonymity towards other users or also towards the service provider. Cf. Schnabel and Freund 2010, 718.

Declaration Not to Commit Any Unlawful Acts

Also the second requirement stipulated by the revised Ministerial draft raises questions. According to the legislative proposal WLAN operators should only grant Internet access to users who have declared not to commit any unlawful acts when using the wireless network. Technically, users would be required to provide such a declaration by ticking a checkbox when logging on to the network.

This provision would probably be ineffectual: how could such a declaration provide any effective protection against any violations of the law? It therefore did not come as a surprise that representatives of the music business, which is suffering from illegal file sharing, fiercely criticised the legislative proposal as insufficient to prevent IPR infringements.[47]

At the same time, the new requirement of a 'user declaration' would have gone beyond the standard of care defined by case law. So far, the courts have only required an 'instruction duty' (*Belehrungspflicht*) in cases where network access is granted to individuals known by name.[48] The new regulation would have gone one step further requiring *consent* by the network user. If the new rules also applied to wireless networks made available to family members—the wording of the revised legislative proposal did not explicitly exclude these cases—the new standard of care would go beyond the requirements set up by the BGH in *BearShare*. As mentioned above, according to the Court, the operator of a private WLAN has no duty to instruct adult family members regarding the prohibition of IPR infringements.[49]

Final Version of the Reform Act (June 2016)

After the publication of the Opinion[50] of Advocate General *Szpunar* in the *McFadden* case[51] the German legislator, in a somewhat surprising turn, changed the pending proposal again in June 2016.[52] All additional requirements—the 'appropriate security measures' and the 'declaration not to commit any unlawful acts'—were deleted from the proposal. As result, the final version of the reform act which entered into force on 27 July 2016,[53] only adds the new paragraph 3 to § 8 Tele Media Act,

---

[47] Drücke 2015, 95. Effective protection against violation of IPRs and personality rights would only possible if the WLAN operators were entitled to record and retain communication data, which, however, would be incompatible with the principle of telecommunications secrecy under § 88 Telecommunications Act.

[48] See the case law cited *supra* at section "Private networks".

[49] BGH, Judgment of 8 January 2014, Case ref. I ZR 169/12, NJW 2014, 2360 (BearShare).

[50] Opinion of Advocate General *Szpunar*, 16 March 2016, C-484/14, ECLI:EU:C:2016:170.

[51] Case C-484 (*Tobias McFadden v. Sony Music Entertainment Germany GmbH*).

[52] Beschlussempfehlung und Bericht des Ausschusses für Wirtschaft und Energie zu dem Gesetzentwurf der Bundesregierung – Drucksache 18/6745, Bundestag-Drucksache 18/8645.

[53] Bundesgesetzblatt I 2016, 1766.

stating that that the liability privilege for access providers under § 8(1) Tele Media Act also applies to operators of WLAN networks.

While some policy makers celebrated the reform act as a breakthrough for free WLAN in Germany and proclaimed the end of *Störerhaftung*,[54] most observers received the revision of the Tele Media Act with reservations.[55] The new law leaves many questions open: Unlike the original Ministerial proposal the final text of the law does not clearly say whether the new liability privilege for WLAN networks also shields the network operator from injunctive claims. A blanket exemption from injunctions would probably be incompatible with Article 8(3) of the Information Society Directive.[56] Some guidance is provided by the explanatory memorandum attached to the reform act, which underlines that the liability privilege shields the network operator from pre-litigation costs (*Abmahnkosten*) and court costs. At the same time, the explanatory memorandum points out that the liability privilege does not shield from 'court orders' (e.g. blocking orders). However, as the explanatory memorandum is not binding for courts, legal certainty has not really been achieved. Interestingly, the wording of the explanatory memorandum closely—sometimes even literally—follows the line of argument set out in the Opinion of Advocate General Szpunar in the *McFadden* case. Apparently the German legislator wanted to avoid any conflict with EU law. However, if this was the intention, it is difficult to understand why the *Bundestag* did not wait a few more weeks until the CJEU's final decision, which in several key points deviates from Szpunar's Opinion. Indeed, as the following analysis of the *McFadden* case shows, the 2016 reform has turned out to be rather overhasty.[57]

The *McFadden* Case – (Some) Clarification from Luxembourg

While the latest reform of the Tele Media Act aimed to create legal certainty for operators of open WLAN networks, the recent decision of the CJEU in the *McFadden* case has called into doubt the new law only a few weeks after its enactment.[58] In September 2014, the Munich Regional Court referred a number of questions concerning the liability of WLAN operators to the CJEU.[59] The referral for a preliminary ruling sought clarification regarding the interpretation of the 'safe harbour' provisions of the E-Commerce Directive.

---

[54] See Flisek, Klingbeil & Held, 'Freies WLAN in Deutschland kommt!' <http://blogs.spdfraktion. de/netzpolitik/2016/05/11/wlan/>

[55] See e.g. Sesing 2016, 507; Spindler 2016b, 2249; Conraths and Peintinger 2016, 297.

[56] Spindler 2016b, 2452.

[57] For an analysis of the *McFadden* case from a German perspective see Obergfell and Thamer 2017, 203; see also Obergfell 2016b, 3492; Mantz 2016, 817; Nordemann 2016, 1103; Bisle and Frommer 2017, 54.

[58] Case C-484/14 (*Tobias McFadden v. Sony Music Entertainment Germany GmbH*).

[59] Landgericht München, Decision of 18 September 2014, Case ref. 7 O 14719/12, GRUR Int. 2014, 1166. The case has been listed as at the CJEU. For a detailed analysis of the preliminary reference see Mantz and Sassenberg 2015b, 85–90; Stögmüller 2014, 542–5.

Facts of the Case

The preliminary reference arose from a case involving Mr. Tobias McFadden, the owner of a sound equipment and events lighting shop, and Sony Music, the record label. Mr. McFadden, who is also an active member of the German 'Pirate Party' fighting for Internet freedom, operates a password-free wireless network at his shop for customers and passers-by. An unidentified person used this WLAN connection for illegal music file sharing. When Sony filed a warning letter against Mr. McFadden, the latter brought an action for a negative declaratory judgment against Sony with the aim of establishing that he is not liable for the alleged copyright infringement. Sony, in turn, lodged a counterclaim requesting injunctive relief, damages and reimbursement of the costs of the warning letter.

In its order for reference to the CJEU, the Munich court explains that it does not assume that Mr. McFadden has personally committed the alleged copyright infringement. Hence, (primary) liability as tortfeasor is ruled out. The court, however, considers (secondary) liability of the shop owner under the doctrine of *Störerhaftung*. In its reasoning the court indicates that it tends to follow the line of argument laid down by the BGH in *Sommer unseres Lebens*.[60] As explained above, the BGH held that private WLAN operators can only prevent liability for third-party infringements by protecting their network through a password. According to the Munich court, however, such liability might be excluded if Mr. McFadden could invoke the 'mere conduit' defence under Article 12(1) of the E-Commerce Directive. Starting from this premise, the court raises a number of questions regarding the interpretation of the 'safe harbour' provisions of the E-Commerce Directive.

The questions raised by the Munich Regional Court essentially address three key issues, which are more or less the same issues discussed in the context of the reform of the Tele Media Act: (1) Can the operator of a free WLAN be qualified as a provider of information society services within the meaning of Article 12(1) of the E-Commerce Directive? (2) Does the 'mere conduit' defence under Article 12 of the Directive cover only monetary claims or also claims for injunctive relief? (3) Which safety measures have to be taken by a WLAN operator in order to avoid any liability?

Conditions for the Application of the 'Mere Conduit' Defence

The first issue raised by the Munich court was the question whether the provider of a free WLAN can qualify for a 'mere conduit' defence. The liability privilege under Article 12 of the E-Commerce Directive only applies to providers of 'information society services'. Pursuant to the definition in Article 2(a) and Recital 17 of the Directive, this notion refers to 'any services normally provided for remuneration'.

---

[60] BGH, Judgment of 12 May 2010, Case ref. I ZR 121/08, GRUR 2010, 633 (*Sommer unseres Lebens*).

The court asked if this means that: (a) the person specifically concerned normally provides the specific service for remuneration; (b) there are, on the market, any providers who provide this kind of service for remuneration; or (c) the majority of providers on the market offer such services for remuneration?[61]

In his Opinion, Advocate General Szpunar takes a rather pragmatic approach to this question. He does not discuss the subtle distinctions raised by the Munich court but merely states that, where the WLAN service takes place in an 'economic context' it falls within the scope of Article 12 of the E-Commerce Directive.[62] This applies also if the provision of WLAN is only ancillary to a business activity and offered free of charge.[63] The CJEU follows this interpretation and in its reasoning refers to *Papasavvas*, a case concerning online defamation in the context of an Internet service which was funded through advertising revenue.[64] Here the Court held that the concept of 'services normally provided for remuneration' under Article 2(a) of the E-Commerce Directive does not require the service to be paid for by those for whom it is performed.[65] Indeed, most services that are advertised as 'free' are either paid for by someone else (e.g. advertising clients[66]) or the user is paying in a different 'currency' (e.g. allowing access to personal data).[67] In the case at hand, Mr. McFadden operated the WLAN network from his business premises to draw the attention of customers of near-by shops and of passers-by to his business. For the CJEU this was sufficient to establish an 'economic context'.

This line of argument raises the question, whether the provider of a—truly—free wireless connection that is not linked to any business activity (e.g. access to private Wi-fi networks offered by members of the *Freifunk* community) also falls within the scope of the E-Commerce Directive. Unfortunately, this question is explicitly left open by the Advocate General who states that 'there is no need to consider whether the scope of Directive 2000/31 might also extend to the operation of such a network

---

[61] Question 1 referred to CJEU by the Landgericht München, Decision of 18 September 2014, Case ref. 7 O 14719/12, GRUR Int. 2014, 1166 at 1169.

[62] Opinion of Advocate General *Szpunar*, 16 March 2016, C-484/14, ECLI:EU:C:2016:170, para. 41.

[63] Ibid. at paras. 42 and 48.

[64] CJEU, Judgment of 15 September 2016, *McFadden*, C-484/14, ECLI:EU:C:2016:689, para. 41.

[65] CJEU, Judgment of 11 September 2014, *Papasavvas*, C-291/13, ECLI:EU:C:2014:2209, para. 29; see also CJEU, Judgment of 26 April 1988, *Bond van Adverteerders*, 352/85, ECLI: EU:C:1988:196, para. 16.

[66] See Recital 18 of the E-Commerce Directive.

[67] A quite similar question arises with regard to the scope of application of the Consumer Rights Directive 2011/83/EU. According to the definition in Art 2(6) of the Directive, the notion of 'service contract' means 'any contract other than a sales contract under which the trader supplies or undertakes to supply a service to the consumer and the consumer *pays or undertakes to pay the price* thereof' [emphasis added]. Based on this definition the German legislator has restricted the scope of application of the provisions implementing the Directive to contracts for a 'paid service' (*entgeltliche Leistung*) in § 312(1) of the Civil Code. It is doubtful, however, whether such a limitation can be justified in light of the nature of the digital economy, cf. Opinion of Advocate General *Szpunar*, 16 March 2016, C-484/14, ECLI:EU:C:2016:170, para. 47.

in circumstances where there is no other economic context'.[68] Unfortunately, in the CJEU's judgment this issue is not discussed at all.

From a policy perspective, a broader reading of Art. 2(a) of the E-Commerce Directive seems preferable. Hence, the concept of a 'service normally provided for remuneration' should be construed as also covering 'sharing economy' services, which are offered not for a profit. It would be a somewhat paradoxical result if the provider of a—truly—free wireless connection which is not linked to any business activity could not rely on the 'mere conduit' defence while the less altruistic provider who uses the WLAN for advertising purposes benefits from the safe harbour provisions.[69]

The notion 'service normally provided for remuneration' has its origin in EU primary law. As Advocate General Szpunar points out this concept 'is taken from Article 57 TFEU and reflects the principle [...] that only services of an economic nature are covered by the provisions of the TFEU relating to the internal market'.[70] In other words, the requirement of 'remuneration' or 'economic activity' is linked to the very concept of 'market freedoms' that is enshrined in EU primary law where it serves to describe the limits of EU competence. From this perspective one could argue that the question of liability of non-commercial WLAN providers does not relate to the internal market and therefore should be left to the Member States. But such a view would ignore the reality of the digital economy where the technological infrastructure of the internal market—in particular WLAN access—is provided not only through commercial but also through non-commercial market operators. WLAN providers linked to an 'economic activity' in the traditional sense coexist with WLAN providers based on a 'sharing economy' model. For the question whether the liability privileges of the E-Commerce Directive apply, this distinction should not be relevant.[71]

Scope of the Liability Privilege for Access Providers

If one accepts the premise that Article 12 of the E-Commerce Directive is applicable to operators of an open WLAN (if it is offered in an 'economic context'), the question that follows is: Does the 'safe harbour' provision only preclude claims for

---

[68] Ibid. at para. 50; see also CJEU, Judgment of 15 September 2016, *McFadden*, C-484/14, ECLI:EU:C:2016:689, para. 37.

[69] Questions 2 and 3 raised by the Munich court deal with the interpretation of the phrase 'transmission in a communication network of information' and the term 'provide' within the meaning of Art 12 of the E-Commerce Directive. It seems rather obvious that the provision of an open WLAN does fulfill these requirements, cf. Opinion of Advocate General *Szpunar*, 16 March 2016, C-484/14, ECLI:EU:C:2016:170, paras. 51–56.

[70] Ibid. at para. 37.

[71] Until the CJEU clarifies this point, free WLAN operators would be well advised to link the provision of their services to an economic activity. For this purpose it might be sufficient if the landing page or the name of the WLAN network is used for advertising a commercial service.

compensation (including ancillary pre-trial costs and court fees) or does it also provide a defence against claims for injunctive relief (including also costs of giving formal notice and court costs related to the injunction)? Based on a combined reading of paragraphs 1 and 3 of Article 12 of the E-Commerce Directive the CJEU argued that Article 12 does not shield the WLAN provider from claims for injunctive relief.[72]

This interpretation of Article 12 is broadly in line with the view expressed by the German Federal Supreme Court on the parallel issue of liability privileges for hosting providers under Article 14 of the E-Commerce Directive. The German Federal Supreme Court, on several occasions, has expressed the view that the liability privilege applies fully to criminal liability and liability for damages, but does not provide any protection against claims for cease and desist orders.[73] Apparently, the BGH considered that interpretation to be an *acte clair* and therefore did not refer the issue to the CJEU.[74] By contrast, the Munich court indicated in its order for reference that it prefers an interpretation of Article 12(1) which precludes not only monetary claims paragraphs 1 and 3 of Article 12 of the E-Commerce Directive.[75]

The Court's view does not really come as a surprise. Some indication as to how the 'normative tension' within Article 12 should be resolved was already given by the CJEU in *UPC Telekabel Wien*.[76] In this decision, which concerned the interpretation of the Information Society Directive, the Court made clear that Internet access providers can be qualified as 'intermediaries' within the meaning of Article 8(3) of the Directive and thus be subject to injunctive relief for infringements of copyright.

More interesting is the question whether the carve-out for injunctions in Article 12(3) of the E-Commerce Directive also applies to ancillary pre-trial costs and court fees. According to the Advocate General, the limitation of liability under Article 12 "extends not only to claims for compensation, but also to any other pecuniary claim that entails a finding of liability for copyright infringement with respect to the information transmitted, such as a claim for the reimbursement of pre-litigation costs or court costs".[77] Such a reading of Article 12 would have far reaching consequences for countries like Germany where the claimant can recover pre-trial costs related to

---

[72] CJEU, Judgment of 15 September 2016, *McFadden*, C-484/14, ECLI:EU:C:2016:689, paras. 76–78.

[73] BGH, Judgment of 11 March 2004, Case ref. I ZR 304/01, MMR 2004, 668 at 670 (*Internetversteigerung I*); BGH, Judgment of 19 April 2007, Case ref. I ZR 35/04, MMR 2007, 507 at 508 (*Internetversteigerung II*); BGH, Judgment of 30 April 2008, Case ref. I ZR 73/05, MMR 2008, 531 at 532 (*Internetversteigerung III*).

[74] Kur 2014, 533; Leistner 2014, 78.

[75] Landgericht München, Decision of 18 September 2014, Case ref. 7 O 14719/12, GRUR Int. 2014, 1166 at 1170.

[76] CJEU, Judgment of 27 March 2014, *UPC Telekabel Wien*, C-314/12, ECLI:EU:C:2014:192. For a critical review of the decision see Spindler 2014, 826–835; for a more positive view see Lehmann 2015, 680; see also Leistner and Grisse 2015, 19–27 and 105–115.

[77] Opinion of Advocate General Szpunar, 16 March 2016, C-484/14, ECLI:EU:C:2016:170, para. 74.

injunctive claims (e.g. legal fees for warning letters).[78] In practice, the fear of pre-litigation costs for warning letters creates a strong 'chilling effect' on potential providers of WLAN and plays an important role many German court cases.[79] Over time a dubious business has evolved with specialised law firms sending out large numbers of cease-and-desist-letters and advertising this business among right holders with the promise to turn 'piracy into profit'. If the CJEU had followed the Advocate General, this could have dealt a serious blow to the German 'warning letter industry'. The statement that the WLAN provider may not be ordered to pay court costs even if an injunction is granted against him could have stirred even more controversy. This would have been a novelty for German law of civil procedure, which, as a matter of principle, adheres to the 'loser-pays-rule (§ 91 Code of Civil Procedure). It is doubtful, whether a reversal of this principle could be deduced from the E-Commerce Directive and how it should be compatible with the principle of national procedural autonomy.[80]

However, these are hypothetical questions as the CJEU preferred not to follow Szpunar's view with regard to ancillary claims for costs related to injunctions. Instead, the Court considered that Article 12(1) of the E-Commerce Directive does not preclude claims for the reimbursement of the costs of giving formal notice nor claims for court costs incurred in conjunction with a claim for injunctive relief.[81] As a consequence, an interpretation of the new paragraph 3 of § 8 Tele Media Act as providing a shield against pre-litigation costs (as suggested by the legislator's explanatory memorandum) would be incompatible with EU law.[82] In summary, to paraphrase *Mark Twain*, the rumours of the death of the German 'warning letter industry' have been greatly exaggerated.

Reasonable Safety Measures Required from a WLAN Operator

If injunctions against the WLAN operator are not generally barred by Article 12 of the E-Commerce Directive, the crucial question is, which 'safety measures' can reasonably be imposed on the network operator in order to avoid unlawful acts by network users.[83]

Before discussing individual safety measures which have been contemplated by the Munich court in its order for reference, the CJEU describes the limitations applicable to injunctions in general terms. Broadly speaking, national courts must, when issuing an injunction, ensure that a fair balance is struck between the fundamental

---

[78] See e.g. § 97a para. 1 German Copyright Act.

[79] For an overview see Schmitz and Ries 2012.

[80] Spindler 2016b, 2451.

[81] CJEU, Judgment of 15 September 2016, *McFadden*, C-484/14, ECLI:EU:C:2016:689, paras. 77–78.

[82] Bisle and Frommer 2017, 61.

[83] Spindler 2016b, 2451.

rights at issue, in particular, Article 17(2) of the EU Charter regarding intellectual property rights, the freedom to conduct a business, which the WLAN operator enjoys under Article 16 of the EU Charter, and the freedom of information of the network users, whose protection is ensured by Article 11 of the EU Charter.[84] The last point echoes the line of argument sketched out in *Promusicae* and *UPC Telekabel Wien*.[85]

On the basis of these general criteria the CJEU examines the three measures contemplated by the Munich court, namely the termination of the Internet connection, the password protection and the examination of all communications passing through that connection.

Regarding the monitoring of all information transmitted via the WLAN, the Court makes clear that such a measure must be excluded from the outset as contrary to Article 15(1) of the E-Commerce Directive, which excludes the imposition of a general monitoring obligation on the access provider.[86] Similarly, a measure consisting in terminating the WLAN connection completely would be manifestly incompatible with the need for a fair balance to be struck among the fundamental rights involved.[87]

The question of password protection is probably the most interesting aspect with regard to the reform of the Tele Media Act. In its opinion, the Advocate General raised several objections against an obligation of the WLAN operator to secure the access to the network. He argued that such an obligation would impose disproportionate administrative constraints on the network operator, potentially undermining the business model of undertakings offering WLAN access only as an extra service to their customers. He also questioned the effectiveness of a password protection. He correctly notes that 'the use of passwords can potentially limit the circle of users, but does not necessarily prevent infringements of protected works'.[88] As a consequence, Advocate General opined that a measure requiring password-protection would not meet the 'fair balance' test.[89]

Yet, again, the Court only partially followed Advocate General's analysis. The CJEU shared Szpunar's view with regard to the first two measures (i.e. termination of the Internet connection and monitoring of Internet traffic), but took a different position regarding the password-protecting of the Internet connection. In particular, the CJEU held that "a measure consisting in password-protecting an internet con-

---

[84] CJEU, Judgment of 15 September 2016, *McFadden*, C-484/14, ECLI:EU:C:2016:689, paras. 81–83.

[85] CJEU, Judgment of 29 January 2008, C-275/06, Promusicae, ECLI:EU:C:2008:54, paras 68–70; CJEU, Judgment of 27 March 2014, *UPC Telekabel Wien*, C-314/12, ECLI:EU:C:2014:192, para. 47. For an example of such a balancing of fundamental rights based on the principles set out in *UPC Telekabel Wien* see now also BGH, Judgments of 26 November 2015, Case ref. I ZR 174/14, NJW 2016, 794 (*Goldesel*) and Case ref. I ZR 3/14, ZUM-RD 2016, 156 (*3dl.am*); see also Hofmann 2016, 769.

[86] CJEU, Judgment of 15 September 2016, *McFadden*, C-484/14, ECLI:EU:C:2016:689, para. 87.

[87] Ibid. at paras. 88–89.

[88] Opinion of Advocate General *Szpunar*, 16 March 2016, C-484/14, ECLI:EU:C:2016:170, at para. 146.

[89] Ibid. at para. 150.

nection may dissuade the users of that connection from infringing copyright or related rights, provided that those users are required to reveal their identity in order to obtain the required password".[90] If other acceptable measures are not available, such a measure "must be considered to be necessary in order to necessary in order to ensure the effective protection of the fundamental right to protection of intellectual property".[91]

Summary and Outlook: Another Reform of the Tele Media Act Ahead?

In the end, several questions remain open and the debate about balancing regulation and innovation is far from settled: First of all, the *McFadden* decision gives no clear answer as to which standard applies to WLAN access points offered by private persons without any link to an economic activity.[92] Second, and more importantly, with regard to WLAN providers operating in an 'economic context', it is still unclear which safety measures will pass the 'fair balance' test and can thus be imposed by a national court. Unfortunately, the CJEU only examined the three measures envisaged by the Munich court in its referral. In particular, it did not discuss alternative measures such as blocking of ports or typically infringing websites. At first glance, such technical measures seem to be preferable to a generally imposed obligation of password-protecting and identification of users.[93] Password-locking and identification of users would be ineffective as a tool of enforcement without retention of the collected data, which would result in a massive interference with users' right to the protection of personal data ensured by Article 8 of the EU Charter.[94] Rather surprisingly, the Court did not discuss the privacy issues involved but focussed its analysis on balancing the WLAN operator's right to conduct business and the protection of intellectual property rights enjoyed by copyright holders.

In the light of these considerations, the focus of the debate could shift from password-protecting to blocking measures. Indeed, a recent Ministerial draft[95] for another reform of the Tele Media Act, which was presented in February 2017, suggests to add a new para. 4 to § 7 Tele Media Act which shall serve as a legal basis for blocking injunctions. It is questionable, however, whether this new approach is the 'silver bullet' that will create the necessary legal environment for a better WLAN

---

[90] CJEU, Judgment of 15 September 2016, *McFadden*, C-484/14, ECLI:EU:C:2016:689, para. 96.

[91] Ibid. at para. 99.

[92] See Nordemann 2016, 1103.

[93] Mantz 2016, 820; Nordemann 2016, 1102.

[94] On the privacy issues see Mantz and Sassenberg 2015b 90; see also Husovec 2017, 123.

[95] *Entwurf eines Dritten Gesetzes zur Änderung des Telemediengesetzes of 23 February 2017.* The draft is available at the website of the German Ministry of Economic Affairs and Energy: http://www.bmwi.de/Redaktion/DE/Artikel/Service/Gesetzesvorhaben/entwurf-telemediengesetz-drei.html.

coverage in Germany. While blocking injunctions against ISPs have been generally accepted by the CJEU in *UPC Telekabel*, it is doubtful, however, whether such measures can reasonably be imposed on undertakings that offers free WLAN as an additional service to their customers.[96] In particular for small business, e.g. restaurants or cafés, such an obligation would impose disproportionate administrative constraints on the network operator. Moreover, the blocking of individual IP-addresses implies the risk of 'overblocking', whereas the blocking of certain domain name servers can easily be circumvented.[97] In such cases it might be more useful to take measures prohibiting the upload or download of large amounts of data in order to prevent file sharing.[98]

References

Bisle, R., and B. Frommer. 2017. EuGH klärt Verantwortlichkeit bei anonym nutzbaren WLAN-Hotspots – Das Ende der Pläne zur "Abschaffung der Störerhaftung"?' *Computer und Recht* 32: 54–63.

Borges, G. 2010. Pflichten und Haftung beim Betrieb privater WLAN. *Neue Juristische Wochenschrift* 63: 2624–2627.

———. 2014. Die Haftung des Internetanschlussinhabers für Urheberrechtsverletzungen durch Dritte. *Neue Juristische Wochenschrift* 67(32): 2305–2310.

Busch, C. 2014. Secondary liability of service providers. In *German national reports on the 19th international congress of comparative law*, ed. Schmidt-Kessel, 765–779. Tübingen: Mohr Siebeck.

Conraths, T., and S. Peintinger. 2016. Der neue § 8 TMG: Kein Wegfall der Störerhaftung von W-LAN-Betreibern. *Gewerblicher Rechtsschutz und Urheberrecht. Praxis im Immaterialgüter- und Wettbewerbsrecht* 8(14): 297.

Drücke, F. 2015. Haftung bei offenem WLAN? *Zeitschrift für Rechtspolitik* 48(3): 95.

Hoeren, T., and S. Jakopp. 2014. WLAN-Haftung – A never ending story? *Zeitschrift für Rechtspolitik* 47(3): 72–75.

Hoeren, T., and S. Yankova. 2012. The liability of Internet intermediaries – The German perspective. *International Review of Intellectual Property and Competition Law* 43(5): 501–531.

Hofmann, F. 2014. Die Haftung des Inhabers eines privaten Internetanschlusses für Urheberrechtsverletzungen Dritter. *Zeitschrift für Urheber- und Medienrecht* 58(8): 654–660.

———. 2016. Störerhaftung von Access-Providern für Urheberrechtsverletzungen Dritter. *Neue Juristische Wochenschrift* 69(11): 769.

Husovec, M. 2016. Accountable, not liable: Injunctions against Intermediaries. *Tilburg Law and Economics Center (TILEC) Discussion Paper 2016*. Available at. http://ssrn.com/abstract=2773768

———. 2017. Holey cap! CJEU drills (yet) another hole in the e-Commerce Directive's safe harbours. *Journal of Intellectual Property Law & Practice* 12(2): 115–125.

Kaeding, N. 2010. Haftung für Hot Spot Netze. *Computer und Recht* 26: 164–171.

---

[96] For a more detailed discussion of this question see Husovec 2016.

[97] Oberlandesgericht Köln, Judgment of 18 July 2014, Case ref. 6 U 192/11, MMR 2014, 832 at 836; on the technical possibilities for the blocking of access to certain websites and the legal implications see also Leistner and Grisse 2015, 19-27 and 105–115.

[98] Ohly 2015, 317; see also Landgericht Hamburg, Judgment of 20 October 2010, Case ref. 308 O 320/10, ZUM-RD 2011, 561.

Kur, A. 2014. Secondary Liability for Trademark Infringement on the Internet: The Situation in Germany and the Throughout the EU. *Columbia Journal of Law & The Arts* 37: 525–540.

Lehmann, M. 2015. Digitalisierung, cloud computing and Urheberrecht. *GRUR Int.* 64: 677–681.

Leistner, M. 2014. Structural aspects of secondary (provider) liability in Europe. *Journal of Intellectual Property Law & Practice* 9: 75–90.

Leistner, M., and K. Grisse. 2015. Sperrverfügungen gegen Access-Provider im Rahmen der Störerhaftung. *Gewerblicher Rechtsschutz und Urheberrecht* 117(2): 19–27 and 105–115.

Mantz, R. 2013. Die Haftung des Betreibers eines gewerblich betriebenen WLANs und die Haftungsprivilegierung des § 8 TMG. *Gewerblicher Rechtsschutz und Urheberrecht – Rechtsprechung Report* 13(12): 497–500.

Mantz, R., and T. Sassenberg. 2015a. Die Neuregelung der Störerhaftung für öffentliche WLANs. *Computer und Recht* 30(5): 298–306.

———. 2015b. Verantwortlichkeit des Access-Providers auf dem europäischen Prüfstand – Neun Fragen an den EuGH zu Haftungsprivilegierung, Unterlassungsanspruch und Prüfpflichten des WLAN-Betreibers. *Multi Media und Recht* 18(2): 85–90.

Mantz, R. 2016. Rechtssicherheit für WLAN? Die Haftung des WLAN-Betreibers und das McFadden-Urteil des EuGH. *Europäische Zeitschrift für Wirtschaftsrecht* 27: 817–820.

Müller, V., and D-K. Kipker. 2016. Der Entwurf eines Zweiten Gesetzes zur Änderung des Telemediengesetzes – Hat die Bundesregierung eine zeitgemäße Angleichung des TMG verfehlt? *Multi Media und Recht* 19(2): 87–91.

Neuhaus, S. 2011. *Sekundäre Haftung im Lauterkeits- und Immaterialgüterrecht*. Tübingen: Mohr Siebeck.

Nordemann, J.B. 2016. Nach TMG-Reform und EuGH "McFadden" – Das aktuelle Haftungssystem für WLAN- und andere Zugangsprovider. *Gewerblicher Rechtsschutz und Urheberrecht* 118: 1097–1103.

Obergfell, E.I. 2016a. Internettauschbörsen als Haftungsfalle für private WLAN-Anschlussinhaber. *Neue Juristische Wochenschrift* 69(13): 910.

———. 2016b. Gerichtlich verordneter Passwortschutz für WLAN-Hotspots – Zur Reichweite der Access Provider-Privilegierung von kommerziellen WLAN-Anbietern. *Neue Juristische Wochenschrift* 69: 3489–3492.

Obergfell, E.I., and A. Thamer. 2017. (Non-)Regulation of online platforms and internet intermediaries – The facts: Context and overview on state of play. *Gewerblicher Rechtsschutz und Urheberrecht, Internationaler Teil* 66: 201–206.

Ohly, A. 2014. *Urheberrecht in der digitalen Welt – Brauchen wir neue Regelungen zum Urheberrecht und zu dessen Durchsetzung?, Gutachten F zum 70. Deutschen Juristentag*. Munich: C.H.Beck.

———. 2015. Die Verantwortlichkeit von Intermediären. *Zeitschrift für Urheber- und Medienrecht* 59(4): 308–318

Schmitz, S., and T. Ries. 2012. Three songs and you are disconnected from cyberspace? Not in Germany where the industry may 'turn piracy into profit' European Journal for Law and Technology, Vol. 3, No. 1. Available at. http://ejlt.org/article/view/116/190#_edn70

Schnabel, C., and B. Freund. 2010. Ach wie gut, dass niemand weiß... – Selbstdatenschutz bei der Nutzung von Telemedienangeboten. *Computer und Recht*, 718–721.

Sesing, A. 2015. Mehr Rechtssicherheit für Betreiber von (kostenlosen) Funknetzwerken? *Multi Media und Recht* 18(7): 423–427.

———. 2016. Verantwortlichkeit für offense WLAN – Auswirkungen der TMG-Reform auf die Haftung des Anschlussinhabers. *Multi Media und Recht* 19(8): 507.

Solmecke, C. 2015. Haftung bei offenem WLAN? *Zeitschrift für Rechtspolitik*, 95.

Spindler, G. 2010. Haftung für private WLANs im Delikts- und Urheberrecht. *Computer und Recht* 25: 592–600.

———. 2014. Zivilrechtliche Sperrverfügungen gegen Access Provider nach dem EuGH-Urteil „UPC Telekabel. *Gewerblicher Rechtsschutz und Urheberrecht*, 826–835.

————. 2016a. Die geplante Reform der Providerhaftung im TMG und ihre Vereinbarkeit mit Europäischem Recht – Warum die beabsichtigte Reform ihr Ziel verfehlen wird. *Computer und Recht*, 48–56.

————. 2016b. Die neue Providerhaftung für WLANs – Deutsche Störerhaftung adé? *Neue Juristische Wochenschrift*, 2449.

Stögmüller, T. 2014. LG München I: Vorlagefragen an den EuGH zur Verantwortlichkeit des Access-Providers eines offenen WLAN. *Gewerblicher Rechtsschutz und Urheberrecht. Praxis im Immaterialgüter- und Wettbewerbsrecht*, 542–545.

Volkmann, C. 2015. Freies WLAN für einen Cappuccino. *K&R*, 289–291.

International Academy of Comparative Law

19th World Congress
Vienna
July 20–27, 2014
Topic VI: (COMPUTERS)
SUBJECT: SECONDARY LIABILITY OF SERVICE PROVIDERS
General Reporter:
Graeme B. Dinwoodie
University of Oxford

Introductory Remarks About the Topic

The topic "Secondary Liability of Service Providers" could be construed extremely broadly; "service providers" is a very broad phrase. However, I have read the term to focus in particular on online service providers because that is the context in which this question has become most acute in recent years. However, this reading should not preclude reporters from giving the term a broader reading, and the more open-ended questions below allow for this. Indeed, a couple of questions specifically ask about secondary liability in the online and other environments.

Within this questionnaire, there are two basic sets of questions. The first seeks information about the doctrinal structure of secondary liability rules: this inquiry encompasses questions about the legal standard to which service providers are held, the safe harbours (or immunities) of which they may take advantage, and the range of remedies that can be secured against a service provider whose services are used by third parties. The second set of questions has a more thematic focus: the source of these rules, whether they are specific to the online environment, the extent to which they are horizontal or limited to particular causes of action, and the practices that have grown up around the rules. These inquiries reflect areas that I expect to be

© Springer International Publishing AG 2017
G.B. Dinwoodie (ed.), *Secondary Liability of Internet Service Providers*,
Ius Comparatum – Global Studies in Comparative Law 25,
DOI 10.1007/978-3-319-55030-5

the focus of the general report. However, national reporters are encouraged to raise both additional doctrinal questions and pervasive themes that have arisen in their country.

Each question has grouped a number of similar questions under the one numbered "Question." However, please do not allow the structure or order of questions to constrain if the legal structure in your country is dramatically different.

Practical Suggestions for the National Reports

1. The International Academy of Comparative Law does not give fixed instructions as to the length of the reports. National reports usually range between 15 and 30 pages; the General Report is typically between 50 and 60 pages.
2. The language of the report must be either English or French.
3. The **deadline** of delivery to the General Reporter is **October 22, 2013.**
4. Publication is envisaged but depends on the circumstances (finances, length, content, etc.). I am speaking to publishers about the options.
5. Please strive to be clear and concise and stay as close to the substance and sequence of the questions as possible so as to ensure the comparability of answers across a large number of countries. If a question does not make sense in the context of your legal system, briefly say so; if other issues are pertinent, please mention them.
6. Please use footnotes for citations and/or references.

Questionnaire

I. Secondary Liability Standards

1. Are there laws in your country creating the secondary (or indirect, or accessorial) liability of service providers for conduct of others using their services? What are the elements required to establish a secondary liability claim? Is there more than one basis on which to establish secondary liability? (For example, in some countries, contributory liability may co-exist with vicarious liability). Do these laws consist of a single horizontal standard applicable without regard to the specific area of law in question, or does the liability standard vary according to the cause of action (e.g., intellectual property, defamation, product liability etc)?
2. If there are laws in your country creating the secondary liability of service providers for conduct of others using their services, what is the definition of a "service provider"? Is the definition limited to internet (or online) service providers? Please provide example of service providers falling within this definition. Does it include search engines and operators of online market places?

3. Was the law developed by courts or created by statute? In what level of detail does the law prescribe the conduct that will give rise to secondary liability? If developed by the courts, from which existing principles (if any) did the courts draw? To what extent does the standard for secondary liability depart from the general standard of secondary liability in tort (or other relevant) law?
4. What is the relationship between the standard for secondary liability of service providers and the relevant standard for primary liability (either of the service providers or third parties using their services)? To what extent have courts assessing the scope of primary liability taken into account the possibility of secondary liability of service providers (and vice-versa)?

II. Immunity from Secondary Liability

1. Are there laws in your country immunizing service providers against liability for conduct of others using their services? If so, is this a single horizontal standard for immunity applicable without regard to the specific area of law in question? Against what forms of liability (e.g., only secondary liability, only damages) is immunity offered? On what conditions is immunity offered?
2. If so, what is the definition of a "service provider"? Is the definition limited to internet (or online) service providers?
3. Was the law developed by courts or created by statute? If developed by the courts, from which existing principles (if any) did the courts draw?
4. Does your law provide for the possibility of remedies against service providers regardless of their being secondarily liable? If so, in which circumstances? What types of remedies will courts consider when the service providers are not secondarily liable (e.g., information disclosure, or DNS and IP address blocking). How do these remedies differ from those imposed in the event of secondary liability being established?

III. Other Questions

1. To what extent have service providers developed best practices or voluntary codes for dealing with conduct by third parties using their services that allegedly amounts to a violation of law? If so, who was involved in the development of such practices or codes? In what form have these been embodied (e.g., a memorandum of understanding with select right holders)? Have courts paid any attention to these practices or codes in deciding questions of secondary liability?
2. To what extent have service providers been subjected to regulatory regimes (e.g., so-called "graduated response" systems) that require them to cooperate in the enforcement of measures against the third parties who use their service for improper purposes?

3. To the extent that the laws referred to above include a so-called "notice and take-down" system, has there been any concern expressed about right-holders abusing the mechanism or service providers being cautiously "over-compliant" with takedown requests? Does your law contain any penalty or disincentive for either such conducts?

4. What mechanism have courts used to balance the need to ensure effective enforcement of rights with the ability of service providers to conduct business? To what extent have fundamental or constitutional rights of service providers or their customers influenced courts' attitudes to secondary liability of the providers (or the award of remedies against them in the absence of liability)?

5. To what extent might service providers be criminally liable for the conduct of third parties who use their services?

6. In disputes involving the laws discussed under Sections I and II above, to what extent have concerns about extraterritorial application of law been considered by the courts? Should they have been?

7. Are there any particular reform of the current law in your country that would you believe establish a more appropriate standard for secondary liability of service providers than currently exists?

8. Are there any on other issues which are not covered by the questionnaire but are a concern in your country or jurisdiction?

Printed by Printforce, the Netherlands